ALLE ZEIT WACH 1842

SJ

Biological Order and Brain Organization

Selected Works of W. R. Hess

Edited by K. Akert

Translated by
K. Akert M. Bornstein V. Bucher R. M. Hess
S. Hess Ph. Levin G. P. Michel

With 91 Figures

Springer-Verlag
Berlin Heidelberg New York 1981

Professor Dr. Konrad Akert
Physiologisches Institut der Universität Zürich
Rämistraße 69, CH-8028 Zürich

ISBN-13:978-3-642-67950-6 e-ISBN-13:978-3-642-67948-3
DOI: 10.1007/978-3-642-67948-3

Library of Congress Cataloging in Publication Data
Hess, Walter Rudolf, 1881 – Biological order and brain organization
Bibliography: p. Includes index.
1. Neurophysiology – Collected works. 2. Brain – Collected works. 3. Biological control systems – Collected works.
I. Akert, Konrad. II. Title.
[DNLM: 1. Neurophysiology – Collected works. 2. Psychophysiology – Collected works.
WL 7 H586b] QP352.H42 612.'8 80-28797 ISBN-13:978-3-642-67950-6 (U.S.)

Softcover reprint of the hardcover 1st edition 1981

2121/3130-543210

Preface

The centennial of his birthday (17 March 1881) prompted the publication of the *Selected Works* of Walter Rudolf Hess. Although English translation of several of his monographs have appeared, none of his original papers has ever been published in the English language. During his scientific career, Hess made pioneering contributions in the field of hemodynamics, pyhsiological optics, oculomotor diagnostics, regulation of circulation, respiration and temperature, and finally on the somatomotor, visceral, and emotional functions of the diencephalon. His concepts concerning *organization and order in physiology* and his views on the important role of the vegetative nervous system in regulating the activity of the central nervous system are of great interest to science and medicine and were in many respects far in advance of his time. These concepts continue a line of thought which was upheld by such famous physiologists as Xavier Bichat, Claude Bernard, and Walter B. Cannon. Indeed, Walter Rudolf Hess has become one of the rare figures in the recent history of physiology willing to carry out an *integrative* analysis of bodily functions and to search for the basic principles of regulation and interaction between regulatory systems. In fact, he anticipated such ideas in biology as feedback control and servomechanisms long before these notions evolved in the field of engineering and electronics. In analyzing the organizational aspects of biomotricity in the central nervous system, he touched upon such concepts on the psychic level which are of considerable value to us today. He soon realized that his approach could bring him near the brink of teleology. This dilemma was most productive in that it forced him to clarify issues such as order and economy as the guiding principles of the functional organization of living organisms.

His concept of vegetative regulation led Hess to postulate that the vegetative innervation of internal organs has its functional and morphological counterpart in the brain. At that time there were few reliable data to confirm the view that the central nervous system and especially the cerebral cortex are innervated by a terminal network of unmyelinated fibers emanating from brain stem neurons which secrete aminergic transmitter substances (see Chapt. 9, p. 155). Yet, Hess' neurophysiologic and neuropharmacologic experiments gave strong indication that the *Leistungsbereitschaft* of higher centers is subject to a regulatory control which resembles in every respect that exerted in the periphery by the *vegetative nervous system.*[1] The analysis beginning in 1962 with the histochemical method of Falck and Hillårp more than justified Hess' notion.

The two most important concrete problems concerned *sleep* and *motor organization*, the former as a paradigm of vegetative regulation and the latter as a paradigm of the animal (somatic sensorimotor) system. A thorough conceptual and experimental analysis of these two problems remained the themes of his life. It seems most significant that as a former ophthalmologist he would always return to eye functions to clarify the relevant problems.

The present selection is naturally somewhat arbitrary. It was our intention to expose the breadth of Hess' interest, and for this reason we have included papers which deal not only with "Brain and Behavior," but also with many other topics which occupied Hess' mind on his long march, such as Neurohumoral (Acetylcholine) Transmission at the Motor Endplate, Hering-Breuer Reflexes in Respiratory Control, Stereovision, and Hemodynamics. The thread that winds through his seemingly heterogeneous topics is the concept of "Biological Order" on the one hand, and that of the *Leistungs- und Erfolgsbezogene Physiologie*, on the other. Hess became aware of the role of biological order in his early work on hemodynamics, and he later recognized the same principle as the governing factor in motor coordination, psychic functions, and even in the behavior of human society.

The translation of his work was not an easy task. Hess himself was conscious of his heavy and rather involved German style and characterized this uneasiness when he mentioned his "constant struggle against a recalcitrant pen" in his autobiographical sketch (p. 309). Also, certain expressions which are important to his thinking are quite difficult to translate, because no equivalent terms are found in the English language. We quote two examples: *Leistungsbereitschaft* (readiness of performance) and *Leistungsfähigkeit* (ability to accomplish work). Led by the conviction that a compromise between the reader's comfort and the translator's conscience was inevitable – at least to some extent and in certain instances – we decided to translate the difficult passages in a liberal way. Fortunately, I was aided in this pro-

1 "Vegetative nervous system" is a term found mostly in the German literature and is preferable to the English expression "autonomic nervous system." "Autonomic" implies independence from higher and voluntary influences and is thus only partially correct. "Vegetative" refers to the functional goal

ject by competent and enthusiastic translators: Dr. Michael Bornstein, Miss Verena Bucher, Prof. Rudolf M. Hess and his wife Dr. Silvia Hess, Dr. Philip Levin, and Mr. Gilbert P. Michel, to all of whom I am deeply indebted. The translated manuscript went through at least three stages of revision and involved many vigorous discussions about the meaning of the original German text and the best way of expressing it in English. My colleagues and friends G. Baumgartner, H. R. Lüscher, and G. M. Yasargil were particularly involved at this stage. The final responsibility, especially for inadequacies which have undoubtedly remained in the text, rests with the editor. I was especially fortunate in being able to enlist the help of Hess' former staff member, Miss Verena Bucher, in this project. She was the morphological collaborator of Hess from 1929 until his retirement, and he credited her contribution by naming her as the coauthor of important original publications. Another former staff member, Max Jenny, head technican of Hess from 1926 until his retirement, generously offered his help by searching for the original illustrations (most of which had been prepared by himself) in the archives of the Physiology Department. This was all the more important because it allowed us to obtain in nearly all instances the highest possible fidelity in reproduction, especially of the halftone illustrations.

The most significant encouragement for this enterprise came from Hess' widow, Mrs. Louise Hess-Sandmeier, who kindly agreed to it and, at the age of 94, took an active and lively interest in the translator's work.

I wish to extend my thanks especially to my secretaries, Mrs. Hedi Hauser and Miss Barbara Meili for their most dedicated collaboration. The preparation of the final manuscript was aided by a videocomputer kindly put at our disposal by my friend, Dr. Alexander Guttmann, of Zurich. The bibliography was prepared by Mrs. Ruzena Biehal, librarian of the Physiological Institute, who suggested our undertaking this work and to whom we owe the first professionally prepared list of bibliographical references of Walter Rudolf Hess.

Finally, I would like to acknowledge the cooperation of the publishers: American Journal of Physiology (Bethesda), Johann Ambrosius Barth (Leipzig), S. Karger AG (Basel), Orell-Fuessli Verlag (Zurich), B. Schwabe (Basel), Schulthess Polygraphischer Verlag AG (Zurich), Springer-Verlag (Heidelberg). The University of Chicago Press (Chicago), and Churchill Livingstone (Edinburgh), who generously granted the copyright and the translation rights of the original articles which appeared in their journals.

A special word of thanks is due to Prof. Richard Jung, Freiburg im Breisgau, who assisted in the selection of the articles.

The translations would not have been possible without the financial aid that was kindly provided by grants from the pharmaceutical houses of

Sandoz, F. Hoffmann-La Roche, and Ciba-Geigy, Basel, as well as from the Zürcher Hochschulverein. We would like to express our sincere thanks to these institutions. Last but not least, the technical skills and generous assistance provided by Springer-Verlag, Heidelberg, are gratefully acknowledged.

Zurich, February 1981 KONRAD AKERT

Contents

Section I

Biological Order and Human Society Function and Neural Regulation of Internal Organs

Section II

Hemodynamics, Physiological Optics, Neurohumoral Transmission

Section III

Vegetative Nervous System and Psychic Functions, Sleep, Hypothalamic Syndromes

Section IV

Organization of Motor Systems

Section V
Biography and Bibliography

Section I

Biological Order and Human Society Function and Neural Regulation of Internal Organs

Chapter 1

Biological Order and Human Society*

W.R. Hess

Contents

The topic of "war and peace" raises a wealth of questions which are related to the problem of *organization*. In fact, organization, in war as in peace, plays a unique and, in many respects, decisive role. Thus, the physiologist, for whom the *central problem of life* resides in the order governing the world of living organisms, is prompted to study human society from the point of view of biological organization. The first obvious question is whether there is any relation between the workings of the human organism and the principles underlying supraindividual systems of the human community. Two different facts should be remembered in this context.

In war and in peace, the acting and suffering element is *man* with all his needs and aspirations, his feelings and moods, his abilities and failings. The fact that these distinctly physiological characteristics also find expression when man lives or acts in a community means that the community has its roots, within certain bounds at least, in the physiological sphere and, hence, has a biological aspect. Furthermore, in the broader field of comparative physiology, supraindividual organizations, i.e., the formation of "states,"

* Hess, W.R.: Kollektive Ordnung in biologischem Aspekt. In: Festschrift Max Huber (Late President of the International Red Cross) – Vom Krieg und Frieden, pp. 151–172. Zürich: Schulthess 1944

are encountered, thereby confirming that collectivity is a general biological phenomenon. The decisive factor here is that the individual is integrated into a society according to definite laws and performs his functions within the framework of a larger unit.

Prerequisite of successful action is the adjustment of individual capacities to common achievements, established in the strive for preservation of the self and of the species. It necessarily follows that the individual is dependent on the whole, a relation which implies a certain contradiction. To overcome this inconsistency, forces which ensure coherence and deliberate cooperation must come into play.

An Example

Speaking of the queen bee in the bee colony implies the concept of the ruling of a central power and – considering the 100-million-year-old history, according to experts – the acknowledgment that the principle of collective order stood the test of time. It is most remarkable that the unique position of the queen bee is expressed in only one single productive achievement: she alone is capable of reproduction and hence is responsible for preserving the line of future generations. In other respects the queen bee is a helpless creature, in no way comparable to the workers and their invaluable abilities. It is also a fact that she does not regulate important activities of the bee colony. However, she fulfills a task of organization in which she is irreplaceable, in spite of individual qualities which, where they do not concern the reproductive sphere, are poorly developed. She keeps the swarm of workers together. A colony without a queen is restless and disorderly, and it easily falls prey to the rapacious appetites of neighboring colonies. It is only by virtue of the cohesive force emanating from the queen bee that the superior qualities of the worker can be concentrated in successful action and come into play, for example, in the delicate organization of food forage, brood care, heating, protection, and defense, as well as tasks of hygiene within the colony. This supraindividual organism is further distinguished by the bond between the individual and the whole which is developed to such a degree that all activity is carried out in *harmonious cooperation*, and rivalries are eliminated by conflict-resolving organizational measures. As a direct consequence, however, the individual has thus lost all freedom and merely functions as an instrument of the whole; his existence has become so dependent on the collective that he is unable to survive by himself for any length of time. Serving the collective has become the exclusive purpose of this life. The behavior of the individual bee is in keeping with these strong ties as soon as the colony is threatened. An extraordinary, active readiness for defense action is constantly present – action which includes the sacrifice of the individual life. The mode of fighting is organized in such a manner that the loss of its defensive instrument must cause the bee to perish. It is remarkable that the same eagerness to do battle and make sacrifices for the state is also seen in aggression against another colony. Militancy for the sake of preserving and developing collectivity is even more pronounced in certain species of ants which organize veritable predatory incursions and wage "wars."

In the context of the general topic, these remarks could be understood as indicating a direct parallel to human behavior. Although such a comparison is not far fetched, we would like to emphasize the disapproval of this way of thinking, which is rooted in the phenomenology of collective life. Accordingly, the willingness of the individual to make sacrifices on behalf of the collective should not be qualified anthropocentrically by the criteria of "good" and "evil." Objective evidence is available but for a certain form and degree of bonding between the individual and the collective, including some indications as to effects upon the parts, the whole and the environment. It marks ways and means to project biological principles onto the level of human collective life, an approach which will stand up to scientific criticism. This involves an abstraction of the facts inherent in collective life itself and of its underlying principles of organization. Both together constitute the essence of a *biological theory of order*.

Two Spheres

For any organism there is an inside and an outside. The physiologist is particularly aware of this fact when studying the phenomena of regulation. He observes how a group of organs is used to create and maintain inner conditions in such a way that every element of the collectivity – the cell in the cell state – encounters the necessary prerequisites not only for subsistence but also for carrying out certain functions.

This includes the supply of oxygen and nutrients essential for life and the removal of carbonic acid and of other waste products formed in the tissues. The entire complex of functions designed to regulate inner conditions is governed by a single organ known as the vegetative nervous system. It has its own mode of operation and functional laws and is largely – but not completely – independent. Still other aspects appear when we consider the human body as an *individual unit*. Its functional targets now lie in the environment: activities undertaken to acquire food, to protect the body (in the widest sense), and to propagate the race lead to interaction with the environment. Muscles are put to work for this purpose while the sensory organs give direction. The activities having such a specific purpose are organized by a control organ of particular characteristics and subject to particular laws: the so-called animal nervous system. Depending on the direction taken by the effects of the two regulatory organs, we speak of an internal or external front or, if you will, "ministry." Fundamentally, the facts are always the same.

Whereas the separation of the two systems was stressed so far in view of the separate target spheres of the regulatory functions, the equally important interrelations must now be emphasized. Here too, an invariable reciprocity exists for every form of collective life. To take an example from physiology, nourishment of the constituent elements of body tissue is only possible if the individual as a whole successfully strives for food. Expenditure of energy, needed for this purpose, in turn depends on sufficient nourishment of muscle tissue by means of internal regulation (including circulation with its well-balanced transport and distribution functions). These interrelations are largely *automatic*, i.e., they play freely. On the one hand, consumption of nutritive material in the tissues induces a feeling of hunger and searching for food. On the other, physical exertion increases those products of metabolism which activate respiration and circulation. This circular concatenation of functions is supplemented by a direct connection of the two regulatory systems, the vegetative and animal nervous system, thus refining their harmonization, especially regarding the temporal aspect.

Two Phases

The body, its organs, and their functional elements, the cells, alternate between phases of *work* and *rest*. In work, energy is generated (by the muscles), or specific substances are produced (by the glands). Rest is apparently defined by a negative criterion: the *absence* of activity. However, the true state of affairs is somewhat different. Rest too has its positive aspect. Expenditure of energy and of substances in the active phase is compensated by supplementary processes. In earlier times, physiology spoke of an assimilatory and a dissimilatory phase. Construction and decomposition were thus set against each other. Today we know that this antithesis does not reach the heart of the matter. "Decomposition" is not a function. It is more accurate to speak of a phase of activity and a phase of restitution, thereby stressing the *productive nature* of rest. The two phases are inextricably linked. Every activity requires the expenditure of nutritive substances and accordingly limits the time during which the activity can still be carried out. At the same pace, restitutive processes take on urgency in order to prevent exhaustion and its catastrophic consequences. Building up new reserves ensures proficiency in the future. Ready reserves are thus identical with potential activity.

Even though the interlinked phases are strictly separated as to functional aim and orientation, their timing may overlap and become intertwined. However, in the periodic alternation of waking and sleeping, there does exist a clear distinction with regard to time.

The quantitative relations between activity (and consumption) and restitution can be represented as a profit and loss account for the matter and energy metabolism of our body. The balance yields the material basis of existence and further development, since all organic life is aimed at outgrowing the level of mere subsistence. A powerful drive for expansion is a feature inherent to all forms of life, expressed in the fact that the species' laws of order are imprinted on assimilated matter and on the energy bound to it. Growth and propagation are named as primary phenomena of life, and they are realized by a surplus-producing metabolism. Of course, the extent to which the vital drive for expansion can be deployed is dependent on the environment, in particular on the presence of other systems which are asserting themselves and their further development.

An interesting insight into these conditions is gained when, due to particular circumstances, the account shows a deficit which eventually turns into an emergency situation. First and foremost, the situation of stress acts as a stimulus for increased efforts to acquire food. If external circumstances are so unfavorable that the effort fails, the organism turns to other tactics. As soon as the reserves are exhausted, it starts to live on its own substance, which normally is required in carrying on active life. Naturally, this gives the signal for the rapidly accelerating loss of performance. Yet the body is exploiting this one last chance to defer final collapse, death from hunger, and to pull through to a possibly better time. It is remarkable that even in the emergency preceding collapse there is no disorganization. On the contrary. The way the body confronts hunger is characterized by typically prospective measures. It might be supposed that organs continually under stress in times of hunger would be quickly worn out. In reality, however, the organs that sacrifice their substance for the whole are those which remain inactive in the crisis. For example, even after a long period of hunger, the heart muscle has lost very little weight, whereas the skeletal muscles have lost a great deal. The central nervous system too is spared to a large degree. In the course of these economy measures, unequal organs are not treated equally; rather, substance is consumed in keeping with *the organ's value*. The criterion is its importance for the preservation of the whole. This stretches the period of endurance, thereby increasing the chances of survival until conditions have improved. The success of such an emergency organization far exceeds all expectations. Not a few people believe they are "starving" if they have to miss one meal. In fact, man can survive for a long time, depending, of course, on his constitution and the strain which he is undergoing. If the body is at rest, periods of hunger of 2 months and, under certain circumstances up to 3 months, have been confirmed. Conditions of relative hunger can be endured for years, of course not without a loss of performance and a gradual decrease in physical and psychic resistance. A similar evaluation of individuals or groups of individuals has been adopted in food rationing in the context of the supraindividual collectivity of our people. However, insofar as it is based on professional categories, rationing employs exclusively causal and not prospective criteria. Nonetheless, there are cases where prospective argumentation does come into play: pregnant women and young people. Here, the higher insight of prospective organization agrees with human feelings. The link of the individual element to the collectivity is not strong enough. Otherwise, there is in this respect an enormous quantitative difference between man as an individual and man as a member of the community. With increasing emergency, however, other forces intervene, which, more and more, shift the weight on the unit as a part of the whole. A glance at those countries waging war confirms the experience that with suitable organization the capacity for endurance of the whole exceeds by far what would ever be expected, judging from peace-time conditions.

Principle of Efficiency and Economy

A direct consequence of the transition from individual to collective life is the onset of a process of differentiation. In its most elementary form it is expressed in the spatial relations of the parts within the whole. At a higher level, an increasing inner differentiation of the individual elements enables them to perform specific achievements. Here is the place to define the concept of efficiency (*Leistungsfähigkeit*). Physiology usually speaks of function. It is of utmost importance for the organism that its functioning is successful, that it produces a result from which the organism benefits in some way or other. However, performance, in the physiological sense, has its full value only when realized in a certain time span. Performance is the decisive factor which, in terms of quality and quantity, positively influences the organism's prospects for existence and "rewards" it with easier conditions of life. If efficiency is the fundamental prerequisite for survival, the second condition is the more or less complete realization of the principle of economy. The optimal physiological achievement is carried out with the minimum expenditure of matter and energy. An impressive example is the peculiar arrangement of the trabecula in the head of the femur, discovered in 1867 by a Zurich anatomist (H. Meyer) in collaboration with a professor at the Polytechnical College (Culmann). The requisite load-bearing capacity of this supportive apparatus is achieved by a system of inner braces in such a manner that there is a minimum expenditure of supportive substance.

Certain laws governing transport and distribution in the circulatory system are less striking, but not less important. The limited scope of this paper prevents us from presenting anything more than a brief and simplified picture of this system. It is all the more worthy of attention as it concerns the energy sector of the organism. Blood is the vehicle, carrying nutrients to the various organs. The circulation performance, i.e., the blood quantities delivered per time unit to a specific organ, is of decisive importance and is dependent on the rate of blood flow, which requires a certain expenditure of energy. The ratio between blood flow and total blood volume (i.e., the transport apparatus) and the circulation time must be considered as the main factors whose mutual relations determine the economy of the circulation system. One of the most interesting problems of hemodynamics bears on these relationships when seen in terms of energy consumption. To deal with but one facet of the question, it may be stated that where great quantities of blood flow jointly in large vessels, the rate of flow is high. A common path means that frictional resistance is small; higher speed can be achieved by a comparatively modest increase in energy. This really pays off when, in the course of distribution, the blood is divided into an increasing number of branching vessels and the growing friction surface implies an increasingly uneconomical function. The distribution system as a whole then benefits from the previously saved transport time inasmuch as the total circulation time is not overly extended by peripheral slowing. This slowing down of the circulation rate associated with distribution is important because it contributes to saving energy expended in circulation. In critical situations, for instance, in diseases weakening the heart, these economy measures assume special importance for the survival of the whole organism and act also as a protective mechanism against the imminent breakdown. The fact that in industry rationalization requires the development of ever cheaper methods of production and distribution makes clear that similar rules are followed in the collective life of man. Here, too, economy is a formative and preserving factor, making itself felt all the more the tenser the situation develops and the more "urgent" the savings are in handling the crisis.

W. B. Cannon, Professor of Physiology at Harvard Medical School, contrasted the principles of economy and *security* in his paper *The Body Physiologic and the Body Politic* by stating "security comes before economy" [1]. In a certain sense, there is indeed an antagonism. Generally, intensified economies in installations and operations jeopardize reliable functioning and endanger unharmed survival. The best solution to such a dilemma might be found in the motto "safety first;" it does not apply, however, in this absolute form in nature, because every living organization is founded on *probability and not on certainty*. Furthermore, at a closer glance, we see that the principle of safety is not in contradiction to that of economy, but it is an integrated and subordinate part of it, the more so the higher the stage of development of the endangered organization. Thus a balance ensues between economizing the parts and safeguarding the whole, which varies from case to case, according to the level of organization.

It has already been pointed out that the elements and relationships under discussion are also encountered in industrial organizations based on collective labor. Their decisive importance is apparent with particular and tragic clarity in warfare, especially so when we look at large operations rather than at narrow sections.

Competition and Coordination

The layman in biology probably tends to believe that under normal (i.e., healthy) conditions everything in our body works in perfect harmony. The fact of the matter is that antagonistic forces very often collide; indeed, this is the rule in absolutely vital *autonomic functions*. The various organs regulating the inner relationships of the organism are subject to a dual influence. If we take the heart as an example, we see that its activity based primarily on autonomic powers is affected to a greater or lesser degree, depending on circumstances, by impulses from the nerves and also by stimulating agents ("hormones") circulating in the blood. Depending on the type of neural or hormonal component, the effect is either an increase in, or inhibition of, performance. It is worthy of note that the two antagonistic mechanisms both operate simultaneously. Thus, the gradual unfolding of the heart's performance is the result of antagonistic regulatory forces. The implication of this organization is most strikingly revealed in situations of great physical effort, when muscular work calls for a greatly increased blood volume and the heart is summoned to an extra effort for the quantitative fulfillment of its task in the interest of the entire organism. It seems particularly strange that even in such circumstances inhibition asserts itself, and this can be detected by simple experiment: if the heart is artificially cut off from inhibitory control, its beat rate rises even higher, which obviously implies the danger of overexertion. *Inhibition* in this context means *protection*. The activating impulses are followed only to a certain degree; any excess is prevented by the heart's own "safety stop" mechanism, inasmuch as this is required for long-term efficiency. Inhibition in this context represents an *organizational factor*, and the antagonism is based on the competition between the part and the whole, found in every collective. It ensures a quantitative balance. Inhibition, by remaining active when the organism is at rest, ensures that the effort required of the heart in a given situation is just what is needed by the body. Once again, we encounter the principle of economy and at the same time learn the organizational measures by which it works and how, in general terms, it operates as a protective function as soon as an organ is strained and comes close to its limits of performance.

Furthermore, the example of the heart has made us familiar with a mode of behavior of utmost importance in biology: the great *adaptability* of the organism's performance to varying degrees of need. This ensures that variations of strain, inherent in the given external circumstances, are compensated for within broad limits. We enjoy a large degree of freedom in our behavior, which is due to our organ's highly flexible potential performance. Rigid standards would be unbiological and would greatly hamper the capability to overcome eventually increased resistance. When drawing conclusions as to the collective life of man, one has to be aware that such an adaptability is based not only on the readiness of the functional elements to temporarily muster increased forces, but also on the reserve capacity which in rest and in normal activity is not fully exploited.

A contest of a seemingly different kind takes place in our body when various organs require a greater volume of blood because of increased activity. A real competition between the organs arises, the total blood supply being limited. The conflict is resolved according to the above-mentioned primary principle of living order, namely the evaluation of the importance of the part for the whole. Where unqualified equal treatment would result in catastrophe for the organism, survival is ensured by the preference given to the most vital organs. The heart, for example, as the central organ in circulation, receives most *favored treatment* in the physiological regulation of blood distribution. Pathophysiology, unfortunately, knows only too well the deleterious consequences of conditions where this order no longer prevails. Incidentally, this shows once again that protection and economy are not incompatible, since an adequate blood supply based on priorities of need results in a life-preserving balance with a minimum of effort. There is no doubt that, mutatis mutandis, the principle of evaluation according to the benefit to the total organism also applies to conflict in human collective life. In this context again the fateful dangers of a superficial comparison become obvious. Collective order is not ruled by a single principle, but by the organic interweaving of various principles which are also involved in a conflict in which the priorities change according to the concrete situation. If we turn to nature for counseling on the problems of collective order, we shall have to analyze the collective phenomenon with respect to the elements involved, and then transfer our insight – adequately evaluated – onto another system. The conception is thus corroborated that abstraction with subsequent concretization, i.e., analysis followed by synthesis, are operations which make a theory of biological order consistent and suitable for application to human collective life.

Coordination as a Creative Principle

Whenever man brings his will and ability to bear, he requires muscle power. This occurs by the muscle fiber becoming tense in response to a nervous stimulus. The tension lasts only a short while; it is over within a tenth of a second. A nonspecialist will be surprised at this statement. He can carry out an experiment on himself in which he successfully holds up a weight for a prolonged period. However, in other circumstances, his movements are smooth in nature, by no means so quick or twitching. The contradiction is resolved when we learn that the nerve cells which transmit the stimuli for contraction to the muscle fibers do not send single impulses but *series of impulses*. Whenever the muscle fiber is about to relax, a new tension stimulus arrives. This goes on until the intended prolonged effect of muscle power is achieved, whereupon the series of impulses is interrupted. It is also characteristic of muscle fiber that, when engaged, it releases all its potential at once. Instead of grading its function, it gives all or nothing. This statement too will come as a surprise when we consider that we can achieve the finest shades of muscle tension and movement. This ability is due to the fact that muscles consist of groups of fibers of varying excitability. Weak or strong tension is obtained by recruiting a greater or lesser number of individual fibers. By summation of brief tiny energy discharges of individual elements and varying their number, new and most valuable qualities are gained as a creation of collective order. This example may not overly impress the nonbiologist. We are, however, as stated before, interested in determining principles of biological collectivity. In this respect the example is of far-reaching significance. It should be added, moreover, that the accomplishments of coordinating synthesis have not yet been fully described. The effort of a single muscle, the duration and strength of which can be regulated, is but the simple form of the complex interplay of muscle groups. The point of interest here is the coordination of individual organs (muscles) to act as functional units. The process is directed by a rigid *order*. This applies both to simultaneous cooperation and to certain combinations occurring in succession. We cannot here describe at length how a flow of successive movements is generated by integration of gradual steps. But a word must be said about the final result: one example among many is human speech, a model of the creative power of coordination. A measured deployment of energy to move the diaphragm and muscles of the chest generates a continuous flow of air, which in turn causes the vocal cords to vibrate. The degree of tension in the muscles of the larynx determines the pitch. The sound is shaped into a vowel by the muscles of the palate, tongue, and lips; consonants are created by interrupting the flow of air at different places and in different ways. Even a word or a short sentence implies an unexpected profusion of coordinated acts which, when they are lent order, develop into a sentence by means of which a thought is transformed into an acoustic symbol comprehensible to someone else speaking the same language. When we are deeply touched by the playing of an orchestra, we are under the influence of an even more complex coordination of sounds produced by the individual artist on his instrument. The same applies to the "dark" side of human collective action when we are profoundly shocked by the "success" of well-coordinated means of destruction, namely, weapons of every type and size on land, sea, and in the air. Our visionary hopes that at the end of the cruel events a better world order might arise can only come true under the condition that the aspirations of ever-competitive nations based on physiological needs and properties be coordinated in a stable and long-lasting balance. Until such times, and if at all enough will, insight, and capacity to install an order based on adequate evaluation of its elements can be mustered, we shall remember the achievement of the International Red Cross which, by coordinating efforts to help the suffering, efforts of every kind and coming from every latitude, has created a worldwide organism and placed its high endeavor at the service of mankind as an all-encompassing collectivity.

Further Problems of Collective Order

One could mention in this connection a number of principles at work in building larger units from individual elements. I am thinking, for example, of the balance between autonomic powers with their own tendencies as opposed to an extremely centralized organization. Some of these topics gain a special significance for the communal life of man. In the human body, the organs possessing (relative) autonomy are only those which regulate the internal conditions of the organism. Where the organism interacts with the environment, behavior is governed by a rigidly centralized order which is achieved in stages, employing to great profit the principle of representation. Muscle groups and entire systems of groups are concerted in the central nervous system; the "representative," through its connection with the executive organs, enables them to develop their potency for highly complex combinations of actions. This process reaches ever-higher stages until a finely shaded and organized interplay of forces is attained which is the expression of a most differentiated creation of coordination. It would be wrong to think that all contest is ruled out when this harmony of action is attained. In fact, it has only been shifted to another level. The study of the motor system as an organizational principle shows, for example, that the automatic stabilization of body posture is based on a balance of antagonistic nervous "tension." The executive organs (i.e., the muscles), by forgoing their autonomy, have relegated the antagonism to the higher nervous organs, i.e., their central representatives. The antagonism now runs its course quietly at this level with an almost symbolic expenditure of energy only. One further element is required to ensure the smooth functioning of the dynamic balance of innervations: each representative fits into the order of the composite play of forces – with its synergetic or antagonistic interactions – according to its real significance. It would be otherwise impossible to maintain an upright posture even physically, let alone mentally. Excessive emphasis on any single component results in an inappropriate expenditure of energy and an inadequate effect, consequences well known to pathophysiology. Relating the abstract contents of such facts to the multifarious strivings within a nation, one is led to recognize the value of *truth*. It is the acknowledgment of a stratified unity. If the balance is disturbed, it is not the antagonism itself nor the tension which is the cause. On the contrary, antagonism and tension are the prerequisites for continuous reordering. The fault lies in the excessive emphasis by means of which one part of the whole gains ascendance. This insight has obvious relation to the theme of "war and peace." Nor should it be ignored in judging scientific information today, in view of the fact that methods of "propaganda" are making headway in the field of science, where truth is a law of necessity.

A number of other phenomena particular to organic collectivity and playing a role in human communal life would also have to be investigated. There is, for example, the innate tendency in every organic order *to comprehensive unity* with increasing subordination of the parts; translated into other terms, this is a claim to totality with its contradiction to individuality. The relation between organization and force is also topical, as is the consequence of organization resulting in power. Another aspect is the "leading" role of the control function in performing a physiological act. It is also interesting that when preservation of life is by and large ascertained, the organizational powers overcome the pure and direct utility principle and continue to develop order for the sake of order, as it were. In the "harmony of order" we encounter *the biological roots of aesthetics* in ornaments and in general. Another characteristic inherent to life is the tendency to use the unity achieved by organization as a building stone for an organization of a higher order: the development of cells into cell complexes to serve as instruments entrusted with certain tasks. The cell complexes are again joined together into organs and the organs into an organism. Our final consideration will be how this biological process of integration progresses from one level to the next, transcending individual life. This reveals the perspective in which order in the human community is perceived as a biological phenomenon.

Evolution to a Higher Level

Our considerations have led us to reflect on the order in the human community, and it may seem as if the physiologist had overstepped his bounds. I do not think so. It is true that his area of study in a narrow sense is limited to describing what organs do and how they work together in the organism. When analyzing the origins of collective life the fluid boundary is demonstrated most strikingly by the unity of mother and child. In purely physical terms, the infant is dependent on the mother for nourishment and other nursing care. The mother, willing to give up personal wishes for the child's sake, is also an example of the strong *binding forces* which are thus used by nature to overcome individualistic and disuniting tendencies. What are these forces and from where do they come? The answer is obvious. The mother is guided in her behavior by a feeling, the instinctive unfolding of which is innate, as is the activation of other drives. The same holds for the "forces" which build up the bonds of the family against individualistic tendencies or divisive influences. Standing amidst the driving forces of a collective order we must realize that these forces impart their subjective and here justifiable anthropocentric stamp. By accommodating ourselves to a higher order we follow unwritten natural laws of human behavior in relation to communities of different types. The inborn emotional structures normally oriented to life in the community were responsible for the development of a suprapersonal organization long before they had become conscious and the principles were formulated in the laws of civilization. Maternal love, the family bond, fidelity in friendship and loyalty toward one's country, and readiness to participate in a constructive ideal are all forms of man's (but not only man's) ability to adopt social behavior. To man alone, however, in his advanced stage of evolution, is given the capacity to consciously account for it, and to seek and find the balance in the conflicts inherent to every collective, the conflicts between self-interest and devotion to the community or between collectives of different alignment. We believe that in this context the development and pursuit of an ethical mode of life, which can be seen as the contribution of the individual to a higher entity and is so difficult to achieve, has its definitely biological implication. It would seem that the physician, researcher, and teacher, *Constantin von Monakow*, so closely bound to our university, means virtually the same thing when he concludes from various observations and from a rich experience with the healthy and the mentally disturbed: "Everything is arranged in such a way that, by fulfilling our vital program, we unawares rise into the spheres of the superpersonal. In the process of maturation, in reaching out quite naturally to ever wider spheres of community, to ever higher realms of the living, the veritable *unio mystica* is attained."

References

1. Cannon, W.B.: The body physiologic and the body politic. Science *93*, 1–10 (1941)

Chapter 2

Function and Neural Regulation of Internal Organs*

W.R. Hess

Honored Councillor, Honored Rector, Honorable Guests

First I would like to thank the Rector for the great honor bestowed upon me by the University and also for his kind words of introduction. I am also thankful that he has given me guidelines for my lecture today. It would be of interest to the audience, he suggested, to learn about the stages of development of the work which has today placed me in the limelight of the scientific and medical community by the award of the Nobel Prize. By accepting with pleasure the Rector's proposition I am spared from the suspicion that the following address awakes the impression of egocentricity, which would be embarrassing for you as well as for me. Thus, I stand here before you not as an individual but rather as a representative of a certain way of scientific endeavor and of a field of knowledge which has profited from it.

When I look back to the time in which I first had contact with persons interested in scientific matters I see myself in the anatomy laboratory of the University of Zurich. The group of students to which I belonged had to dissect the arteries of the foot. One day, an anomaly was discovered which was called to our attention by our teacher Prof. Ruge. He mentioned the fact without further elaboration. Yet, in my mind the question automatically arose as to "why." Such a striving toward causal understanding is not uncommon though it varies remarkably from person to person. Since I felt strongly motivated in this direction, I turned to my knowledge of hydrodynamics for consultation. This induced me, in the following semester, which was spent in Berlin, to make a side trip to Halle where Wilhelm Roux, the pioneer of so-called developmental mechanics, had the chair of anatomy. Following the discussion which he obligingly granted me, he sug-

* Hess, W.R.: Funktion und nervöse Regulation der inneren Organe. Vierteljahrsschr. Naturforsch. Ges. Zürich 95, 249–264 (1950)

Lecture held on 1 November 1949, in the Main Lecture Hall of the University of Zurich on the occasion of the ceremony celebrating the Nobel Prize for Medicine and Physiology

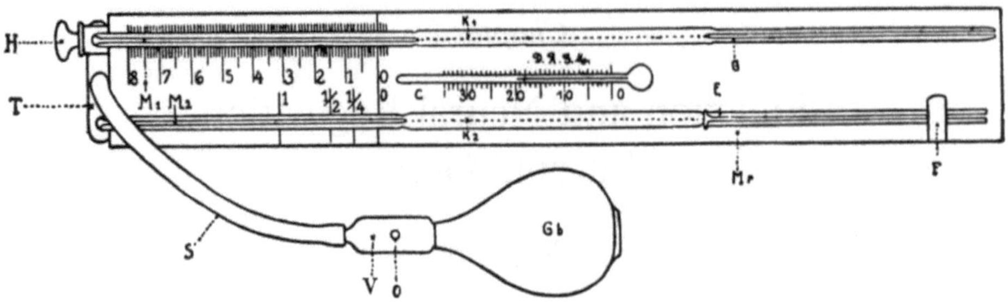

Fig. 2.1. Blood viscometer with direct reading. One to two drops of blood are sufficient. (From Hess [2])

gested that the dissection findings should be reported in writing. Thus came my first work: *A Mechanically Induced Conformity in the Structure of the Vascular System* [1].

But the interest awakened in hemodynamics was not thus quieted. On the contrary! It exercised a strong attraction, particularly because the tasks of the circulatory system are transparent, and because further, the laws which govern the flow of liquids, including the blood, were well known to me. An important energetic factor is the *internal resistance* of the fluid. Because there was no apparatus available to measure viscosity on freshly obtained, unaltered blood before coagulation occurred, a suitable instrument had to be designed. The measurement should be possible with only a few drops of blood and, hopefully, without complicated calculations. The first application of this new instrument [3] with direct reading (Fig. 2.1) was to help define the relationship between blood viscosity and the driving force of the blood flow, that is, the heart. Thus in a concrete case, one of the biologically important relationships between *performance* and *expenditure of energy* was made the object of an experimental investigation. The report of the findings and their evaluation was the brief content of my dissertation *Viscosity of the Blood and Cardiac Work*. In this connection I mention a little episode which belongs not only to that time, but also certainly to the present day, and characterizes certain situations which are not exactly encouraging. I sent the ready-to-print dissertation confidently to the editor of *Pflügers Archiv für die gesamte Physiologie des Menschen und der Tiere*. The notice of receipt followed in the form of a question as to under whose direction the work had been carried out. From my chief at that time, the well-known surgeon of the Thurgau Canton Hospital in Münsterlingen, Dr. Conrad Brunner, I could understandably not take either substantive or formal reference, but rather I had to truthfully admit that I carried out the investigation on my own initiative and under my own direction and critique. The not-to-be misunderstood answer to this admission was that my work was returned in a few days without further explanation. I must admit that this first experience with authoritarian behavior in the field of science depressed me. In my naiveté I believed that I was entitled to at least be given the reasons. The considerate help provided by Prof. H. Zangger, later my faculty colleague, brought the necessary correction: he arranged publication in the *Vierteljahrsschrift der Zürcher Naturforschenden Gesellschaft* [2].

The next step in the development of scientific interest was determined by my economic situation. Although even during my studies, physiology was for me the most attractive goal, I turned to a practical specialty and became an ophthalmologist. Decisive for this choice was the transparent relations offered by the eye which are due to direct accessibility, especially with respect to dioptrics. In any case there was no break in continuity in turning to this discipline. I later saw in the career of well-known scientists such as Donners, Hering, and C. v. Hess that a close kinship seems to exist between physiology and ophthalmology to which predisposed individuals feel attracted.

The Eye Clinic of Zurich at that time was under the direction of Prof. O. Haab, an investigator of highest exactitude and a surgeon of extraordinary skills. There the conditions were such – as they were, by the way, also under Conrad Brunner – that one could learn what is important in experimental physiology. Thus I believe that many experiences from my clinical activity proved fruitful in my subsequent activity in the field of my true choice. Here belongs, among others, the impulse toward painstaking work which arises from the sense of responsibility for the patients. Even if occasional pressure came from the chief – especially from Conrad Brunner – I realized even at an early time that thanks were due to my teachers.

In the course of ophthalmological activities I became aware of an interesting scientific problem: It may be said without impairing my good remembrance of Professor Haab that, with his predominantly morphological orientation, he had little understanding of oculomotor paralysis, which evidently required a dynamic approach. This, on the other hand, stimulated his assistant, because disturbances in the dynamics of muscle actions led directly to the problem of dynamic equilibrium of innervation and, indeed, in a very instructive way. The attempt to analyze motor disturbances in a quantitative manner prompted me to design an apparatus which could directly demonstrate the functional deficit (Fig. 2.2). The "Coordimeter" [4, see translation in this volume, p. 55], an apparatus widely used in practice for the diagnosis of oculomotor paralysis, produced a concept of motor coordination that was immediately applicable to other problems as well. It was thus on this basis that the analysis of the experimentally induced motor effects from the diencephalon could be taken up. With the aid of geometrical models representing the vectorial diagrams of forces, it was possible to base the analysis and interpretation of experimental data on a more solid and useful foundation.

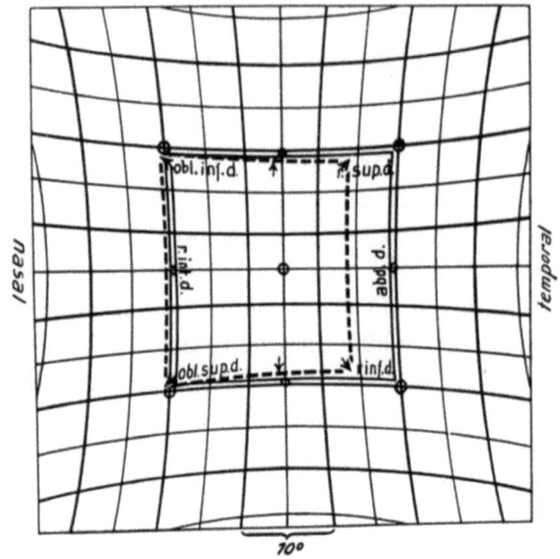

Fig. 2.2. Graphic demonstration of oculomotor paralysis. In the present case the external rectus muscle is involved. Comparison between the doubly traced *spherical-type square* and the *dashed rectangle* shows the limited motility of the paralyzed muscle

Our publications concerning the contribution of the diencephalon to automatic motor behavior have clarified matters extensively. Beyond this, insight into the general organization of the motor system has deepened. Such progress is not based on experimental findings alone. A no less important role is also played, as has just been said, by the dynamic concept of motor innervation arising from the analysis of oculomotor disturbances. It should be emphasized that this development leads directly to the understanding of the functions of motor centers and their mode of operation, to mention only integration of reflexes and their proprioceptive control; the higher centers of body posture (Fig. 2.3); synergistic coordination between goal-directed voluntary movements and reflex-regulated postural mechanisms; and motor coordination by primary central association and by secondarily mediated proprioceptive input. Furthermore, the use of the not yet widely exploited theoretical models was shown to enhance our understanding of the biomotor system [11].

In the preceding account of my scientific development I have limited myself to the problems of the motor system. In reality the interests at each phase had a wide range involving questions whose treatment partly collided but partly also could be pursued along the same line. The hemodynamic studies should be mentioned at this point. With the viscosity of the blood, only one of the essential factors governing the energetics of circulation is considered. Equally decisive is the organization of the vascular tree, particularly of the arterial system, in which by far the greatest part of the en-

Fig. 2.3. Motor effect: raising of the head induced by stimulation of the diencephalon. After cessation of the stimulus the head returns immediately to the normal position. The substrate stimulated is clearly recognizable from the sections of the brain (caudal diencephalon, region of the rubrospinal tract)

ergy produced by the heart is consumed. It should be noted that the function assigned to the arterial system is not only to conduct the blood flow but also includes the problem of distribution by dividing the compact flow in the aorta. This dispersion is accomplished by progressive branching until an extremely fine diameter of the capillaries is attained and the necessary intimate contact between blood and tissues can occur. With respect to cross-sectional relationships, anatomy and physiology teach us that the sum of the cross sections of the branches arising from one trunk is larger than the cross section of the trunk itself. Unanswered and, as far as I know, not expounded, is the question of quantitatively defined proportions. In this connection it can be shown that the number of the steps for the increase in cross section which occurs with each branching has considerable energetic consequences. Taking into consideration that there are many examples of nature's preference for the most economical utilization of material and energy, the question arises as to which vascular cross section is the most favorable for the transport of blood from the heart into the periphery. The calculations based on the fact that an increase of the contact surface necessarily goes together with each division of a fluid column, gives us the criterion of the optimal cross-sectional proportions along the pathway. As a consequence the way was opened for control observations on suitable biological objects, and a series of works resulted. The treatment of this problem was eventually presented as my *Habilitationsschrift* (docent's thesis) under the title *The Principle of the Minimal Dissipation of Energy As Related to Hemodynamic Research* [5] and was accepted, even though, as I soon noted, with suppressed enthusiasm. The topic was interpreted as speculative by some faculty members. One had at that time not regained the right balance after the excessive swing of teleological speculation. Captured in a rigid mechanistic attitude one did not dare to approach the question of a purposeful design in nature, not even when the occasion was compelling. In medical circles there was an unwillingness to accept the symbolic (mathematical) formulation of quantitative relationships, and a static form of thinking predominated. Today I understand this intuitive rejection somewhat better than at that time; the empirical approach provides the largest part of the basis of medicine and, further, morphology is more convenient for static thinking.

However, I saw no objective reason for me to abandon my preferred approach at that time, especially in view of the memorable demonstration before the *Zürich Naturforschenden Gesellschaft* of the orderly structure of the trabeculations in the spongiosa of the bones by the anatomist Meyer (we called him "bony Meyer") and Professor Culmann from the Polytechnic High School in Zurich. They explained that the arrangement of the bracing material in the bone corresponded exactly to that which an engineer would propose if he wished to attain a certain load capacity with a minimal amount of supporting material. Thus, their considerations were on the same level as the investigations related to the arterial system which we mentioned before. The only difference is that the arrangement of the bony trabeculations can be easily visualized and impresses even one who has no precise concept of the technical requirements. Unfortunately, the hemodynamic events cannot be illustrated in an analogous way, but can only be expressed by mathematical formulations. However, even though this form of presentation cannot be easily grasped, the same general principle prevails. Actually, the energetic economy of circulation has a higher priority in that a system which economizes the burden of the heart daily, hourly, or even per minute is relevant indeed. In an emergency it can mean at once the balance between life and death.

It must be admitted that our problem evolved from an extrapolation of observations made elsewhere. On the other hand the idea of an economical operation of the cardiovascular system came as an *incentive for quantitative research*. Furthermore, preoccupation with the cross section of vessels automatically led to a further theme which was connected with the fact that the arteries undergo luminal changes by virtue of the musculature embedded in their walls and thereby determine the blood supply of the individual organs. The concept of a quantitative coordination of functionally related arterial segments was based on a vascular innervation which is by nature also organized according to quantitative principles. In referring this principle to pathological conditions, concepts for the understanding of functional disturbances in the arterial circulation have been developed in terms of "neurovascular ataxia."

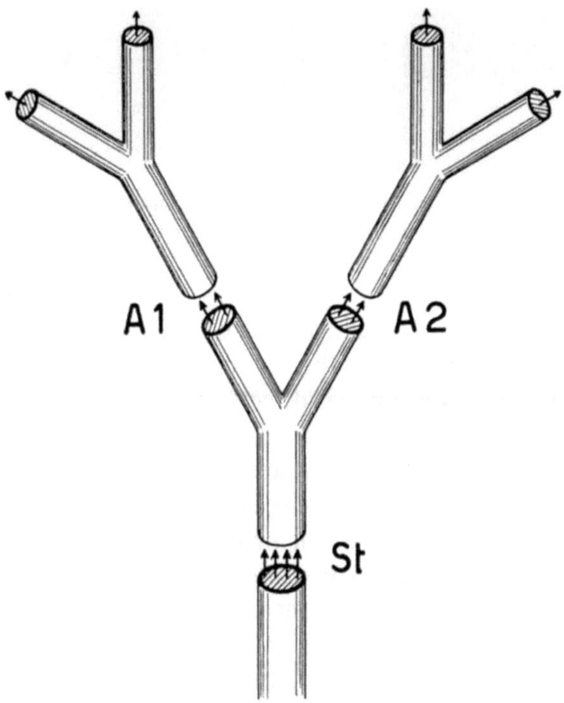

Fig. 2.4. The sum of the cross sections (A_1 and A_2) is larger than the cross section of the trunk *St*. The blood flow in the branches is correspondingly slower. Under optimal energetic conditions the cross-sectional proportions are $1:\sqrt[3]{2}$ (From Hess [5])

First attention was drawn to the mechanism by which the cross-sectional quotient from the trunk vessels to the branches, from the branches to the twigs, etc., was maintained automatically and under changing conditions of blood flow. This concept is outlined in Fig. 2.4. Although certain assumptions were made and experiments designed, this matter was not followed further experimentally. Only unhappily did I give up this plan, for it is clear that a principle could be studied here which provides automatic optimation of a function. Closely related to this is that a minimal effort is required in the presence of a balance of antagonizing forces, provided that the oscillation of the system is guided by a stimulus from one as well as from the other side that may be expressed as a (differential) quotient [6]. With this concept of the "regulatory valence" we touch upon a basic biological law which is also encountered in the physiology of sensation.

Considering the difficulties which could be predicted for truly reliable experiments, the research program reduced itself to the investigation of the properties of living arteries. In fact, since that time arteries have been for me a particularly interesting experimental object. At the same time they have been a *subject*, as it were, actively engaged in the mechanisms regulat-

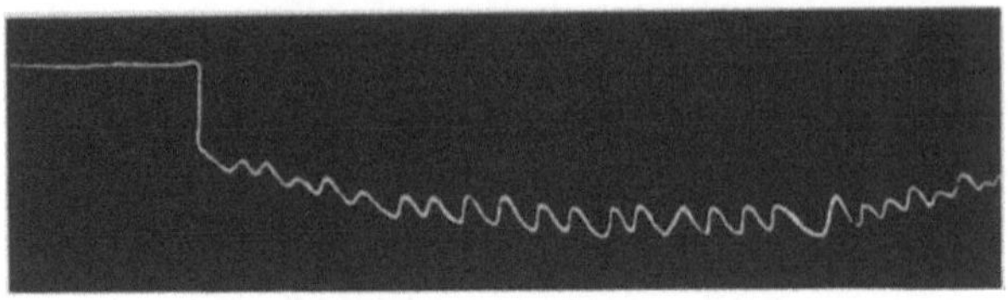

Fig. 2.5. Rhythmic variations in tone of the isolated arterial segment. The variations are induced by a slight stretch

ing circulation. Numerous investigations were devoted to this topic by my collaborators (Fig. 2.5). Besides the investigations on the isolated, freshly dissected artery, our attention was directed very soon to the entire arterial tree, that is, to the integration of individual arterial segments, with their autonomic potency, into the arterial system. From this work, fruitful knowledge resulted. Even today, in some circles one is still prepared to categorize an innervated system in terms of the neurally or humorally induced influence on the individual target organs; interest is limited to whether the effect is activating or inhibiting. In the frame of hemodynamic considerations one must clearly keep in mind the end effect that the mosaic of manifold single functions may exert on circulation. In a concrete case the activation of vasoconstrictors means an inhibition of the blood stream in the same vessel because the artery becomes narrower. Slackening of the arterial constriction permits, on the other hand, an increased blood flow in that internal pressure widens the artery. Thereby, a decreased activity (i.e., a decrease of vasoconstriction), if related to the performance of the arterial systems (that is, in regard to blood flow), appears as positive. This view became the point of departure for considerations regarding efficiency *(leistungsbezogene Betrachtungsweise)* which, in all areas of automatic regulation processes, provide the key to the understanding of functional integration. The designation "ergon" for work, success, or result of a function leads to the understanding of functional organization in terms of synergetic coordination. To this belongs the idea that, in biology, "being" is conceived as the result of an equilibrium. Thus and only thus is one in a position to relate organized systems to the results of causal analysis.

Before a consistent use of such a method of interpretation was employed, the study of the close functional connections between respiration and blood circulation – the latter as carrier of respiratory gases between lung and tissue – was the next step for me. Thus, after the *Regulation of Blood Circulation* [8], the *Regulation of Respiration* [9] was also treated, especially with respect to neural mechanisms, in monograph form. In each of these systems the autonomic potency of the peripheral apparatus forms the logical point of departure. As the next step, elementary and still peripheral mechanisms are established. The organization of complex regulatory

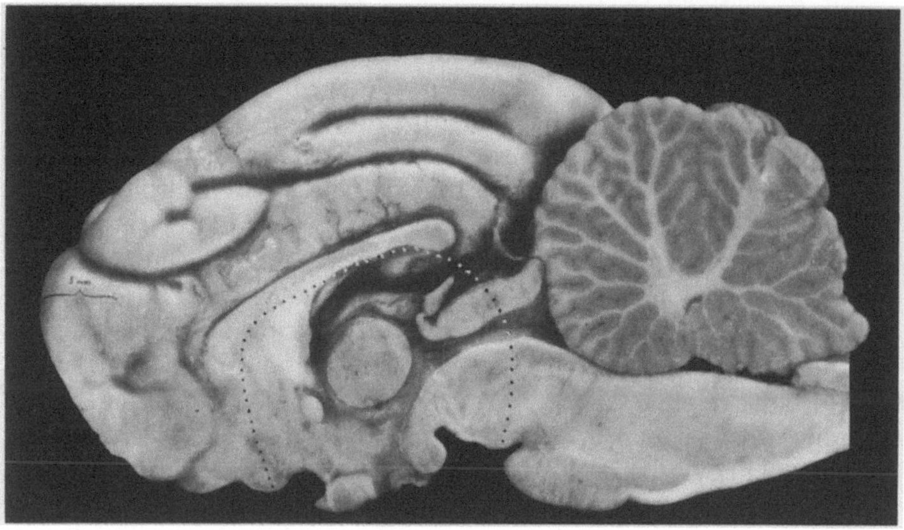

Fig. 2.6. Midsagittal view of the cat diencephalon and the bordering areas. The *dotted line* indicates the area of experimental exploration

mechanisms leads into the central nervous system with its specific arrangements constituting the physiological equivalent of a so-called center.

In regard to circulatory and respiratory regulation I initially followed the then classical concept of the leading role of the medulla oblongata, that is, that region of the central nervous system which is directly contiguous with the spinal cord. This interpretation is supported by the experience that experimental transection immediately below the medulla oblongata severely impairs the circulatory system and makes spontaneous respiration absolutely impossible, while a transection above the area remains without such consequences. Still, one was already aware through occasional investigations that lesions and stimulation in the region of the diencephalon could influence the function of these systems. Correspondingly, the extent of these descending influences had to be more closely examined with regard to the regulation of other vegetative functions.

Relevant observations were already available in our laboratory. In fact for various reasons we had already been prompted to look at the diencephalon [7, see translation, p. 71] for a long time. As Fig. 2.6 shows, because of the overlying cerebral hemispheres, this area of the central nervous system is difficult to approach experimentally. This circumstance may have been the reason that the knowledge of this region, compared with experience of the relatively easily approachable cortex, was very limited. This gap in information was most likely responsible for the fact that at about the same time – independent of our own program – analogous plans were drawn by Ranson (Chicago) to study the diencephalon experimentally.

While Ranson followed a conventional technique, it appeared urgent to me to begin the development of an appropriate method which would be optimally adapted to the situation. The development of this method took a good 2 years and was published under the title *Method of Localized Stimulation and Destruction of Subcortical Areas of the Brain* [10]. Utilizing all of the available technology, the most recent experiences regarding the importance of the wave shape of electrical impulses in regard to their differential effect on heterogeneous neural elements were also taken into consideration. The induction coil which had served physiology well for 100 years had to give way, for our purposes, to another principle of stimulation, namely, the use of rhythmically interrupted direct current with a special wave form of single impulses. Great weight was placed on the dimensions of the electrodes which, with their diameter of just 0.25 mm, could be inserted into the brain without damaging, to any disturbing extent, the fiber tracts which are of importance for the stimulation experiment.

In carrying out the experiments we were not looking for specific effects. Considering the complexity of the diencephalon and the great lack of knowledge about the biological significance of the various areas that it includes, we preferred to follow a systematic scheme by which as many points as possible could be electrically stimulated in the diencephalon. So the procedure extended point-by-point from caudal to rostral, from dorsal to ventral, and finally from the midline in the lateral direction. One of the first series of investigations yielded so many results that it was not possible to determine any relationship between stimulation sites and effects. For this reason a second, third, and finally a ninth series were added. Only then were the observations so extensive and the effects so often confirmed that justified conclusions could be drawn. Naturally time also passed – indeed, a great deal of time – so that the projected 2–3 years finally turned out to be 25 years! However, from time to time interim reports about the accumulating observations were written and other work also filled in certain periods. The interruptions were mainly caused by the fact that each brain investigated had to be histologically examined in order to determine precisely the site of stimulation. Here our modest situation, particularly the lack of personnel, was a markedly retarding factor that could only be overcome by patience which is, in any case, a prerequisite for research work. Also, the final overview was not possible without fulfilling certain additional requirements. Thus, in order to determine the precise localization of the points of stimulation and to form relationships to neighboring structures, large photographic atlases had to be prepared. These were done on the basis of preparations with fiber as well as cell staining; in addition, they were prepared in horizontal and sagittal sections. All of the results obtained had to be continuously registered and the stimulation points had to be plotted in so-called stimulation point atlases, each with the corresponding case num-

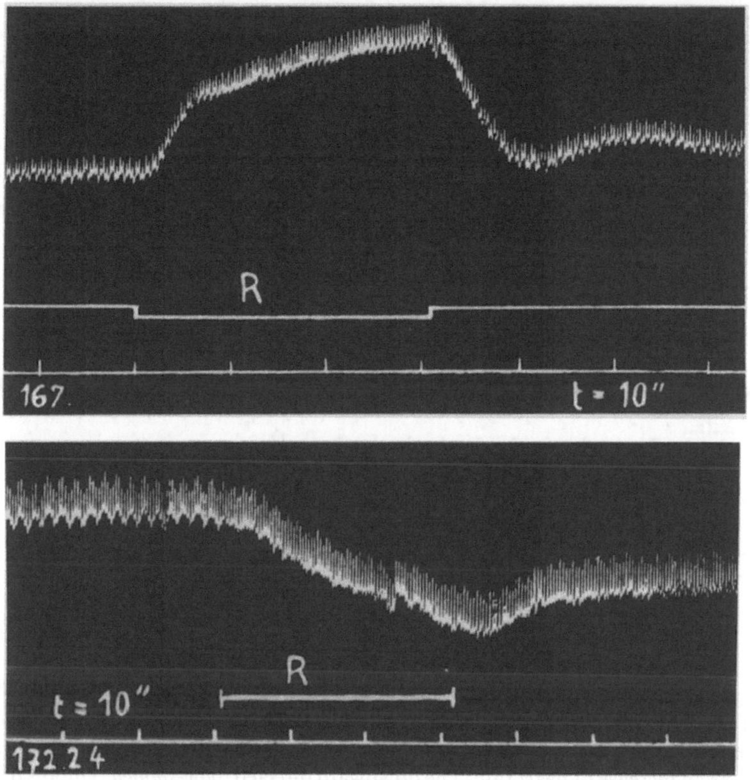

a

b

Fig. 2.7a, b. Vegetative effects. *a* Increase in blood pressure elicited by electrical stimulation of the caudal region of the ventral diencephalon (hypothalamus). *b* Decrease in blood pressure elicited by stimulation of the rostrolateral region of the ventral diencephalon (hypothalamus)

ber of the animal. Furthermore, it was necessary to demonstrate single fiber tracts by a special technique (the Marchi stain) in order to investigate the distant connections of certain particularly effective points of stimulation. Parallel to this was the determination of the physiological condition due to the destruction of certain areas which had been immediately preceded by their stimulation. The effects were documented by verbal description and by cinematography.

A high-frequency current of 40 mA and 10–12 sec duration was sufficient to make a small coagulation at a defined point. Where deficit occurred, the stimulation results complemented this in an instructive way. Finally the stage was reached where the relationship between points of stimulation and the effect, which could not be interpreted in the beginning, could clearly be correlated. The way to accomplish this was the plotting of individual symptoms in separate maps; through the choice of suitable symbols certain function-specific areas became apparent. Naturally, the representation on a

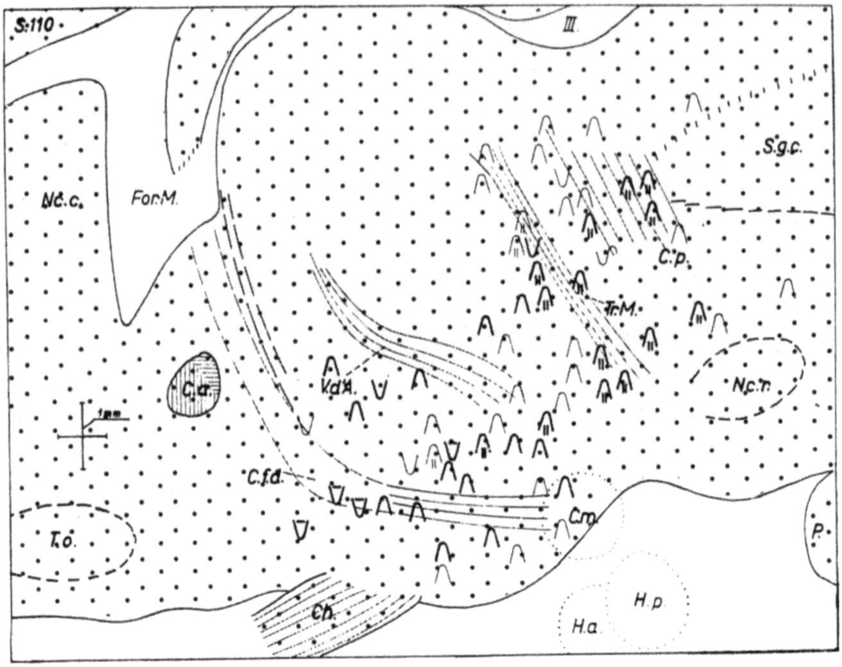

Fig. 2.8. Localization of the blood pressure effects elicited by electrical stimulation. Sagittal plane of the diencephalon projected slightly laterally from the midline. The *bow-shaped* symbols directed upward correspond to an increase in blood pressure. The insertion of *parallel lines* means an increase of heart rate. The downward directed, *bow-shaped* symbols designate blood pressure decrease. The insertion of *horizontal lines* means slowing of the heart rate. *Hatched field* in S. 130 (*Fig. 2.9*) represents zone from which adynamia was elicited. Abbreviations: *C.a.*, anterior commissure; *Ch*, optic chiasm; *C.f.d.*, fornix; *C.m.*, mammillary body; *C.p.*, posterior commissure; *For. M.*, foramen Monro; *H.a.*, adenohypophysis; *H.p.*, neurohypophysis; *N.c.*, caudate nucleus; *Nc.r.*, red nucleus; *P*, pons; *S.g.c.*, central gray substance; *T.o.*, optic tract; *Tr.M.*, habenulo-interpeduncular tract; *V.d'A.*, mammillothalamic tract; *III*, third ventricle

single plane was, by far, not sufficient. The three-dimensional plot required the utilization of an entire series of sections and, for facilitating the spatial concept, planes in different directions were also necessary. Such accounts on the individual phases in the progress of work are not trivial here, because all these efforts required, temporally and materially, very great expenditures, which were carried to the largest extent by private donors and, further, through support from a well-known insurance company, from the Rockefeller Foundation of New York, from the Foundation for Scientific Research of the University of Zurich, and from the Swiss Academy of Medical Science.

The motor symptoms we obtained have already been mentioned. The demonstration of vegetative syndromes remains to be discussed; however,

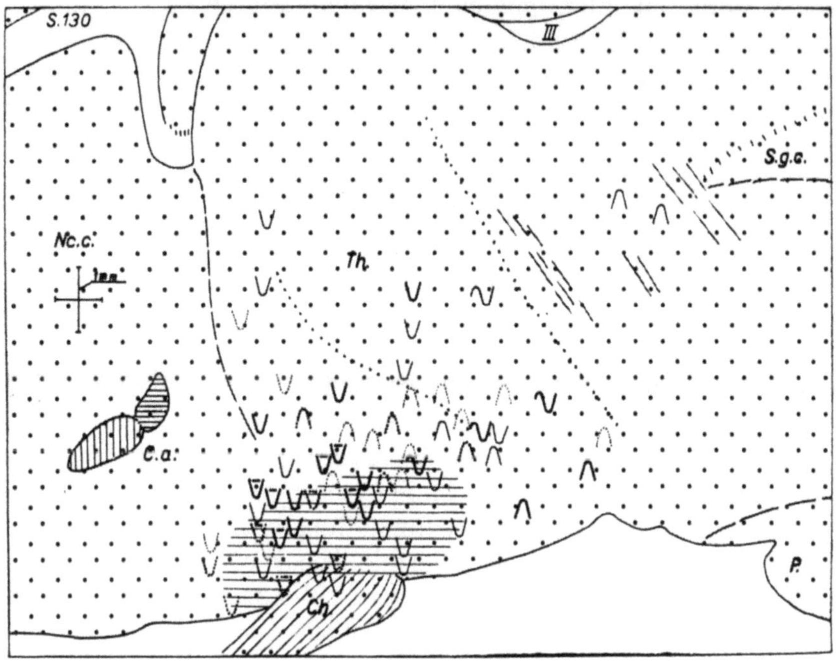

Fig. 2.9. Legend as in *Fig. 2.8.* The plane of the section lies only 1 mm more lateral. Here the decrease of blood pressure predominates, particularly in the rostral area of the hypothalamus

in the time remaining, this can only be summarized (Figs. 2.7–2.9). It was shown that the triggering of vegetative effects was limited to the most ventral parts of the diencephalon, that is, to the so-called *hypothalamus* and caudally to the central gray matter of the midbrain. From the area so described one obtains effects of stimulation analogous to those produced by stimulation of the sympathetic trunk. The decisive difference is that, in the latter case, according to the position of the stimulus, this or that effect limited to an individual organ is produced, whereas one obtains *a generalized syndrome* from the diencephalon. Obviously, at this level all the effector organs of the sympathetic system are associated, including respiration and circulation. Of particular interest is that the region which can be designated as the "ergotropic," respectively, "dynamogenic" zone is not exclusively limited to the collective control of the sympathetic nervous system. Often there was a very pronounced change in the mood of the animal so that, for example, a previously very tame cat acted as if it were placed opposite a dog, i.e., it exhibited a defensively oriented affect. There were clear criteria that the induced condition, rage, is perceived in a subjective way. The cat not only displayed piloerection, growling, and hissing, but under some conditions, it also went right into a directed attack and remained irritable for some time after, apparently becoming entirely peaceful following discontinuation of the stimulus.

From the anterior section of the hypothalamus a completely different syndrome is elicited. It concerns effects which are mediated by the parasympathetic innervation and which are related to protection, economizing of energy, and restitution. Accordingly, we speak of the *endophylactic-trophotropic* region of the diencephalon. In this area (in contrast to the ergotropic zone), the differing functions are not necessarily elicited in a global fashion. There is a clear-cut differentiation of representation areas with the pathways leading to or away from them. Similar to the effects triggered from the ergotropic zone, the influence of the diencephalon goes beyond the functional "repertoire" which can be released by peripheral stimulation. In fact, compact functional syndromes may develop from areas in which the activity of the vegetative peripheral organs are associated with somatomotor functions. Urination and defecation in the appropriate posture are impressive examples. The same is true for sniffing, licking and, finally, sleeping: From a circumscribed area of the middle part of the diencephalon the attentiveness of the animal decreases more and more, and the reactivity to external stimuli visibly diminishes; the pupils narrow, the lids become closed, the animal lies down and rolls up, developing a condition which, as far as we can see, does not differ from physiological sleep. Hence, this condition induced by stimulation of the diencephalon expresses itself in the form of an inhibition of activity. Here, too, the subordination of certain cortical functions becomes manifest; this concerns, for example, consciousness. Thus, the key position of the diencephalon in the coordination of vegetative and somatomotor functions and in the integration of sensory and associative performance is confirmed.

One result especially to be mentioned is only appreciated when one follows the earlier concepts of relating definite functions to specific nuclei of the brain (visible in the brain sections). The symptom-for-symptom, point-for-point correlation analysis of functional and morphological aspects permits clear relationships to be established only to the extent that functionally specific *areas* appear. These areas overlap, at times considerably, and for certain syndromes the attempt to provide a topographical differentiation ends in complete failure. With respect to the functional significance of nuclear complexes, one can only venture that they may represent *trophic centers* consisting of more or less wide-reaching fiber connections, of which the latter are the actual site of coordination.

At this point the monograph *Vegetative Functions and the Diencephalon* [12] was prepared. Further, in direct connection with this and with an overview of the entire organization of the vegetative nervous system, the monograph *The Functional Organization of the Vegetative Nervous System* [13] and the monograph in which finally all the data and also the methods are summarized, *The Diencephalon* [14], appeared. The results described therein were those which were recognized by the authorities of the Nobel Institute for Physiology and Medicine (Karolinska Institute, Stockholm) with the words "for the discovery of the functional organization of the diencephalon for the coordination of internal organs."

The strong emphasis which I have placed upon these works and their development in my account should not be, as far as I am concerned, a final exclamation point, for there is still much to be done along the same path. In fact, the next step is already being taken. Through numerous papers by predominantly American researchers the concept has been popularized that the vegetative effects elicited from specific sections of the forebrain may be considered an expression of still higher-ranking vegetative centers. I myself tend toward another concept according to which the cortex is not the highest level of vegetative functions. Rather, I hold it to be probable that the vegetative effects evoked from the cerebral cortex involve a response of the vegetative nervous system which is normally associated with emotions. Besides this, an uncoordinated irradiation may also take place. In any case it is necessary to seek an explanation by experimental means. For this reason the Physiological Institute of our university has been greatly involved with the frontal lobe for more than 1 year. Thus, a theme has been touched upon which, entirely unplanned, has led to a joint venture with Egas Moniz, who shares with me the Nobel Prize for his "Discovery of the Therapeutic Value of Prefrontal Lobotomy in Certain Psychoses." His work also involves, eventually, the functional significance of the frontal cortex. Only in one case is the analytic and, in the other, the therapeutic approach concerned.

Ladies and Gentlemen!

Permit me to add a final word to the conclusion of this lecture. It is directed especially to the young scientists who will be called in their time to advance a step in their field by committing their talents and utilizing their available resources. The way I have outlined seems to be so consistent that one is led to think that it was the realization of a preconceived plan. It was, however, not so. Leading me from step to step were only the priority of my interests and a genuine disposition for logical induction. Conscious will was involved only to the extent of rejecting opportunism and of not giving up the fight with a recalcitrant pen. Valuable support meant, in all phases, the considerate and ready help of good collaborators. I close with my thanks to them.

References

1. Hess, W.R.: Eine mechanisch bedingte Gesetzmäßigkeit im Bau des Blutgefäßsystems. Arch. Entwicklungsmech. Org. *16*, 632–641 (1903)
2. Hess, W.R.: Zum Thema: Viskosität des Blutes und Herzarbeit. M.D. Dissertation. Vierteljahrsschr. Naturforsch. Ges. Zürich *51*, 236–251 (1906)
3. Hess, W.R.: Die Bestimmung der Viskosität des Blutes. Münch. Med. Wochenschr. *54*, 2225–2229 (1907)
4. Hess, W.R.: Eine neue Untersuchungsmethode bei Doppelbildern. Arch. Augenheilkd. *62*, 233–238 (1908)
5. Hess, W.R.: Das Prinzip des kleinsten Kraftverbrauches im Dienste hämodynamischer Forschung. Habilitationsschrift. (Docent thesis). Leipzig: Veit 1913
6. Hess, W.R.: Über die periphere Regulierung der Blutzirkulation. Pflügers Arch. Ges. Physiol. Menschen Tiere *168*, 439–490 (1917)
7. Hess, W.R.: Über die Wechselbeziehungen zwischen psychischen und vegetativen Funktionen. Schweiz. Arch. Neurol. Psychiatr. (Neurol. Psychiatr. Abh.) *15*, 260–277 (1924); *16*, 36–55, 285–306 (1925)
8. Hess, W.R.: Die Regulierung des Blutkreislaufes. Leipzig: Thieme 1930
9. Hess, W.R.: Die Regulierung der Atmung. Leipzig: Thieme 1931
10. Hess, W.R.: Die Methodik der lokalisierten Reizung und Ausschaltung subkortikaler Hirnabschnitte. Leipzig: Thieme 1932
11. Hess, W.R.: Die Motorik als Organisationsproblem. Biol. Zentralbl. *61*, 545–572 (1941)
12. Hess, W.R.: Vegetative Funktionen und Zwischenhirn. Helv. Physiol. Acta Suppl. *IV*, 1–65 (1947)
13. Hess, W.R.: Die funktionelle Organisation des vegetativen Nervensystems. Basel: Schwabe 1948
14. Hess, W.R.: Das Zwischenhirn. Syndrome, Lokalisationen, Funktionen. Basel: Schwabe 1949

Section II

Hemodynamics, Physiological Optics, Neurohumoral Transmission

Chapter 3

Blood Circulation and Finality*

Academic Inaugural Lecture held on 15 June 1918

W.R. HESS

In 1866, Meyer [11], the then Professor of Anatomy of our university, presented to the *Naturforschende Gesellschaft Zürich* (the Society of Natural Sciences) his results of investigations which made him widely known on the inner architecture of the bones. He had made the observation that a very pronounced regularity and order is present in the arrangement of the osseous lammellae and trabeculae and suggested that this phenomenon should be brought into relation with the mechanical forces acting within the bone and between bones. Culmann, the renowned expert in *graphische Statik* (graphical statics) of the Federal Institute of Technology, gave this correlation a more exact form by proving that the orientation of the osseous lamellae followed lines that static analysis identified as curves of *maximum pressure* and *maximum tension*; in other words, lines that indicate the directions in which the resistances to pressure and tension are bound to be most heavily concentrated. Thus, the arrangement of bony lamellae respresents a materialization, or incarnation as it were, of the pressure-tension lines and is therefore adapted especially to resist the mechanical stresses that arise from the load on the skeleton.

The relationships become especially apparent where the stress on the skeleton is one-sided, i.e., where it is exposed to a stress of a specific kind in its purest possible form. This occurs for example, at the femur, whose architecture has become the most readily known object for the elucidation of Meyer and Culmann's discovery. The correspondence between the course of the trabeculae and the lines which have been reconstructed on the basis of pressure and tension measurements in Culmann's laboratory, using an artificial model with the form and mechanical load of the femur, is very striking indeed. As in the thigh, Meyer could show an architecture obeying the laws of statics for an additional number of parts of the skeleton. Since then, Fuchs [4] has demonstrated analogous correlations with respect to the structure of the roof of the skull.

* Hess, W.R.: Die Zweckmäßigkeit im Blutkreislauf. Akademische Antrittsvorlesung, gehalten am 15. Juni 1918. Basel: Schwabe 1918

The discoverers knew clearly the significance of the above-mentioned aspect of the structure of the bone: It allows the bony resistance to be fully effective in the most accomplished fashion. As a consequence, the skeleton is built to bear a certain load with a minimal amount of bony substance. From this point of view we designate the inner architecture of the bone as "purposive."

When I use this expression I am conscious that the danger of eliciting controversy is connected with it. In many places we encounter a pronounced shyness in considering biological arrangements from the standpoint of their "purpose." There is a tendency to deny expressly the term "purpose" or "finality," and to reject the thought that in the shaping of the organic world the attainment of a goal or, as may be, of a purpose, plays a role in any way! To connect such a teleological meaning with our use of the concept readily suggests itself; however, it would be a misunderstanding! Jensen [10] in his essay *Biological Purpose, Evolution and Heredity from the Point of View of Physiology* has emphasized that the investigations pertaining to purposiveness frequently suffer from a lack of an adequate differentiation between more or less related concepts and through this give occasion to confusion.

It seems advisable therefore, that we define the sense in which the expression should be used here. According to Jensen, "purposive" is defined in common language in two ways: (1) that which is the effect or result of a directed action (one thinks of the *realization* of a plan), and (2) that which serves to maintain and further the organism. Here one takes its *properties* and functions into account.

For our considerations the second interpretation is pertinent. We shall hold to the conceptual analysis which Jensen summarizes in the statement: "In the organism the purposive can be characterized purely objectively, without consideration of an effective principle of purpose that is perhaps extant at the origin of the organism." This objective criterion of organs and biological processes concerns the maintenance and furtherance of the organism. In other words, it increases the probability of survival.

We could have circumvented this definition of the concept "purposive" had we used another expression in its place. Roux [14] suggests the designation *dauerfördernd* (prolonging existence). Ribbert, [12] in his treatise *Zweckmäßigkeit in der Pathologie* (Purposiveness in Pathology), prefers the term *dauermäßig* (apt for endurance). I adhere to the expression "purposive" because it is most widely used, and the danger of being misunderstood can exist only so long as one neglects to give a precise definition. With this we return to the object of our introductory discussion and what has been said about the architecture of the bone. This example of purposive organization is not unique. On the contrary, we know a large number of other typical cases – I recall the dolphin fin which was examined by Roux, the founder of developmental mechanics, with regard to the relation between function and structure. His examination determined that this organ has an outer form and an inner structure which allow adjustment to the normal mechanical stress with *a minimum of material and force*.

Another example is the body of the bird, from whose organization one can recognize, stage for stage, the principle of economy. The same is to be said of the typical form of the fish's body which distinguishes itself by its unsurpassed fitness for cutting through the water. We find the same design in the domain of shipbuilding, where the ship's hull is given the form that is most advantageous for its passage.

Still more cases of purposive organization could be cited if we only follow Spengel [15] on the subject of finality. Yet our interest today goes in a direction other than toward further naming of already known examples. Our intent, rather, is to seek that which is not yet recognized as such.

Among the different functions of the organism we find that those operating in the interior of the body do not play a less significant role than those which are effective toward the exterior. We need to mention only the blood circulation, and here we raise the question as to whether this system, too, operates with *the smallest possible expenditure of energy*. This is the problem which we should like to approach today!

To this question we can give at first only a very fragmentary answer. Fuchs [4] recognized that the histological structure of the vascular wall corresponds in its detailed arrangement to the functional demand. Roux [13] examined the ramifications of the arteries and found that the configuration of the branching vessel follows, in its finest detail, the currents of a fluid. Through the close fit of the wall to the contours of the stream the danger of turbulence is precluded and with this the squandering of energy.

Besides these findings, additional properties, real, and only believed purposiveness in the structure of the circulatory system have become known; all, however, represent very special cases and are of secondary significance (e.g., [1]. Uncertainty prevailed until recently with respect to the *principal factors* upon which the magnitude of the circulatory work of the heart depended. In an attempt to gain some information we had to realize that the situation is quite complicated and by far not as simple as, for example, in the skeleton. There the relation of structure to function is obvious for the observing eye. An accidental view of the bone tissue offered to an expert in static analysis was enough to give a full understanding of it. With circulation, conversely, there is a need for more or less cumbersome research and calculations *to establish the criteria of "purposive."* To know them will now be the next goal in our discussion.

Doubts will perhaps exist here and there as to whether the interest in the question at all justifies an investigation – especially its treatment before this audience. It could hardly be justified were it not for the deepening of our understanding of the mechanisms of circulation. By choosing the way of the *"purposive"* we can recall the example of tried investigators. Mach, for example, points out that the concept of purposiveness has led to the most important physiological discoveries. Through it Kepler was led to the discovery of accommodation of the eye. The question of the purpose of the valves of the heart and of the veins led Harvey to the recognition of the essence of blood circulation.

It is the striving to widen our knowledge and understanding that motivates us to seek the relationship between function and purpose of circulation. What are the distinctive criteria of the purposeful? The function of circulation is the mediation of exchange of materials between the tissues and the external world, on the one hand, and the various tissues among themselves, on the other. This function is bound to a continuing displacement of the blood, the vehicle for the materials that are to be transported. The motion of the blood requires continuous work which must be produced by the heart and which, according to its rate, burdens the energy metabolism more or less. This energy consumption is determined by (1) the quantity of blood which the heart must expel in a certain time, for example in 1 min, to satisfy the tissues' need for blood, and (2) the pressure necessary to keep this quantity in circulation against the *resistance* of the vascular system. We know these resistances to be dependent essentially upon the length of the stretches over which the blood must flow, upon the cross sections and ramifications in the vascular tree and, finally, upon the viscosity of the blood.

With this the significance of the form of the vascular tree for the cardiac work is obvious. Still unclear, however, is how the formation must be in order to keep the resistance at a level which sets the demand for the pressure exerted by the heart to the smallest possible amount.

There seems to be little doubt that the minimum amount of energy is utilized when the distance over which the blood must flow is kept as short as possible by avoiding all possible detours in conducting the blood to the organs that are to be supplied.

A glance at Fig. 3.1 will give us further clarification. It represents a schema of the course of two vessels directed to two separate territories of tissue (represented by intestinal segments). Let C be a principal arterial trunk. The demand for the shortest possible path would be satisfied by a straight vascular course as indicated by the solid lines. In any case, the way over which the blood must travel is momentously shorter than with the course of the vessels as marked by the dashed lines. *Yet, the shorter conduit need not be unconditionally the one that represents the most favorable conditions of resistance*, for the following reason: The second of the given possibilities allows the blood first to travel over a common stretch of way, which is in reality a detour for each area that is to be supplied.

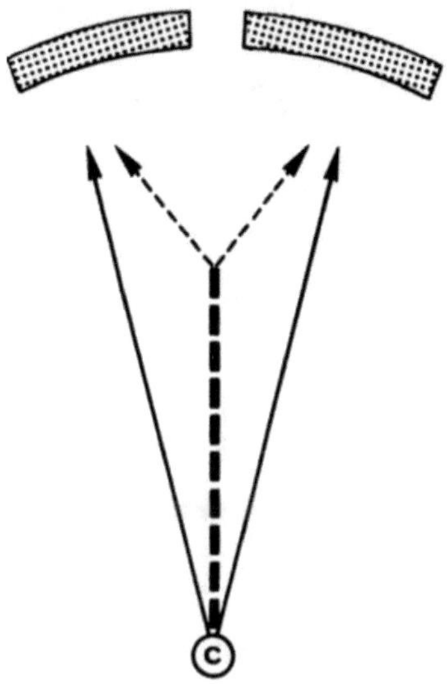

Fig. 3.1. Branching of an artery. For explanation, see text

However, the common stream bed offers essential advantages from the point of view of energy, because *the surface of contact between the streaming fluid and the rigid wall is diminished.* In fact, the flow through one pipe requires only half the energy needed to drive the same volume per minute through two separate pipes when the total cross section is the same in both cases (see Addendum 2).

Of the two possibilities indicated in Fig. 3.1 the advantages of the shortness of the total way, combined with the disadvantages of an early separation of the two streams, are contrasted with the advantages of the common conduit combined with the disadvantages of a detour. In which of the two situations can one find the more compelling advantage? In other words, which solution is the more favorable, *the more purposeful* from the point of view of economy?

The answer is given on the basis of a quantitative estimation of the determining factors. To be included in the calculation are conditions which prevail when unequally large areas are to be supplied by dividing a vascular trunk into two unequally large branches.

The calculation (see Addendum 1) can be explained as follows:

The most favorable site and angle of branching (that is, the angle between the direction of the trunk and that of the branch) are dependent upon the ratio of the size of the branch to that of the trunk. The splitting off of a separate path, in the sense of its shortening, is less energy saving with a decreasing quantity of blood to be conducted in the branch in proportion

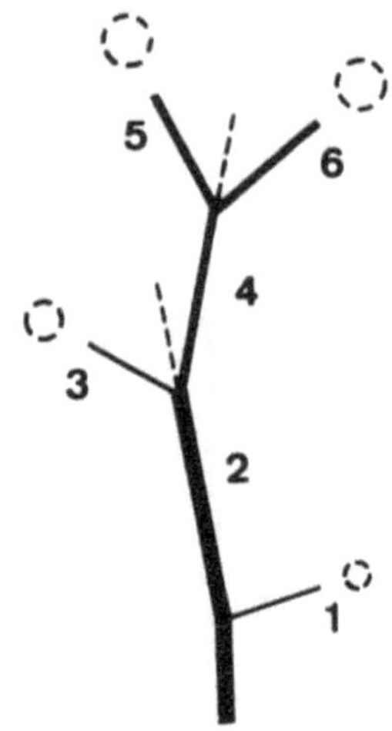

Fig. 3.2. Angle of branching and vascular diameter. For explanation, see text

to the total amount in the trunk. For thin branches, the most advantageous point of branching is that which exhibits the longest possible common path; such is attained with a branching of the vessel at the nearly right angle (branch 1 in Fig. 3.2).

The larger the blood flow, the less advantage of the common path has import and the more readily the significance of the path comes to the fore. Early branching, i.e., distant from the organs to be supplied, is realized with more acute angles of branching.

They are favorable for the larger branches, especially those that are nearly as large as the trunk (see branches 2 and 4 in Fig. 3.2).

With the asymmetrical branching, the larger of the two branches emanates at the more acute angle (see branches 3 and 4 in Fig. 3.2). With the partition of a trunk into two equally large branches, the branching at both sides is at the same angle. What has been explained now in words is found precisely formulated in the result of the calculation. It shows that *the ratio of energy losses between the trunk and the branch is identical with the cosine of the angle of branching* (see Addendum 1). It is this relationship from which the preceding statements presented are derived.

And now the question: How do the branchings of the vessels in the natural vascular tree present themselves with respect to this cosine law? An examination of this problem is made easy for us. Twenty-five years before its theoretical establishment, intensive empirical investigations had been made by Roux [14] on the branching of the human vascular tree. It is true that he directed his attention to the angle of branching more distal to the supply area. Nevertheless, his findings give a basis for examining our question.

These results were summarized in a series of rules, of which those that interest us here are as follows:

1. "If a trunk divides itself into two branches of equal diameter, both stand at the same angle to the direction of the trunk."
2. "The deviation of the arterial trunk (i.e., the principal branch) at the branching site is always smaller than that of its branch (the side branch)."
3. "Those branches of the aorta, of the brachial artery, of the femoral artery, and of the coronary artery, that are so small that the trunk (the principal branch) does not show any deviation at the branching site, emanate mostly at large angles of more than 70 °." (Thus with decreasing relative size of the branch an approximation of an angle of branching to the right angle is observed.)
4. "Those branches that are so large that the trunk, the principal branch, deviates considerably at their dividing point, emanate mostly at angles of less than 70 °." (Thus with an increasing relative size of the branch a decrease of the branching angle is seen, and with great regularity those angles of branching seem to prevail which we have learned to characterize as most favorable. As in the bone with respect to the spongiosa structure, we find in the vascular tree with respect to the branching angle *an obvious tendency toward the optimal configuration.*

As in the structure of the bone, deviations from the strictly formulated law may be expected, since a series of additional factors has an effect on the direction of the course of the blood vessels; such deviations from the rule are most likely due to latent alterations in the vascular tree which are caused by growth.

Among the factors upon which the resistances of a hemodynamic system chiefly depend we have mentioned the cross sections of the blood vessels. This raises the question whether characteristics of purposiveness may be detected in this regard. Here, we have to proceed from the fact that the *pressure gradient* in a tubular system with a given flow per minute *decreases with increasingly large cross section.* The circulation of blood requires a correspondingly smaller pressure to be produced by the heart, the more dilated the vascular network. The largest possible cross section of the vascular tree minimizes resistances.

Something special is present in the living organism to the extent that for different reasons *certain limits are set for the widening of the cross-sectional area* of the vascular tree. The following consideration should illuminate this point: The enlargement of the vascular tree is bound to the condition that the blood volume is adequate to fill the space contained in the whole system. Enlargement of the vascular system is, for example, conceivable only if an increase of the absolute blood volume keeps pace with it. A doubling

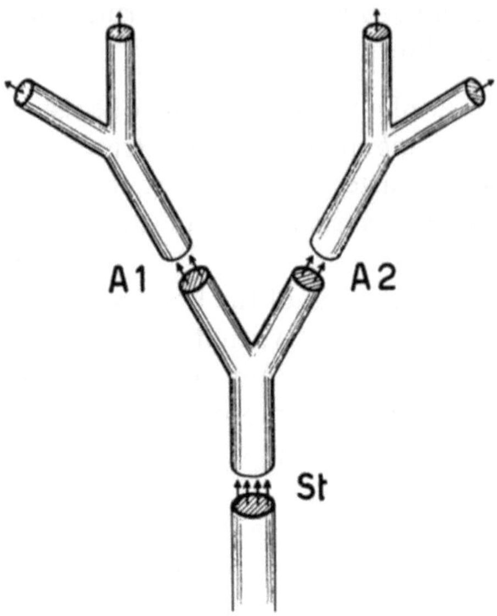

Fig. 3.3. Geometry of the vascular tree. For explanation, see text

of the total cross-sectional area would be possible only with an accompanying doubling of the blood volume.

That such an increment of mass, viewed apart from the resistance factor, has a further consequence is clear after what has been said in our previous discussion on the bone structure. It contradicts the demand for the minimization of substance. This postulate pertains even more to circulation than to the bone tissue. Every increment not only interferes indirectly with the economy of material and energy of the organism, e.g. by burdening the locomotor apparatus, but it also imposes particular stresses on the circulation system, all the more since blood itself is a tissue with a considerable, genuine metabolism; it consumes nutrient substances and oxygen as every other living tissue does. Blood corpuscles continually perish through wear and must be replaced. The more blood there is in the body, the larger the demands of its own nutrition and of replacement. Herein lies a factor (though not the only one) which imposes limitations upon the dimensions of the vascular tree.

From this we need only to draw the right conclusions to recognize clearly a criterion of purposiveness which relates to the cross section of the blood vessels: The evaluation of the blood vessels from the position of their functional significance allows us to differentiate *vascular segments of different rank orders*, namely trunks, branches, twigs, and so forth. These relations are schematically represented in Fig. 3.3.

Theoretically, such a system could be designed to partition the limited content uniformly among the vessels of different rank order, that is, to set the diameter in proportion to their minute volume. Another solution is that with the general rationing of cross section – let this way of expression be permitted today – the vessels of different rank order would be considered differently. I think, for example, of a relative preference of the lower ranks. This would be the case if with every ramification the sum of the cross sections of the branches exceeded the cross section of the trunk from which they emanate so that *a widening of the path is obtained.* The degree of widening is measured by the quotient formed by the sum of the cross sections of the branches as numerator and the cross section of the trunk as denominator. This quotient is designated as the cross-sectional coefficient. It is essential to know that with determined branching angles and established volume of a conduit system, its resistance is still a function of the *cross-sectional* coefficient provided that a systematic relation exists between diameters of trunk and branches. The details may be ascertained by calculation (Addendum 1). The outcome is demonstrated in the curve of Fig. 3.4, in which the abscissae represent the cross-sectional coefficient and the calculated resistances are entered as ordinates.

If the sum of the cross sections of the branches just corresponds to the cross section of the trunk, if then q, the cross-sectional quotient $= I$, a total resistance as measured in the ordinate R_1 results. An increase of the cross section of the trunk on the account of that of the branch, i.e., a decrease of the cross-sectional quotient below the value I, has as its consequence, as the curve shows, an important increase of resistance (the onset of the curve is rising toward the left). Conversely, an increase of the quotient brings an improvement of the conditions of resistance, with the curve sinking toward the right. It is true only as long as a certain value, M, at the ordinate is thereby not exceeded. Otherwise, an increase of resistance is inevitable, since the curve rises again in the direction of ordinate R_2. We have a special interest in the point of reversal at M, the minimum of the curve. For a symmetrical branching it lies at the cross-sectional quotient of 1.26. Under these conditions, then, the vascular tree presents the lowest resistance to the blood stream that is consistent with the total available volume.

We have found a new characteristic sign of purposiveness. Do we meet the same in the natural vascular system? This question is not easily answered. Accurate measurements of the cross sections of vessels in the living encounter the greatest difficulties, because various kinds of stimuli can effect an alteration of the diameter. Direct measurements of the vascular diameters under fully normal conditions can be made by photographic records of the ocular fundus, for example, according to Dimmer. Yet, the small dimension of the objects increases the errors beyond the range that is desirable

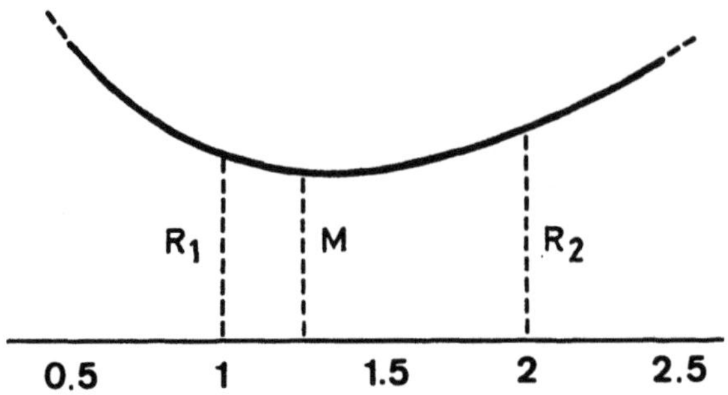

Fig. 3.4. Vascular resistance and cross-sectional coefficient. For explanation, see text

for the exactitude of the measurements. Of equally limited value are the measurements of dye-injected vessels of any organs; since the pressure relationship in the living organism cannot be reproduced with the injection, we have to reckon with abnormal distension.

A new method is based upon the determination of the *mass of the vascular wall.* Under the assumption that this mass is proportional to the mean load, the cross-sectional coefficient can be extrapolated. From Blum's [2] recently completed investigations on a series of mesenteric vessels of the horse, values may be calculated that *approximate* the optimal cross-sectional coefficient, namely between 1.23 and 1.4 (see Addendum 3). A glance at the curve in Fig. 3.4 shows that the corresponding ordinates, as measures for the resistances, differ only slightly from *M*, the minimum of resistance. We must accept these figures with some reservations because they are obtained indirectly, but they do allow the conclusion that *important deviations* from the optimal cross-sectional coefficient are certainly *alien* to the natural vascular system.

Among the factors that are of decisive significance for the magnitude of the resistance, besides those that are related to the vascular parameters, we have named one concerning the physical constitution of the blood. Specifically, I mean the property that is measured in the coefficient of internal friction, in the *viscosity of the blood.* There is a proportional relationship between viscosity and pressure gradient within certain limits, other conditions remaining the same.

This relationship implies that the physical property of the blood has an effect on the magnitude of cardiac work. This influence, however, is not expressed in terms of a simple proportionality, for example, in the sense that a low viscosity is equivalent to a low level of cardiac work! On the contrary, we shall learn that under certain circumstances a significant *increase of cardiac load* can be connected with the lowering of blood viscosity! This seeming contradiction resolves itself when we again view the finality of circulation: the supply of the tissues with nutrients and oxygen and the withdrawal of cellular metabolites. This aim is fulfilled only if bidirectional transport occurs in adequate measure. This depends for the one part upon the blood flow that irrigates a certain territory of tissue. For the other part, the functional capacity of the blood, i.e., its role as a vehicle for the substances that are exchanged, is decisive. Its adequacy may vary according to its composition; we may think, for example, of the dependence of the oxygen capacity upon the absolute hemoglobin content.

The central point of the problem concerning the relation between cardiac output and blood viscosity lies in the fact that principally *the same substances that are essential constituents of the blood also determine the level of blood viscosity.*

This fact implies that every change in blood composition influences cardiac output in a dual manner: Decrease of the hematocrit is connected with a decrease of the coefficient of friction. Cardiac work, if calculated to the *unit volume of blood,* becomes smaller thereby, but due to its limited oxygen capacity, a relatively large cardiac output is needed for the transport of a *useful load.* The demand placed upon the heart for an increased flow can make the advantage that lies in the lower viscosity fully or partly illusory. The increased concentration of blood constituents and the more effective transport function are connected to a relatively *small minute volume;* diminution of the transported volume means limitation of cardiac work. The saving finds its counterbalance, however, in that the blood with the larger transport capacity, the "superior blood," is also more viscous and its minute flow requires an increased pressure in order to overcome the heightened resistance.

From the point of view of energetics we would like to know which of these reciprocally acting factors is the more effective in a given case. This can be determined if the relationship between viscosity and specific transport capacity of the blood is known. Such data are available from the investigations of Blunschy [3] and Ulmer [8, 16], and from our own findings [6]. The following facts can be extracted (see Addendum 3):

The amount of work necessary to transport a certain amount of useful load through resistive channels varies with the concentration (viscosity) of the blood. Of all possible concentrations one distinguishes itself by accomplishing the transport with the smallest possible work. Again, there is an optimum! Optimation of blood viscosity is therefore set for *the minimal work that may be imposed upon the heart in order to fulfill the purpose of circulation within a given configuration of the vascular tree.*

According to the present experimental data the rather wide optimum lies in the region of viscosity from 3.5 to 5.0, a striking correspondence with the values that we know as normal [7] and which vary between 3.0 and 5.5, depending on age, sex, etc.

With this a third criterion of purposiveness in the blood circulation is revealed

We recapitulate: The angle of branching of the vessels, the cross-sectional relationships between trunk and branches, and the coefficient of friction of the blood are the principle factors that determine circulatory resistance. Our analysis shows that, in relation to every single factor, conditions prevail that allow the work to approach an absolute minimum. With no other organization of the circulation of the blood could the same efficiency with respect to nutrition and ventilation of the tissues be accomplished. Herein lies *the purposiveness of blood circulation*. To have recognized it in the individual characteristics is of interest in two respects: In addition to those that were already known we have recognized new examples that show us how the striving nature fulfills certain tasks with the smallest possible means – here especially from the aspect of energetics. The success of these strivings is in no way to be considered as small, since the daily work of the heart under the described optimal conditions amounts to 18,000 mkg, thus equivalent to a transport of 10 kg from sea level to the Rigi mountain. In this order of magnitude must lie the quantities of energy of which the organism is defrauded when it deviates from the criteria of purposiveness with respect to structure and function of the vascular system or in the composition of the blood. It is obvious that such waste of energy must significantly lower the probability of survival!

It is not, however, the attainment of this insight that has motivated our analysis of the geometry of the vascular tree and the physical properties of the blood. What is of primary concern to us is to have shown that the consideration of finality can lead to new insights in the domain of circulation.

Addenda

The following are provided for the clarification of the calculations referred to in the preceding text and from which the distinguishing criteria of "purposiveness" have been derived.

Addendum 1: Calculation of the Optimal Angle of Branching [5]

In Fig. 3.5 below let

$$CD = DE = EF = l,$$

further

$$D_1P = DP;\ E_1P = EP;\ F_1P = FP.$$

In addition let the specific loss of energy that the blood suffers in traveling a certain length in the trunk vessels be G, and in the branch vessels, G'.

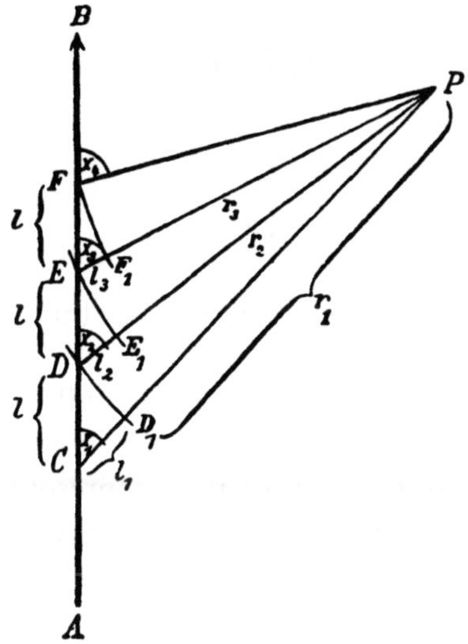

Fig. 3.5.

We shall show that every angle smaller than x_2 or larger than x_3 is more unfavorable than x_2 or x_3, when these are chosen so that the same loss of energy is suffered by both.

Let then:

$$l_2G' + r_2G' = lG + (l_3 + r_3)G' .$$

Since

$$r_2 = l_3 + r_3 ,$$

it follows that $lG = l_2G'$.

If we compare x_1 with x_2, that is, segment CD with segment CD_1,

$$l_1 > l_2 ,$$

thus

$$l_1V' > l_2V'$$

from which follows

$$l_1G' > lG ,$$

that is, CD_1 is more unfavorable than CD, and thus the angle x_1 is more unfavorable than angle x_2.

We then compare x_4 with x_3, that is, segment EF with segment EF_1:

$$l_3 < l_2 ,$$

and thus

$$l_3G' < l_2G',$$

from which follows

$$l_3G' < lG,$$

that is, EF_1 is more favorable than EF, so that the angle x_3 is more favorable than the angle x_4.

If we choose to designate l as infinitesimally small, x_2 and x_3 then differ from one another only infinitesimally little; they can then be set equal to one another and therefore may represent the most purposeful angle. If, indeed, l is infinitesimally small, we have in DE_1E a right-angled triangle and $l_2 = l\cos x_2$. Since according to the construction

$$lG = l_2G',$$

$$lG = lG' \cos x_2,$$

from which follows

$$\cos x_2 = \frac{G}{G'}.$$

The result is that the most favorable branching angle is that whose cosine is equal to the ratio of the loss of energy that the blood suffers in the trunk vessels to the loss of the energy that it suffers in an equally long segment of the branch.

Addendum 2: Calculation of the Optimal Cross-sectional Quotient [9]

a) *Relation Between Cross Section and Resistance of an Undivided Fluid Path Which is Represented by a Single Tube*

Let the resistance of a fluid path be defined by the equation

$$V_s = \frac{P}{R}, \tag{1}$$

where V_s is the volume per second, P is the pressure gradient, and R is the resistance.

The factors that determine a volume per flow per second are fixed in the Poiseuille equation:

$$V_s = \frac{\pi}{8\eta l} Pq^2, \tag{2}$$

wherein l is the length of the fluid path; η, the viscosity of the fluid; and q, the circular cross section of the tube forming the fluid path.

From formulas 1 and 2 it follows that

$$\frac{P}{R}=\frac{\pi}{8\eta l}Pq^2\,, \qquad R=\frac{8\eta l}{\pi}\cdot\frac{1}{q^2}\,.$$

For the relations for which we interpret the Poiseuille formula, the only variable is q; the length of the test tube as well as the viscosity of the fluid are constant.

These two factors we include with the other constants in the factor K. For the quotient between cross section and resistance it follows (for a simple type with circular cross section):

$$R=\frac{K}{q^2}\,. \tag{3}$$

b) *Relation Between Cross Section and Resistance in Test Tubes with Identical Dimensions and Arranged in Parallel*

Under the premise that the pressure at the beginning and at the end of the two tubes is identical, the resistance represented by the two tubes may be calculated from the single resistances according to the formula:

$$\frac{1}{R}=\frac{1}{R_1}+\frac{1}{R_2}\,, \tag{4}$$

where R_1 is the resistance of one tube, R_2 the resistance of the second tube, which is set in parallel with it. R is the resistance of the two tubes that are united into a common tube.

According to Eq. (3)

$$R_1=\frac{K}{q_1^2}\,, \text{ from which there follows } \frac{1}{R_1}=\frac{q_1^2}{K}\,,$$

$$R_2=\frac{K}{q_2^2}\,, \text{ from which there follows } \frac{1}{R_2}=\frac{q_2^2}{K}\,.$$

Equation (4) can now be substituted as follows:

$$\frac{1}{R}=\frac{q_1^2+q_2^2}{K}\,,$$

or

$$R=\frac{K}{q_1^2+q_2^2}. \tag{5}$$

But we set the precondition that the two tubes have the same cross section,

$$q_1=q_2, \quad \text{so with this,} \quad R=\frac{K}{2q_1^2}.$$

Numerator and denominator multiplied by 2 yield $R=\frac{2K}{(2q_1)^2}$.

Because $q_1=q_2$, it follows that $2q_1$ is the sum of the cross section of both tubes set in parallel; we define this sum as $q_n=2q_1=q_1+q_2$.

The relation between resistance and the sum of the cross section of two tubes of equal width set in parallel is given in the following equation:

$$R=\frac{2K}{q_n^2}. \tag{6}$$

c) *Equality of the Differential Quotient of the Resistance with Respect to the Cross Section in All Serially Arranged Vessels is the Criterion of a Test Tube with the Smallest Possible Resistance*

Proof: A test tube of given length but arbitrary configuration of cross section is to undergo a change of cross section, without changing length and mean cross-sectional area, i.e., with constant volume such that the total resistance is diminished.

The attainment of this goal presupposes that there exists somewhere in the system a section whose dilatation reduces the resistance more strongly than an equally large narrowing of any other equally long section would increase it. Such a section fails to exist when the differential quotient of the resistance with respect to the cross section in all successive sections of the test tube is the same. Under these conditions there exists no possibility with the given test tube of producing another cross-sectional configuration characterized by a smaller resistance. The minimum amount of resistance is already attained.

d) *Determination of the Relation of the Cross Section of the Trunk to the Sum of the Cross Section of the Branches with Equality of the Differential Quotient in the Zone of the Trunk and in the Zone of the Branches*

Equality of the differential quotients obtains when

$$\frac{d}{dq_a}\left(\frac{K}{q_a^2}\right)=\frac{d}{dq_n}\left(\frac{2K}{q_n^2}\right) \quad \text{[from Eqs. (3) and (6)]},$$

$$\frac{-2K}{q_a^3}=\frac{-2\cdot 2K}{q_n^3}, \quad q_a^3=\frac{q_n^3}{2}, \quad q_n=\sqrt[3]{2q^a}.$$

Result: For a composite test tube with q_a as cross section of the trunk and q_n as the sum of the cross section of two equal branches, equality of the differential quotients obtains when the sum of the cross sections is 1.26 ($=\sqrt[3]{2}$) times the cross section of the trunk.

Addendum 2 a

Indirect determination of the cross-sectional quotient by Blum [2] is based upon the premise that in immediately neighboring sections of vessels, trunk, and branch, the quantity of material invested in the wall is fitted to the average tension that is to be borne; in other words, correspondence with respect to specific load of the wall material is postulated.

If this is the case, there exists in vessels of different diameter, if related to equal internal pressure, direct proportionality between the mass of the wall and the cross section. The mass of the wall is determined by direct weighing of excised pieces of vessels, and the pressure gradient for the purpose of correcting the internal pressure is determined by measuring the reduction of the mass of the wall from one section to the next.

Addendum 3: Calculation of the Work Done to Transport the Mass of Dissolved or Suspended Substance per Unit Time Though a Resistive System

The work expended by the flow A, is given from the pressure gradient, p, and the volume of flow, V, as

$$A = pV. \tag{1}$$

In relation to the unit of time (for example, 1 min) whereby

$$\frac{A}{t} = A_m \quad \text{and} \tag{2}$$

$$\frac{V}{t} = V_m, \quad \text{it follows that} \tag{3}$$

$$A_m = pV_m. \tag{4}$$

Calculated for a transportational effect per minute, of a unit, N_m, of the useful load,

$$\frac{A_m}{N_m} = \frac{pV_m}{N_m} \tag{5}$$

The pressure gradient p is a function of the time in which the volume V is driven through the resistive system. For laminar flow the equation, derived from the formula of Poiseuille, is valid:

$$V = \frac{Kpt}{\eta}. \tag{6}$$

Whereby η is the viscosity and K, a constant which is dependent upon the configuration of the test tube.

From this and Eq. (3) there follows

$$V_m = \frac{Kp}{\eta}; \quad \text{thus} \quad p = \frac{V_m \eta}{K} \tag{7}$$

and from this and from Eq. (5) there follows

$$\frac{A_m}{N_m} = \frac{V_m^2 \eta}{N_m K}.$$

If C designates the concentration gr/cm^3 of dissolved or, as there may be, suspended substance in the fluid, then the useful load that is contained in the volume V is $N = VC$ and the transportational effect per minute, $N_m = V_m C$.

From this there follows

$$V_m = \frac{N_m}{C}. \tag{8}$$

It is our goal to bring all the work that is consumed in the flow into relation with the effect, per minute, of transported useful load; we seek then the amount of work $[A_m]$, which under given circumstances transports the unit of mass of the dissolved or, as may be, suspended substance in the unit of time through the resistive system.

$$[A_m] = \frac{A_m}{N_m}. \tag{9}$$

From this and Eqs. (7) and (8) there follows

$$[A_m] = \frac{N_m}{K} \cdot \frac{\eta}{C^2}. \tag{10}$$

The evaluation of this formula allows the construction, upon the basis of the experimentally established relationships between composition and viscosity of the blood [6], of a curve which presents the minimum shown on p. 44 (Fig. 3.4).

References

1. Bärner, M.: Über den histologischen Bau der Arterien in der Brust- und Bauchhöhle des Pferdes mit besonderer Berücksichtigung der Anpassung dieser Gefäße an die Umgebung. M.D. Dissertation. Jena: Fischer 1905
2. Blum, E.: Über die Querschnittsbeziehungen zwischen Stamm und Ästen im Arteriensystem. Pflügers Arch. Ges. Physiol. Menschen Tiere *175*, 1–19 (1919)
3. Blunschy, F.: Beiträge zur Lehre der Viskosität des Blutes. M.D. Dissertation, University of Zürich. Einsiedeln 1908
4. Fuchs, R.F.: Zur Physiologie und Wachstumsmechanik des Blutgefäßsystems. Master's thesis. Jena: Fischer 1902
5. Hess, W.R.: Eine mechanisch bedingte Gesetzmäßigkeit im Bau des Blutgefäßsystems. Arch. Entwicklungsmech. Org. *16*, 632–641 (1903)
6. Hess, W.R.: Zum Thema: Viskosität des Blutes und Herzarbeit. M.D. Dissertation. Zürich: Zürcher & Furrer 1906. Vierteljahrsschr. Naturforsch. Ges. Zürich *51*, 236–251 (1906)
7. Hess, W.R.: Die Viskosität des Blutes bei Gesunden. Dtsch. Arch. Klin. Med. *94*, 404–408 (1908)
8. Hess, W.R.: Über Relationen zwischen Blutbeschaffenheit und Hämodynamik. In: Festschrift zur Eröffnung des gerichtlich-medizinischen Institutes der Universität Zürich, pp. 91–108. Berlin: Hirschwald 1912. Vierteljahrsschr. Gerichtl. Med. 3. Fol. *44*, 93–108 (1912)
9. Hess, W.R.: Über die periphere Regulierung der Blutzirkulation. Pflügers Arch. Ges. Physiol. Menschen Tiere *168*, 439–490 (1917)
10. Jensen, P.: Organische Zweckmäßigkeit, Entwicklung und Vererbung vom Standpunkt der Physiologie. Jena: Fischer 1907
11. Meyer, H.: Die Architektur der Spongiosa. Arch. Anat. Physiol. Wiss. Med. 615–628 (1867)
12. Ribbert, H.: Zweckmäßigkeit in der Pathologie. Bonn: Cohen 1906
13. Roux, W.: Über die Verzweigungen der Blutgefäße des Menschen. Eine morphologische Studie. M.D. Dissertation. Jena. Z. Naturwiss. *12*, 205–266 (1878)
14. Roux, W.: Gesamte Abhandlungen über Entwicklungsmechanik der Organismen. Leipzig: Engelmann 1895
15. Spengel, J.W.: Zweckmäßigkeit und Anpassung. Akademische Rede. Jena: Fischer 1898
16. Ulmer, A.: Die Bestimmung des Volumens der Blutkörperchen auf viscosimetrischem Wege. M.D. Dissertation, University of Zürich. Zürich: Leeman, M.D. 1909

Chapter 4

A New Examination Method for Diplopia*

W.R. Hess

I was prompted by the rather complicated nature of the methods hitherto used in examining diplopia by the relatively great degree of dependence on the intelligence and goodwill of the patient and, above all, by the difficulty of obtaining precise records, to find a method which would be as free as possible of these practical drawbacks. This paper presents the results of my efforts.

Four broad green lines are drawn on a blackboard. Two of them intersect with the other two to mark the limits of a rectangular surface. The patient is seated in front of the board and is wearing spectacles of which one glass is red and the other green. The examiner has a black stick in his hand which bears a red mark at its end.

The patient perceives the green lines and the red mark held against the blackboard. He does not see both with each eye. He perceives the red mark only with the eye covered by the red glass. The eye covered with the green glass sees the lines. The complementary color appears black in the colored glass and thus disappears against the black background. The same principle is used as when testing for malingering color blindness with red-green spectacles and red-green letters.

The patient is told to follow the red mark which the physician moves across the board and to say whenever the mark crosses a previously designated green line. A patient seeing correctly with both eyes will also state the right moment. The diplopic patient will perceive the red mark as displaced in relation to the green line when it is in fact on the line. On the other hand, the mark will appear to him to be on the line when the two in reality are displaced in relation to each other. The angle of displacement will correspond to that of the diverging visual axes of the two eyes. The eye following the red mark receives its impression on the macula. As a result, the image of the green line must pass through the macula of the other eye if the patient is to perceive the mark on the line.

* Hess, W.R.: Eine neue Untersuchungsmethode bei Doppelbildern. Arch. Augenheilkd. *62*, 233–238 (1908)

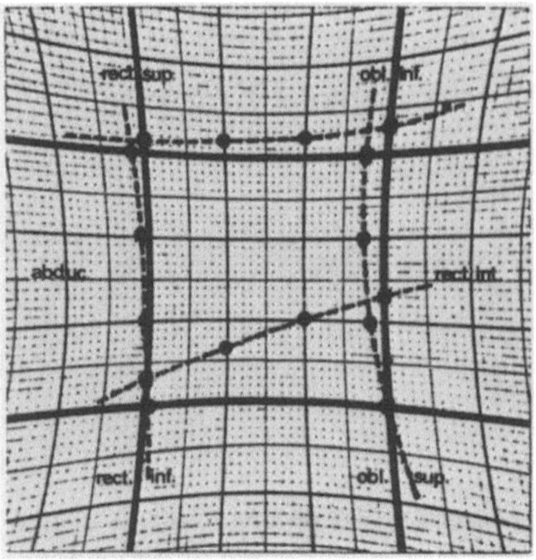

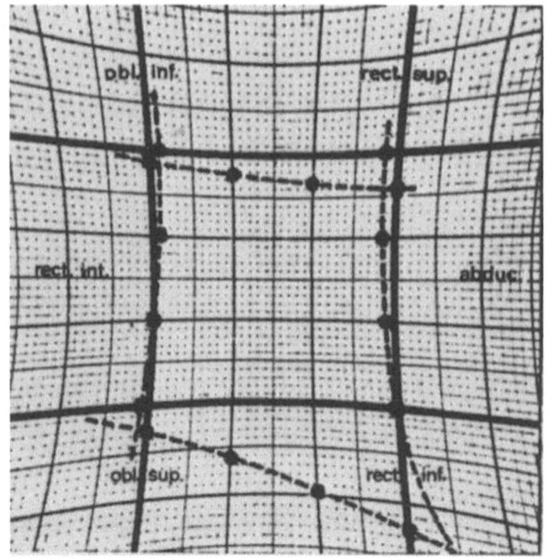

Fig. 4.1a.b. The lines are drawn with a red pencil on a diagram, preprinted in green, which corresponds to the coordinates on the black examination board. This reproduction has been reduced by half. The red lines have been replaced by black dashes (diagnosis: paresis of the left superior oblique muscle)

If the mark is moved across each line at various points the patient's comment will inform the examiner about the respective positions of the eyes in the various directions of gaze, and this gives a direct indication as to how the muscles are working.

A concrete example will illustrate this (Fig. 4.1): the patient has indicated the spots marked with a dot as the points of intersection of the red mark with the green lines. The dots are connected by a line. The lines obtained in this way connect all the fixation points seen through the red glass while the eye covered with the green glass sees the points of the corresponding green lines. Furthermore, these lines delimit an area which contains all the fixation points of the former eye when the latter is seeing points which fall in the area encompassed by the green lines. The mutual relation of red and green areas corresponds in size, form, and direction to the areas of gaze of the two respective eyes. The recordings are thus a direct graphic representation of the positions and movements of both eyes within a given field.

Two different records are obtained from each patient when, during the examination, the red glass is in front of the right or the left eye. Depending on which eye is affected by paresis, the red mark represents the function and position of the diseased or healthy eye, while the green lines reflect the functions and positions of the healthy or respectively diseased eye. If red is related to the diseased eye, the recording is termed primary (Fig. 4.1 a). The regular green figure then corresponds to the healthy eye. The red figure is smaller than the green because of the diseased eye's limited movement. If the glasses are exchanged, we obtain the secondary recording (Fig. 4.1. b).

The green field is related to the diseased eye, and the red field must be larger because the healthy eye assumes a larger role. We can use both recordings to establish a precise diagnosis of the affected muscle or muscles. However, both recordings must be made in every case because the examination will then show which eye is paralyzed: the paretic eye is the one to which the smaller field in both recordings is related. Furthermore, comparison of the two tracings must be opposite in all respects. For instance, in the given example the red field is cut off at the bottom right and rotated to the left, whereas in the other tracing the bottom right is enlarged and the tracing rotated to the right.

In follow-up examinations, e.g., to determine any recovery, it is of course sufficient merely to repeat the primary recording, i.e., the one in which the red glass is situated in front of the paretic eye.

Now that the method has been explained, some practical details follow. Blackboard, green lines, and red mark are made of felt and wool in order to avoid reflection. This ensures that the green lines and the red mark are effaced as much as possible by the complementary color of the glass. The board measures 120 × 120 cm. I did not think it advisable to make it any smaller, because an eye with a relatively strong deviation would project the mark beside the board, where it could hardly be perceived. Furthermore, the results are more precise when the area encompassed by the green lines is large. The board, if larger than suggested, might be impractical, although very useful in individual cases.

The division on the board corresponds to the projection of right angles on a plane, the perpendicular distance of which is 1 m from the vertex. In order to be able to keep this distance easily at all times, a string of the desired length is fastened at the midpoint of this distance. Stretched, the string should just reach the root of the patient's nose. Under these conditions, the distance between the green lines is 30°, and between the lines used to determine the location of the red mark, 10° and 1°, respectively. Four points of convergence per line should be tested. They should be entered on the diagram exactly as the patient indicates and then be interconnected. More than four points would be of value only in theoretical investigations. One can easily do with only two points, especially if the patient is undergoing a follow-up examination to determine any improvement or deterioration in his condition. Two different diagrams correspond to the two different types of recording. One of them is used when the red glass is on the right, the other when the same glass is on the left. The patient should try not to move his head during the examination. This is made easier if he leans his head against a headrest fitted to a chair, such as is used by photographers. It is not really necessary to hold the head tightly because, when it is leaning (but not free), both physician and patient perceive the slightest movement. The patient can then be told to remain still and the physician can check the patient.

Before adjusting the red-green spectacles, the purpose of the test is explained to the patient and he is told that the moment the mark and line intersect, he must say "now." To check whether the instructions have been properly understood, the test is conducted a few times without the spectacles and with one eye closed. Of course, the patient's perception and reality must coincide. If the patient does not readily see the red mark with the spectacles, his attention is directed to it by moving it back and forth across the blackboard. Then he is instructed to follow it with his eyes, whereupon the mark is moved along the lines of the coordinate system (starting at the top or the bottom, on the left or the right). It is easiest to use the middle four lines lying 10° apart. The intersections can be made at any other place, but they must always be entered on the diagram at the corresponding place. In some cases, the patient will be unable to state the position of the mark on certain parts of the board. This is so when the movement of one eye is greatly impaired, so that it cannot follow the mark or the lines to the periphery. In this case, the examiner will be satisfied with points of intersection lying within the field of gaze accessible to both eyes. The figure reconstructed from these points will be incomplete, as a direct consequence of oculomotor dysfunction which can go right up to complete paralysis. Knowledge of muscle function helps to decide which muscle or muscles are paretic, for the primary recording always exactly points to the restricted field of gaze and, at the same time, it reveals a displacement of the entire field in the direction in which the antagonists force the diseased eye. For the diagnosis of any muscle paresis, it is enough to know the action of the muscles contributing to horizontal and vertical movements. The action of the rectus superior and inferior and of the obliquus superior and inferior can easily be distinguished by keeping in mind the following rule: the former group moves the eyeball up- and downward in the temporal field of gaze, while the latter performs the opposite function in the nasal field. The red field, therefore, will be restricted at the top or at the bottom outside, and at the bottom or at the top inside, respectively. Paresis of the rectus externus and internus causes a straight restriction from the outside or the inside. To simplify the diagnosis, the name of each muscle in its functional area is printed on the diagram. The diagnosis can, therefore, be read off directly. Once familiarity with the recordings has been acquired, a single glance will inform the examiner about the quantitative and qualitative aspects of the impairment, even if multiple deficits are present.

The method can also be used in paretic strabismus, provided, of course, that the acuity of vision in each eye is not too greatly impaired. The new tool to obtain a precise picture of the position and movement of the eyes, e.g., before and after an operation for strabismus, will yield important practical and theoretical results. It is very simple and instructive to carry out the objective demonstration of the false projection produced by prisms and of the convergence which accompanies accomodation when minus lenses are put on. The advantages of the method may be summarized briefly as follows:

A simple and rapid method has been described for obtaining a graphic representation, both precise and direct, of the position and movement of both eyes. Errors due to the fusion tendency, or to the patient's imperfect understanding or bad will are virtually ruled out. At any rate, they can always be recognized. The light stimuli to which the patient must pay heed fall on the fovea of both eyes. Hence, the information provided will be more precise than when he has to react to a peripheral stimulus. Furthermore, information can still be obtained even if the image has been diverted or already suppressed to a large degree.

Chapter 5

Direct Stereoscopic Photographs*

W.R. HESS

Man's ability to perceive depth is basically due to binocular vision. The physiological process responsible for this remarkable achievement is in principle based on the fact that the disparity between the two eyes provides two slightly different images of the external world which arise from different viewing angles. These two images are then fused in our mind into one three-dimensional picture. Thus the image which is created displays the optical properties of an object when viewed at two different angles.

Prior to the invention of the stereoscope, three-dimensional vision existed only when real, three-dimensional objects were viewed. A pictorial representation of space and corporality was then not possible. Thus, the stereoscope (Wheatstone's mirror stereoscope of 1838 and Brewster's lens stereoscope of 1849) was a genuine innovation in the art of pictorial representation.

In spite of the great and to some extent unique advantages of the stereoscopic over the two-dimensional picture, the former has been unable to achieve the wide distribution of the latter. The main reason is that the advantage of the stereoscopic picture can only be appreciated when an instrument, the stereoscope, is used. This limitation blocks the advance of the stereoscopic picture wherever it could be of use.

This insight forces us to ask whether it is not possible to create a photographic picture which can express space and corporality immediately, i.e., without a stereoscope.

The desire to produce such pictures is not new, and there have been innumerable attempts to do so [1]. Various procedures have been reported by Ives, Rigl, Lippman, 'tHooft, Friedmann, and Reiffenstein. However, none of these has hitherto proved suitable for pictorial representation.

The practical results which I have achieved in my own attempts give some reason to hope that we are now finally able to produce useful stereo-

* Hess, W.R.: Direkt wirkende Stereoskopbilder. Z. Wiss. Photogr. Photophys. Photochem. *14*, 33–38 (1914)

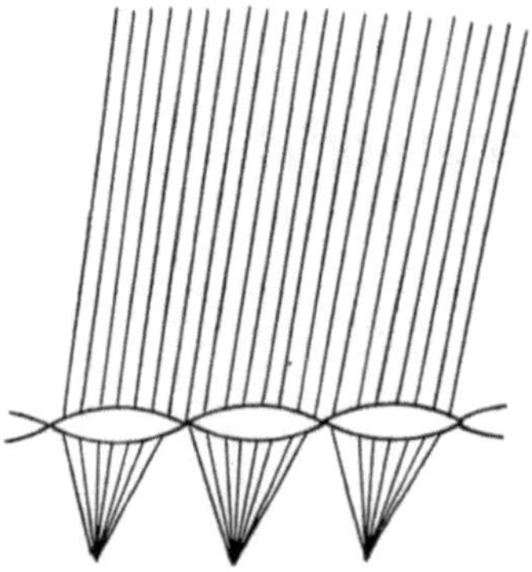

Fig. 5.1

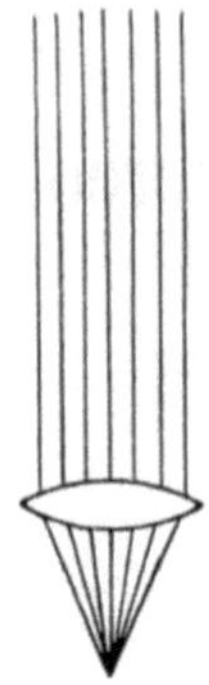

Fig. 5.2

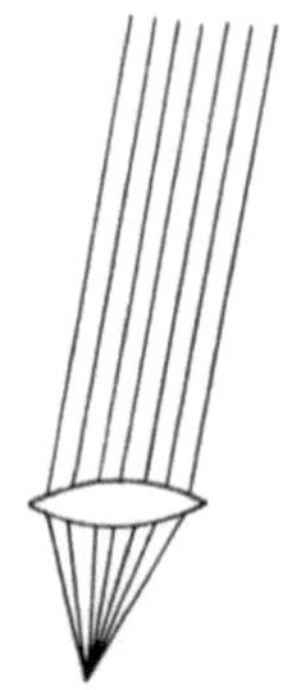

Fig. 5.3

scopic photographs which can be viewed directly. The photographs can be produced in such a simple manner, as can be seen from the process depicted in the diagrams, that there is no reason why mass production cannot be undertaken[1].

My starting point was the effect of a lens which forms parallel light beams, provided that the pointed light source is at the focal distance from the lens (Fig. 5.1).

If its position corresponds exactly to that of the focal point, the rays passing through the lens are parallel to the lens axis. If the light source is somewhat to the side of the focal point (i.e., eccentric to the lens axis), the parallel rays are at an oblique angle to this axis (Fig. 5.2).

If the light source is to the right of the focal point, the rays, once they have passed through the lens, are directed to the left; if the light source is to the left, the rays are directed to the right (Fig. 5.3).

The opposite process occurs because of the reversibility of the rays' direction when parallel light falls on the lens. The rays are collected at one point. This point is congruent with the focal point when the incident rays are parallel to the lens axis; it is to the left or right of the focal point when the incident rays fall on the lens at an oblique angle from the left or the right.

We now construct a system of contiguous, microscopically fine lenses which are joined together in a continuous surface, much like the surface of an insect's eye, the difference being that the surface of our lens system is not curved but flat.

If we direct parallel light onto this compound lens surface, each individual lens influences the rays as described above. The parallel rays of light are split up into individual bundles and converge to single pointed light sources (Fig. 5.4).

---

[1] Editor's note: At the time Hess wrote this article, the firm Stereo-Photographie AG in Zurich manufactured and distributed such pictures

Fig. 5.4

The position of these light sources is determined by the focal length of the individual lenses and also by the angle of the incident light rays. For each angle there is a corresponding point of convergence. A light-sensitive emulsion is inserted behind the lens surface at a distance corresponding to that of the focal length (where the light spots are projected by the lenses).

If we examine the finished copy with the lens system inserted in front of it, we notice the following phenomenon: the points marked on the light-sensitive layer are visible through the lenses, but not from all directions. They are visible only from one specific direction. The lenses are effective in both directions in exactly the same way, i.e., when viewing and when copying.

The next step for achieving our goal is rather simple. We send the copying light through a photographic negative before it strikes the lens surface. Depending on whether they are situated behind bright or dark spots on the negative, the various individual lenses receive much, little, or no light, and the spots are made intense, weak, or nonexistent. Seen as a whole, these individual spots can be combined into a continuous positive image, which, as for each individual spot, can be perceived only from the angle of the incident light.

It should be emphasized that this point matrix-like character cannot be perceived from the lens side. As Fig. 5.4 shows, once the light has passed through, the parallel light rays of each individual lens immediately, i.e., without a gap, join the rays of the next individual lens. This is important because it creates perfect continuity of the observed image.

For producing the copy, we have chosen the right half of a stereoscopic pair and have copied it by means of light which fell on the lens from the right. We now allow light to strike the same surface a second time, this time, however, at an oblique angle from the left. Before it strikes the surface, it is sent through the left half of the same stereoscopic pair. Once again a continuous positive is obtained which can be distinguished from the first one by the fact that it can only be seen when looking from the left onto the plate.

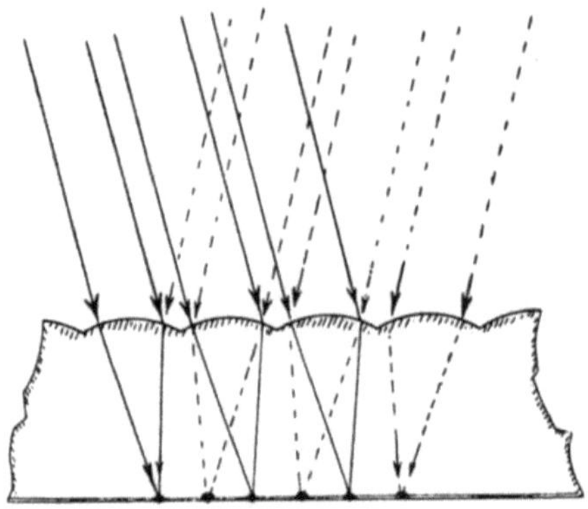

Fig. 5.5

Now let us examine this double copy from the lens side by looking straight onto the plate! The right eye, because of its position, looks to the right of the median line and from the right side onto the image surface and thus sees the right stereoscopic picture. At the same time, however, the left eye, looking more from the left, perceives the left stereoscopic picture and only the left one. Since, however, each eye sees its image at the same spot, the observer is forced to fuse and blend the images to a single picture. By virtue of this synthesis of two stereoscopic images, the unified picture has the optical properties which the depicted object displays when viewed at angles. Thus, the picture has three-dimensional character.

This is what we observe from the front. From the rear we see a multipoint screen, each pair of points belonging to one individual lens. One point of each pair lies to the left of the intersection between lens axes and image plane, while the other is to the right (Fig. 5.5). The total of all the points generates a hazy and partially double image because, without the aid of the lenses, the two images are seen by both eyes at once.

So much for the theoretical description! Now to the practical application, which is easier than it seems at first glance: a film with a thickness equivalent to the focal length of the individual lenses and with a joined lens system imprinted on the side opposite the emulsion fulfills all the optical requirements (Fig. 5.5). For technical reasons, cylindrical rather than spherical lenses are used, but this does not affect the image if care is taken during copying that the cylinder axes are viewed at a 90° angle.

This "stereo film," differing from conventional film only by its special thickness and the imprinted lens system, is suitable for taking a stereoscopic pair of pictures so that the result is seen as a genuine stereoscopic photograph. Care must be taken that the copying light does not fall randomly onto the film. As has already been said, for each stereoscopic pair, the light must come at the angle at which the image is to be subsequently viewed.

This special requirement of the copying process can be fulfilled in a number of ways, e.g., by use of a copying apparatus with corresponding windows through which the copying light falls correctly onto the plate.

In conclusion, a few words about the final result. It has already been mentioned that the photograph, as will be evident from what has been said above, must be viewed from the side on which the lens system has been imprinted. Whether the light originates from the observer's side (incident light) or from the reverse side (transparent light) is basically immaterial, apart from the general rule that lantern slides require a finer resolution of the point matrix.

Practical experiments, however, show that the stereoscopic effect is far stronger with transparent light. Another factor influencing the effect is the distance at which the photograph is held before one's eyes. The absolute dimensions of thc individual lenses should correspond to those used at normal reading distance.

References

1. 'tHooft, G.O.: Das Parallax-Problem. Z. Wiss. Photogr. Photophys. Photochem. *9*, 1–12 (1910)

Chapter 6

The Effect of Acetylcholine on Skeletal Muscle*

W.R. Hess

The investigations reported here and conducted together with van Neergard [1] were carried out to study the question of autonomic innervation of skeletal muscle. Our extirpation experiments, published by Takahashi [4], came to a completely negative conclusion about the influence of sympathetic nerve on skeletal muscle tone. No influence on the muscle tone of the thigh musculature at rest was observed in the guinea pig after unilateral extirpation of the abdominal sympathetic nerves, and in spite of the sympathetic extirpation the trophic state of the musculature remained unchanged.

In the present experiments, we looked for an influence of the autonomic nervous system by means of peripheral application of substances stimulating the sympathetic and parasympathetic nerves, respectively.

The experiments were carried out on the gastrocnemius muscle of the frog which was artificially perfused from the aorta with Ringer's solution saturated with oxygen. At regular intervals the muscle was made to contract by means of an exactly dosed electric shock. The contraction was recorded from the mobilized Achilles tendon.

Adrenaline in concentrations of 1:100,000 to 1:1,000,000 had no effect on the muscle, provided that secondary influences due to vasoconstriction were prevented. Occasional relaxation seemed to occur, but its appearance was quite uncertain and may be due to secondary factors.

Acetylcholine caused prompt contraction. The experiments conducted to establish exactly the nature of this contraction partially confirm the findings of Riesser and co-workers [2, 3] and also partially contradict them. A clear-cut effect was observed at a concentration of 1:50,000,000, i.e., in the concentration range effective in the intestinal tract. Muscle contraction appears and disappears with the appearance and disappearance of the stimulant in the nutritive fluid. The result is reproducible. Individual twitches

---

* Hess, W.R.: Über die Wirkung von Acetylcholin auf den Skelettmuskel. Q. J. Exp. Physiol. Suppl. Vol. Proc. XIth Physiol. Congress, Edinburgh 144–146 (1923)

elicited under the influence of acetylcholine were more pronounced than normal twitch contractions, but did not correspond to the degree of shortening of the unstimulated muscle. Under acetylcholine, the descending slope of the contraction curve falls beneath the baseline, not as a consequence of a mechanical artifact, but as the expression of an inhibition of acetylcholine contraction. There is no change in the duration of the contraction or of the latency period. An enhanced responsiveness of the muscle to electric stimulation is likewise not observed. Tetanic contraction of the muscle cannot be further increased by acetylcholine; on the other hand, at the peak of the acetylcholine contraction, tetanization does not produce a higher amplitude. After a strong acetylcholine stimulation, additional stimulation by acetylcholine or electric tetanization produces a weaker response.

Atropine concentrations, e.g., 1:1,000,000 (a ratio which has a pronounced influence on intestine and heart) have no effect. Increasing the concentration until it affects the acetylcholine contraction also reduces the amplitudes of contraction. This phenomenon occurs at atropine concentrations in the range between 1:10,000 and 1:1,000.

The irrigation fluid derived from skeletal muscle stimulates an intestinal preparation (guinea pig, rat) similar to the effect obtained with a highly diluted acetylcholine solution. This effect subsides following continuous irrigation, while it is considerably increased by tetanization. Atropine in a 1:2,000,000 solution inhibits intestinal contraction.

The overall result indicates a strikingly close relation between muscle contraction produced by acetylcholine and that produced by an electric stimulus. In both cases, the *same* muscle fiber elements appear to bring about the contraction. In the interpretation of our atropine experiments, the acetylcholine effect has no relation to the parasympathetic innervation of skeletal muscle.

References

1. Hess, W.R., Neergard, K.v.: Die Beziehungen der Acetylcholinverkürzung des Skelettmuskels zur Einzelzuckung und zum Tetanus. Pflügers Arch. Ges. Physiol. Menschen Tiere *205*, 506–517 (1924)
2. Riesser, O., Neuschloss, S.M.: Physiologische und kolloid-chemische Untersuchungen über den Mechanismus der durch Gifte bewirkten Kontraktur quergestreifter Muskeln. Arch. Exp. Pathol. Pharmakol. *91*, 342–365 (1921)
3. Riesser, O., Steinhausen, W.: Über das elektrische Verhalten des Muskels bei Einwirkung von Acetylcholin. Pflügers Arch. Ges. Physiol. Menschen Tiere *197*, 288–299 (1922)
4. Takahashi, N.: Untersuchungen über die tonisierenden und trophischen Funktionen des Sympathicus. Pflügers Arch. Ges. Physiol. Menschen Tiere *193*, 322–328 (1922)

Section III

Vegetative Nervous System and Psychic Functions, Sleep, Hypothalamic Syndromes

Chapter 7

The Reciprocal Relations Between Psychic and Vegetative Functions*

W.R. Hess

Contents

I. General Physiology of the Vegetative Nervous System and its Relationship to the Animal Nervous System

Research progress in the field of the vegetative nervous system has aroused an ever-increasing interest in the interactions between psychic and vegetative functions. The significance of this relationship has been emphasized in particular by psychiatrists. For example, Specht [49] in his work en-

---

* Hess, W.R.: Über die Wechselbeziehungen zwischen psychischen und vegetativen Funktionen. Schweiz. Arch. Neurol. Psychiatr. *15*, 260–277 (1924); *16*, 36–55, 285–306 (1925)

titled *Vegetative Nervous System and Psychopathology* stated that the understanding of psychopathological conditions derives considerable advantage from current knowledge of the vegetative nervous system. Küppers [36, 37] also pointed out that the relationship between psychic and vegetative mechanisms is in pressing need of clarification. The concept developed by Küppers of the structural organization of the nervous system and the relationship between vegetative and psychic centers is very attractive, although lacking experimental support. Our own point of view is mainly based on organic mechanisms as opposed to the aim of psychiatric and psychological research, which is entirely oriented toward mental activity. Since every problem should be approached from different sides, it would appear that our own way of research is fully justified. Beginning with the study of the regulation of blood circulation [24, 27, 28], we were prompted to pay more attention to the mechanisms which regulate vegetative functions in general. This led eventually to the problem of the mutual relationship between psychic and vegetative functions.

The Nature of Vegetative Functions. All behavioral manifestations of an organism seem to fulfill two purpose, one of which is to establish relationships between the individual and his environment. The individual as a whole is concerned with the external world: his orientation is aided by sensory organs, and his adaptive responses are mainly performed by skeletal muscles. Hence, the nervous system serves as a connecting link between sensory and motor mechanisms, thereby providing aim and guidance to the individual's active interactions with the environment.

A second group of organic functions is devoted to the regulation of the vital conditions of tissues and cells. Respiration and blood circulation take care of the oxygen supply to the milieu which remains in direct reciprocal relation with the cells. Digestion, supported by blood supply, provides the humoral and energetic agents to the nutritive medium from which the cells satisfy their needs. Excretion by the kidneys and lungs evacuates materials which are no longer needed.

The functional division into two systems of organs which we have just described corresponds to the well-established dichotomy between so-called *animal and vegetative physiology*. We shall therefore adhere to these terms and define the vegetative nervous system as regulating the vital functions required for the maintenance of the functional milieu of the cells.

It seems logical to designate that part of the nervous system which regulates the behavior of the individual with respect to its environment as the "animal" nervous system, although remaining well aware that Müller rejects this term as "unfortunate" [49]. This author maintains that "animal" means the same as beast. Low forms by this definition belong to the "animalia," although in his opinion they possess only a vegetative innervation. However, here we disagree. For even in lower forms we can distinguish between behavior that is directed toward the external environment and processes within the organism which provide the suitable conditions for the cells.

In his arguments against the term "animal," Müller obviously uses a topographical-anatomical criterion. Coelenterates with the nervous system limited to the coelom are destitute of an animal nervous system in his opinion. We, on the other hand, consider "vegetative" and "animal" in functional and not anatomical terms. For this reason, there is no inconsistency in the assertion that animal nerve elements may coexist with vegetative nerves in the coelom of these animals, just as they do in the spinal cord of higher forms.

The Relationship Between the Vegetative and the Animal Nervous System. Vegetative and animal functions appear as coordinated mechanisms as long as they are considered at the level of separate organs within the organism. With respect to the aims and purposes of the organism as a whole, however, their position is a different one. Metazoic organisms are evidently organized at different levels of complexity. An organism, in principle, consists of cells, each cell representing a basic unit. Together, the cells are united in a collective whereby certain cell groups undergo various degrees of differentiation and are brought together to form new units. Such units, in the hierarchy of organic evolution, operate at a higher level than the simple cell. It follows that vegetative and animal functions can exist in organisms of lower as well as higher rank order.

The functional milieu of the cells lies within the interior of the individual. Thus one may consider the mechanisms regulating the internal milieu of the cells as *internal factors*, in contrast to the *external factors* by which the individual remains in reciprocal connection with the surrounding world. It is noteworthy that, according to our concept, the dichotomy of vegetative and animal nervous systems appears in phylogenetic evolution the moment metazoic organisms arise, i.e., when organisms consist of two kinds of entities: the cell as a structural and functional unit and the individual representing the integration of all cells. Each kind has its specific vital realm. Hence two kinds of regulatory mechanisms which function side-by-side are from now on required. One is the more ancient and regulates the factors constituting the internal milieu of the cell. The second is added with the development of the multicell organism. It is responsible for the integration of the new unit into the *vital realm* of the individual. There is reason to emphasize this conclusion because there is generally little understanding of the relationship between vegetative and animal nervous systems in lower animals. This is particularly important because even competent investigators have tacitly assumed that the innervation of lower animals is exclusively vegetative in nature. Occasionally we also encounter the opposite concept whereby the vegetative nervous system is supposed to be second in order besides appearing at a relatively late phase of metazoic evolution.

Admittedly, the functions of animal regulation systems are more closely linked to visceral regulation in organisms at lower than at higher levels of differentiation. This is because at an early phase of evolution the factors constituting the *milieu interne* of the cell and the vital range of the individuum have many aspects in common. This is particularly true for low organisms living in marine water. Nevertheless, the animal principle exists even at that level, since adaptive behavior which serves the interactions between the individual and its surroundings is already recognizable. A good example is the locomotion of metazoic organisms, one of the phylogenetically oldest forms of animal behavior, which results from spatial and temporal coordination of the musculature. The action of single muscle fibers is not significant for the internal milieu of the cells or for the adaptive behavior of the individual. Yet, it contributes to a mechanical act which is clearly responsible for a directed response to environmental stimuli. The coordination of muscle fibers is undoubtedly of animal nature even in cases where only the most elementary forms of locomotion are involved.

The successful operation of animal functions depends on specific conditions in the internal milieu of the muscle fibers. Regardless of the species – mollusc or human being – these conditions determine whether the muscle cells are able to function properly and lead to a directed response. In other words, the vegetative component provides essential conditions for the execution of an integrated performance.

The Antagonism Between Animal and Vegetative Factors. The aim of biological regulations is to improve the living conditions and to ensure the survival of the organism and the species. Since these regulations concern the internal milieu of the cell as well as the external environment of the individual, one may conclude that in this respect functional orientation is the same for the vegetative and the animal nervous systems. The survival of the cell is a prerequisite to the survival of the individual, and vice versa. The requirements of the individual (and their fulfillment) essentially condition the survival of the cells.

In spite of this synergism between vegetative and animal functions which may occur by simultaneous as well as by successive interactions, there is a sharply defined antagonism between the two. This antagonism exists at the cellular and organismic levels. Let us recall the example of locomotion in order to illustrate this antagonism. Locomotion serves the interaction between the individual and its surroundings; it is achieved by the action of muscle fibers. With increasing locomotor activity oxygen consumption increases, and so does the CO_2 production. Thus, important factors of the functional milieu of the cells are changed in the adverse direction. Regulatory mechanisms come into play which are designed for compensation. However, such corrections within the skeletal muscles are only achieved by displacing the adverse conditions to other tissues, for instance, by increasing circulation and respiratory activity.

A similar but reversed situation with respect to the antagonism between vegetative and animal functions exists during high activity of the digestive tract when the blood supply to the glands and resorptive epithelia is increased. Because of the centralization of the blood supply on behalf of gastrointestinal functions, the nutritive conditions of the somatic system are jeopardized and its efficiency is immediately decreased, i.e., before new regulations come into play.

The Balance Between Factors Governing the Internal Milieu of the Cell and the External Surroundings of the Individual. Changes of survival of the individual depend both on the general living conditions and on the functional conditions at the cellular level. For this reason the antagonism between the vegetative and the animal must never reach an extreme situation in which one overpowers the other. For should the vegetative predominate, the organism would be in danger, and vice versa. The optimal state is the result of a balance which tends to maintain the needs of both vegetative and animal regulations, and this in turn depends on the functional interaction of both regulating systems. If, for example, during intensive bodily exertion the functional state of the muscle tissue is changed in the adverse direction, specific mechanisms come into play which prevent further deterioration by reducing animal activity. If, in contrast, dangerous situations within the sphere of the individual require prompt measures, for example, defense by flight, maximal efficiency of animal and vegetative systems must be effected by their anticipatory release.

Physiology tells us that reciprocal interactions between vegetative and animal functions exist. The influence on the vegetative sphere by the animal system is exemplified during physical exercise when circulation and respiration are activated. The reverse is seen during fatigue, which corresponds to a protection of the animal system induced by vegetative regulations, and not primarily to a manifestation of exhaustion. Another paradigm of vegetative influence upon the animal system is provided by temperature regulation which depends to a considerable degree upon the external environment. At a high environmental temperature, the heat loss and cooling mechanisms via the vegetative system (for example, vasodilatation and sweating) may be insufficient. In such instances the cells of the body may be threatened by overheating.

This danger can be opposed by suppressive actions on mechanisms which normally participate in heat production, i.e., somatomotor activity. Indeed, when the danger of overheating prevails, one finds the individual in a state of motor and mental inactivation. These regulations occur automatically and they are certainly not the consequence of heat damage, since under similar conditions, for instance during fever, quite intensive bodily exercise can be consistent with even higher body temperatures. On the contrary, the suppression of animal activity and the low tonus of skeletal muscles are clearly the result of a mechanism serving temperature regulation. The animal system, under these conditions, is governed by the vegetative.

This example shows once more that there is no separation between vegetative and animal regulation systems, and that no one-sided predominance of one system occurs over the other. Thus, the antagonism between vegetative and animal is kept in balance by their mutual interdependence.

The Principle of Antagonistic Innervation. The decisive aspect in the balance between antagonistic factors is a quantitative one. Oppositional factors interact with forces whose intensity can be measured; their interrelation is analogous to that existing between tension and countertension.

A good example of the balance between antagonistic innervations is present in the regulation of cardiac activity. Let us consider the heart as an organism within the organism, in order to illustrate our point of view. The rest of the bodily organs represent the range of cardiac influence, i.e., the periphery with which the heart remains reciprocally related. In other words: influences resulting from the interaction between heart and periphery are considered to be external, whereas factors arising from within the cardiac muscle and determining its functional conditions (i.e., internal milieu) are considered to be internal.

It is known that cardiac function is dominated by two separate innervations: one which tends to increase circulation, and the other, a competing one, which tends to diminish it. The balance between these two innervations determines the cardiac output.

The physiological significance of each of these two forces can be stated as follows: the activating system eventually improves the blood supply of the body, for instance of the skeletal muscles during physical exercise. Without this system the muscle tissue would suffer from the increasing discrepancy (a "strain" of a sort) between blood supply and energy requirement. This "strain" affects the peripheral organs adversely. However, under the influence of the activating system the heart intensifies its output and thereby reduces the "debit" (Spannung) in the peripheral tissues. However, with increasing cardiac activity, the functional state of the heart itself is jeopardized. The oxygen consumption increases, and metabolites accumulate in the internal milieu of the cardiac musculature. This creates a debit which increases with the functional load. Here again, compensating mechanisms are called into action, which we shall not deal with in order to avoid complications. Suffice it to say that reducing the debit in peripheral tissues leads necessarily to an increase of debit in the cardiac muscle. It follows that even in the case of a single organ there exists an antagonism between external and internal factors. This conclusion is reinforced if we examine the consequences of cardiac inhibition. Since cardiac work is reduced with decreasing load, the condition of the internal milieu of the heart is improved. Decrease of debit, however, affects the blood supply to the bodily organs, for example the skeletal musculature, and a discrepancy between consumption and supply results, i.e., an increased debit in peripheral tissues. In this situation, however, the heart does not follow exclusively either the activating or the inhibiting commands, but tends to harmonize the antagonistic influences and thereby equilibrates the tension between external and internal factors. Owing to the antagonistic innervation, the autonomous activity of the heart operates at a level ensuring the quantitative equilibration. In other words, the adjustment of antagonistic nerve impulses leads to the removal of debits arising within the heart itself and within the tissues of bodily organs that depend on cardiac activity.

The Organization of the Vegetative Regulation System. Cardiac activity exemplifies the fact that the antagonistic components of the vegetative nervous system are represented by the sympathetic and the parasympathetic system. The sympathetic nervous system – with regard to the heart – is oriented towards the external factors while the parasympathetic nervous system subserves the internal milieu. Thus, the functional significance of the sympathetic system for the heart is analogous to that of the animal nervous system for the individual. At the same time we recognize that the sympathetic nervous system enables the animal system to penetrate into the domain of vegetative regulations.

This occurs by summoning those functional components of the vegetative system that are best qualified to momentarily strengthen the efficiency of the animal apparatus. This role is exerted by the sympathetic system not only via cardiac innervation. Vasoconstriction, especially of the splanchnic vessels (as reported elsewhere [24, 27, 28], should be considered within the same context since a redistribution of the blood supply is achieved in favor of the animal system. The reduction of blood supply during physical exercise in those organs which are not directly or indirectly involved brings a relative advantage to the animal apparatus (skeletal musculature), including its vegetative auxiliary elements. Furthermore, functional inhibition of the digestive tract by the sympathetic nervous system is consistent with the same interpretation, since the corollary attenuation of digestive functions leads to a reduction of those activating stimuli which would otherwise arise from hyperemia in the corresponding tissues.

The sympathetic system plays a role as a connecting link between animal and vegetative functions in still another area: by reducing the resistance within the respiratory tract when oxygen demand is rising. Excitation of the sympathetic nervous system induces bronchodilatation and this – in connection with circulatory activation – may be considered to be another example of a most effective synergism.

The role which we attribute to the sympathetic system is consistent with Cannon's concept [10], according to which the release of adrenaline in emotional states is to be considered a regulatory mechanism assuring maximal efficiency of bodily reactions. The effect of the parasympathetic system is in every respect the reverse of the sympathetic: attenuation of cardiac activity on behalf of the internal conditions of the heart. For the animal system, the parasympathetic influence means a reduction of effectiveness. The parasympathetic nervous system therefore favors the interests of the functional elements of the heart. Corollary effects of the parasympathetic system exhibit the same tendency: by activating the vasodilatators an increased blood supply (and with it, improved nutrition) is provided.

Thus a synergism is established which counteracts functional overload by diminishing wear and by augmenting the supplies. The enhancement of digestive functions exerted by the parasympathetic system is consistent with the same concept since the main significance of the gastrointestinal tract lies in the improvement of the internal milieu of the cells. The same is true for the role of the parasympathetic system with respect to excretory functions, which relieve the functional milieu of the cells by eliminating noxious products. According to Cannon's concept the parasympathetic is the nerve of assimilation, the sympathetic the nerve of dissimilation and catabolism. Our own concept goes deeper. As regards the antagonism between sympathetic and parasympathetic innervation, we see the competition between the interests of the individual and those of the tissues and cells, interests which need continuous equilibration.

The regulation of vegetative functions is organized in still another direction. Aside from nervous impulses, chemical stimuli play an important role in the control of vegetative organs. The heart especially is under the influence of adrenaline as a physiological factor. With the aid of the method of the denervated pupil, Hartmann et al. [21] have demonstrated that adrenaline is released into the blood stream during physical exercise. The influence of adrenaline upon the heart corresponds to that of the sympathetic nervous system. Little information exists on the possible role of the antagonist of adrenaline, acetylcholine, in the regulation of cardiac activity. However, we know from the investigations of Geiger and Loewi [18] and from our own experimental data [25, translated in this volume, p. 67] that in the contracting skeletal muscle a heart-inhibiting principle, similar in effect to choline and acetylcholine, is formed. In our experiments, we have also seen that the cardiac-inhibiting principle is released from skeletal muscle into the perfusate. We therefore suggest that chemical regulation contributes also by inhibiting agents to the regulation of cardiac activity. Of particular importance is the fact that hormonal and nervous regulations participate not only in parallel arrangement but also in serial order. Release of adrenaline into the blood stream is due to sympathetic nerve excitation; thus, the nervous action precedes that of chemical stimuli. The opposite relationship is encountered, according to Loewi, in the mechanism of transmission of nerve impulses to the heart, since the nervous impulses are not acting directly but only by mediation of specific mediating principles which are mobilized by the nerve impulses. Thus, nervous and hormonal influences are not to be considered mutually exclusive. On the contrary, they represent two mechanisms of regulation which support and complement each other. The nerve impulses mediate sharply differentiated and rapidly effective changes, while the hormonal signal is characterized by its long-lasting and more generalized effects.

II. Sleep as a Manifestation of the Vegetative Nervous System

In view of the eminent role played by sleep in the maintenance of an optimally functioning organism and indeed for survival, it is understandable that research on the nature of sleep has been under way for a long time. It is not surprising, therefore, that a large number of theories exist today although not one has gained general acceptance. It is not the purpose of this review to discuss all the various theories, because numerous accounts on the subject are already available. Instead, we refer primarily to the monograph of Trömner [53], in which the problem of sleep is extensively covered, and also to the various communications on the problem of insomnia presented at the German Congress of Internal Medicine in 1914 [17] in which the problem of sleep has been treated in detail. Additional information is found in the contributions by Reinhart [47] and by Stern [50], which are contained in the monograph *Epidemic Encephalitis.*

To summarize the current status of the sleep problem: the notion that sleep is not a manifestation of a functional failure, a deficit phenomenon, but rather a positive and purposeful manifestation of the organism, has gained general acceptance. The main characteristics consist of a suppression of receptor mechanisms and a failure of the organism to execute animal functions in general. The main criterion of this negatively oriented behavior seems to be protection against exhaustion. The credit for bringing this concept to general acceptance belongs to Claparède [11, 12]. However, the mechanism responsible for the induction of sleep is still an unsolved problem. It does not help the physiologist to say with Claparède that sleep is an instinctive phenomenon. More concrete indications are given by Verworn [54], whose theory we shall briefly mention because it is typical of the idea that the sleep-inducing mechanisms may be located within the animal domain of the central nervous system. Verworn's hypothesis is based on the fact that sensory stimuli are blocked and that the nervous centers which undergo increasing fatigue during wakefulness fall into a state of refractoriness. A special role is attributed by Verworn to the central representation of the levator palpebrae muscle. When the excitability of this center decreases owing to fatigue, the subjective feeling of drowsiness from the ocular region arises, together with the urge to close the eyelids. Once the organism is deprived of the most important sensory stimuli, sleep ensues automatically on the basis of the prevailing fatigue. According to Verworn, the absence of sensory stimulation leads inevitably to sleep. In contrast, Kreidl and Herz [35], who investigated mentally deficient human subjects, pointed out that the role of sensory stimuli in the sleep mechanism should not be overestimated.

Verworn's concept is somewhat similar to that of Trömner who likewise emphasized the role of sensory deprivation. One difference between the two consists in Trömner's idea that sensory deprivation may be achieved by central inhibition of sensory afferents. The inhibitory function of the corresponding center would be limited to sensory stimuli entering consciousness, while unconscious activities would not be affected. Since the thalamus opticus is a relay station for corticopetal sensory excitations (olfactory excitation excepted), Trömner believed that the thalamus contains this center. Trömner's contention is obviously based on the assumption that the sleep center is identical with a relay station for sensory excitations. This is particularly evident in Trömner's argument against Mauthner's theory. Mauthner [41] suggested that the cells lying in the gray matter surrounding the third and fourth ventricles may represent the sleep organ, because clinical observations suggest that inflammatory processes in these areas are possibly correlated with the clinical symptom of somnolence.

A special position with respect to the sleep theory is taken by Küppers [36, 37] and concerns the basic organization of the nervous system and localization of mental functions. Küppers localizes the sleep-inducing mechanisms along the entire cerebrospinal axis and mentions the term "sleep dissociation." In his opinion the nervous system consists of three main divisions: (1) the "animal system" shielding the organism from the external world; (2) a "mural system" constituting the internal frontier; and (3) the *Binnensystem* (milieu interne) as a connecting link between the two. "When falling asleep, the *Binnensystem* dismisses the somatic and the mural systems and dissociates itself from them. During arousal the relationships are reinstated and contacts which have been interrupted during sleep are reestablished with the external world."

This brief survey suffices to characterize the current situation with respect to the sleep problem. We believe that the four theories which have been mentioned contain all the points of view which need to be seriously considered by a physiologist. The histological theory of Mathias Duval and the notion that sleep is a consequence of vasomotor reactions or of CO_2 accumulation will not be considered here.

In our general physiological discussion we have established important foundations which lead almost inevitably to an explanation of sleep, and this in a very concrete form.

We shall begin with the notion that unconditional independence of the animal from the vegetative regulation system is incompatible with, as pointed out, their mutual relationship. Hence, we conclude that the conditions for survival are fulfilled only if vegetative regulations cooperate with the animal and if, on the other hand, animal functions are coordinated with vegetative needs. We consider this interaction as the main foundation for the equilibrium between the internal milieu and the individual, keeping in mind that the vegetative regulation system harmonizes vegetative functions, while the animal regulation apparatus – conceived as an isolated entity – integrates animal functions into a continuing succession of purposeful acts and eventually into goal-directed behavior, thus securing the individual's adaptation to the external world.

The reciprocal relationships between vegetative and animal regulations, finally, are responsible for the harmonious integration of the two systems.

In this context we consider sleep to be a *specific phenomenon* which consists of a reduction of animal activity (*Leistungsbereitschaft*) induced by the vegetative nervous system. Thus, we seek the cause for the induction of sleep within the internal milieu of the cells. Of primary but not exclusive importance, we consider the conditions in the animal division of the central nervous system, inasmuch as *during sleep the elements of the animal regulation system – including cortex and senses may become target organs of vegetative regulations.*

According to our basic concept concerning the organization of the vegetative nervous system, we believe that the attenuation of animal activity represents a shift in the balance between the parasympathetic and the sympathetic system. For the moment we shall ignore possible changes in the hormonal balance which may also play a role. Owing to sleep-induced attenuation within the animal realm, animal functions are reduced to a minimum. The opposite extreme is encountered in affective states which lead to maximal activation of the organism, for example, during outbursts of rage. In this instance, we believe that there is a shift in favor of the sympathetic nervous system, and the result is maximal animal activity. Again we refer to the agreement between our concept and that of Cannon [10].

We are aware, of course, that our sleep hypothesis is based on assumptions which must be experimentally confirmed. We shall return to this point in detail.

We do not ignore the fact that our concept is likely to arouse dispute, because the assumption that the highest levels of the central nervous system may be influenced by vegetative innervation is not consistent with conventional thinking. Instead of an independent animal nervous system, we postulate one which is governed by a vegetative innervation. We also realize that our concept with its implications far exceeds the limits of a sleep theory. We have already alluded to the possibility of similar mechanisms coming into play in affective states. However, extreme and even intermediate levels in the balance between enhancing and attenuating factors of excitability are important, and one should be aware that not only the *level* of activity but also its *direction* is essentially dependent upon vegetative innervation. Hence, we consider sleep not isolated from other psychic processes which are influenced by the vegetative nervous system. And it should be added here that the regulation of animal functions by the vegetative nervous system is obviously important for the pathogenesis of psychic disturbances. New explanations may be arrived at if we consider the failure of psychic functions to be a manifestation of a disturbed balance between the parasympathetic and sympathetic systems. The causal factor may be remote from the substrate of psychic functions. This may explain the "functional" nature of these disturbances. The psychic disorder appears as a symptom which parallels the impairment of functions normally regulated by the vegetative nervous system. The significance of hormonal influences in the pathogenesis of psychoses (von Monakow) will not be touched upon here, since we have elsewhere pointed out the intimate connections between hormonal and nervous signals and their significance as carriers in the realm of vegetative regulation mechanisms.

Let us now return to the sleep problem. Sleep is one of the most elementary and most impressive manifestations of vegetative influences acting upon psychic activity and therefore represents the first opportunity to examine our concept experimentally. In order to do this we must be more specific about certain points in our concept.

The target of vegetative regulations is preferably not identified in terms of a circumscribed localization. On the contrary, we must consider the entire interphase between sensory and motor representations. The question, for example, of vegetative innervation of skeletal musculature which is widely discussed today represents only part of our problem. The vegetative innervation of the iris is another special case illustrating the fact that the intensity of animal stimuli is being regulated. The influence of the vegetative nervous system upon the transmission of somatomotor activity within the central nervous system, e.g., the various efferent pathways, concerns additional points of the same problem.

It follows from our concept that the influence of vegetative regulations upon the animal system is not mediated by impulses that merely pass from one system to the other. The influence involves rather the graded regulation of excitability *(Leistungsbereitschaft)*. The animal apparatus maintains its *autonomy* with respect to the exchange of excitatory and inhibitory impulses, but this autonomy is a *limited* one because excitability of the animal elements is under the influence of the vegetative nervous system. On the other hand, the vegetative nervous system itself is not autonomous, but depends on the animal system. Via the sympathetic nervous system the animal apparatus may increase the level of activity, e.g., of circulation and respiration. However, the animal apparatus by way of the sympathetic nervous system may also influence the excitability of its own elements in the sense of a general increase of its level of activity. This closed circuit counterbalances the sympathetically mediated attenuation of excitability. In this way external sensory stimuli may regulate the level of activity of the animal system, and we can thus understand the sleep-disturbing role of sensory stimuli.

Now, let us examine how our concept on the nature of sleep may be supported by direct observations and by experiments. We would prefer not to concentrate on details, but devote our attention to principal matters concerning the fundamental change of animal and psychic functions.

Three principal phases can be distinguished in the organization of animal functions: stimulus reception, stimulus processing, and stimulus response. In reflex activities we can encounter each of these three phases in a relatively simple form. At higher levels, in so-called reactions, we are confronted with more complex organizations because both sensory as well as effector mechanisms are characterized by built-in reflex chains which function as auxiliary devices. During the phase of stimulus processing we find that excitation spreads from the afferent to the efferent limb in the horizontal and vertical direction. The patterns of excitation within these pathways may become conscious in terms of psychic processes. Even where psychic activity seems to appear on its own, it may be valued only as a link within a closed physiological circuit. It is true, however, that the temporal linkage with receptor and effector elements may be rather weak in such instances. We shall emphasize the general physiological functions of sleep by extending the analysis not only to psychic phenomena, but also by including the three principal phases of animal functions.

The Narrow Pupil During Sleep. The narrowing of the pupil during sleep has already been discussed by Brown-Séquard [9] in his interpretation of the nature of sleep. He considered it one of the symptoms characterizing sleep as an inhibitory state. Pupilloconstriction, as well as the contraction of the orbicular muscle of the eye expresses an activity which clearly contradicts the interpretation of sleep as a simple suspension of functions.

Braunstein [8] has put forward a theory concerning the special mechanism of pupilloconstriction during sleep. According to this theory the cerebral cortex is supposed to exert an inhibitory influence on the oculomotor nuclei during wakefulness. During sleep this inhibition subsides. The oculomotor activity can now develop with full intensity. It would therefore appear that the constricted pupil is a consequence of *cortical sleep*. This mechanism has also been implicated in explanations regarding the narrow pupil after the administration of morphine and other pupillomotor drugs with sleep-facilitating action [42].

Braunstein's hypothesis is of particular interest to us because it represents the crystallization of the corticocentric concept of sleep which prevails today. However, we must confess that this theory may satisfy only to a relatively limited degree. The physiological significance which should be attributed to the disinhibition of oculomotor centers remains to be clarified. At the most, this mechanism could be part of a more general inhibitory influence which the cerebral cortex exerts upon lower levels, and sleep would in consequence appear as a manifestation of functional decortication. Yet, the reflex behavior during sleep clearly demonstrates that this is not the case. We shall return to this point later.

More recently, Wieland and Schoen [59] have postulated that the narrow pupil during sleep (and during morphine action and general anesthesia) should be attributed to an increased CO_2 concentration in the blood which leads to an increased excitation of the pupilloconstriction center.

According to the data of Straub [52], Gollwitzer-Meier and Kroetz [19], Enderes [15], and Bass and Herr [86], the CO_2 concentration in the blood rises during sleep. Numerous investigations by Wieland and Schoen prove furthermore that a parallelism exists between the size of the pupil and the factors responsible for the CO_2 accumulation. Nevertheless, the conclusion that pupilloconstriction and CO_2 concentration are causally related cannot be accepted. As a matter of principle it can be assumed that these two symptoms appear in parallel, following a synergistic change within the relevant regulation centers. It is noteworthy in this context that a sudden arousal stimulus may instantly relieve pupilloconstriction. In its place there is an enormous pupillodilatation within only a few seconds. On the other hand, the change in CO_2 concentration is characterized by considerable inertia. This change will outlast the phase of pupillodilatation and continue well into wakefulness. Thus, the temporal discrepancy between the two mechanisms fails to support the hypothesis of Wieland and Schoen.

The speed with which pupilloconstriction follows an arousal stimulus contradicts, in addition, the assumption of a hormonal mechanism or one induced by a so-called hypnotoxin [45], at least in the sense of a decisive factor. The prompt reaction of the pupil to arousal stimuli, even to stimuli which remain below the waking threshold, suggests a mechanism which may be analogous to the one governing the normal oscillation of the pupil during wakefulness.

In considering that the iris normally attenuates the excessive intensity of light stimuli, we are confronted at first not only with an inexplicable but also a paradoxical situation with respect to the narrow pupil during sleep, because the eyelids are closed during sleep, and they thus reduce considerably the amount of light penetrating the eyeball. Furthermore, the pupil is likewise narrowed when sleep occurs in a completely darkened room, where stimuli which could possibly induce pupilloconstriction are notably absent.

One might argue that pupilloconstriction during sleep is perhaps related to accommodation and convergence. Indeed, Johannes Müller has mentioned this relationship. However, intensive investigations of Pietrusky [46] confirmed that the narrow size of the pupils during sleep is completely independent of eye position. Furthermore, the eye is not in a state of accommodation, and the recovery felt by the overworked eyes after sleep indicates that lens accommodation has been at rest during sleep.

Since our search for a satisfactory explanation and interpretation of the narrow pupil during sleep has been unsuccessful, we may end up with the conclusion that our question was perhaps wrong from the beginning. This is indeed the case if we apply the same criteria of pupillary functions to the sleeping and the waking state.

The situation changes when this special symptom is related to the physiological significance of sleep. In the first section of this monograph we pointed out the contrast between animal and vegetative functions. Animal activity challenges cellular functions. Inhibition of animal function means relief of cellular functions, in the sense of an enhancement of vegetative restitution. In considering the eye, we can readily see that the act of pupilloconstriction diminishes the functional load of retinal elements and of those nerve elements which carry retinal excitations. In the waking state, the oscillations of the pupil regulate the access of light to the retina and enable the eye to fulfill its task of visual orientation without overstressing receptors and subsequent visual pathways. The same mechanism which regulates the light intensity during wakefulness is clearly able to bring rest to the retina whenever such rest is required for vegetative reasons.

This is evidently the case when the need for rest calls for sleep. The narrow position of the pupil during sleep performs the same function as a plug in the ears: it is a device for stimulus attenuation. It seems not improbable that the sound-conducting mechanisms are attenuated during sleep. The increased tension of the eardrum and the increased tonus of the tensor tympani muscle may perform this task. Unfortunately, the examination of these mechanisms is not as easily carried out as that of the pupil, and experimental data on this point are not known to me. The significance of the pupillomotor apparatus in ocular function should not be underestimated. From the diameter of the pupil we can calculate the range for the regulation of stimulus intensity. The pinhead-size pupil in deep sleep permits hardly more than $^1/_{10}$th or $^1/_{20}$th of the light to penetrate the eyeball as compared with the fully open pupil.

This means that a 10–20 times stronger visual stimulus is required to reach the intensity threshold of the retina. Consider also, that during sleep the eyelids are closed, thus affording a protective mechanism against mechanical insults which could harm the cornea. During wakefulness the lid closure reflex is seen only when the eye is exposed to extremely bright light. During sleep the light stimulus must be completely eliminated. Thus, we find in sleep the same synergism between iris and eyelid as during wakefulness.

In the latter state, we know that a third mechanism exists which attenuates visual excitation. It is localized within the retina and consists in a change of retinal excitability: the light-adapted retina is far less responsive than the dark-adapted one.

To our knowledge, the problem of retinal adaptation in the sleeping person has not been raised until now. Yet, the question seems quite relevant here. If eyelid, iris, and retinal adaptation are mechanisms which cooperate synergistically in the waking state by regulating visually induced excitation, it is tempting to assume that retinal adaptation takes place also during sleep in order to support the effort made by lid closure and pupilloconstriction in attenuating the effect of visual stimuli.

Unfortunately, it is difficult to answer this question. The usual determination of retinal threshold is based on responses of the subject. Such responses cannot be obtained from a sleeping person. Perhaps the examination could take place immediately after arousal. However, the results may not be convincing, because the change in adaptation of the adult retina could be very rapid indeed. Although retinal adaptation after light exposure requires considerable time, the adaptation which might take place during arousal may be much shorter, because in the sleep-adapted retina, the metabolic products which have accumulated in the light-exposed retina as a result of intensive activity are lacking. Consequently, no time may be required for the elimination of such products from the sleep-adapted retina. In spite of this difficulty we are not completely left without criteria for the evaluation of retinal sensitivity during sleep. It is known that the size of the pupil is not exclusively dependent upon the amount of light falling upon the retina. For a given light exposure we find that the higher the light intensity to which the retina is adapted the greater the pupillodilatation.

We have just mentioned that the pupil immediately after arousal is abnormally wide in spite of the increased light exposure of the eye. Only secondarily does it react to light by the normal reflex constriction. This behavior may indicate that immediately after arousal, retinal excitability is still reduced.

Finally, the reaction of the pupil during sleep should be discussed with respect to innervation. Analogous to pupillary oscillations during wakefulness, the narrow pupil of the sleeping person represents a functional state which is regulated by vegetative innervation. We take this for granted and yet a peculiarity exists which is worthy of closer consideration, because the vegetative nervous system exerts an enhanced action upon the animal system by controlling the diaphragm which is inserted in the pathway to be taken by the animal stimulus. The narrow pupil signals inhibition. Wide pupils enhance the effect of visual stimulation; pupillodilatation signifies facilitation. A special feature of this inhibition and facilitation is that they act directly on the stimulus and not on the substrate of the subsequent nervous excitation. This peculiarity, however, is not very significant. In the end, both mechanisms are equivalent: the vegetative nervous systems turns out to be a regulator of animal functions.

We believe that the concept we have exposed in the first section of this review provides an adequate background for comprehending this specific relation. The narrow pupil during sleep clearly demonstrates that the resting state of the sleeping person is taking part in the restitution process of the whole organism. Resting is thus a manifestation of vegetative regulation. Attenuation of light exposure as a consequence of pupilloconstriction is a means to inactivate significant parts of the animal system. Thus, the eye appears as a target organ of the vegetative nervous system. Because of the transparency of the optic apparatus, the dependency of animal functions on vegetative regulations can be directly visualized. Let us point out that the same dependency exists in the waking state, the difference being that vegetative orientation does not invariably predominate, but remains in keeping with the functional requirements of the animal apparatus. The latter presupposes a sufficient visual contact with the external environment. Pupillodilatation during arousal, therefore, demonstrates that the retina is now governed by sympathetic innervation, known to enhance the function of the animal system. At the same time, because of this, the parasympathetic innervation which prevails during sleep is dominated by the sympathetic system, but not unconditionally so. For, as soon as the light exposure, by exceeding a certain level of intensity, threatens to overload the nervous elements, there will be a return to parasympathetic prevalence, and reflex pupilloconstriction ensues.

The question arises as to whether the pupillomotor apparatus is the only (or only visible) target of vegetative regulation governing animal functions. The preceding discussion prompts us to include the mechanism of retinal adaptation. There was no reason until now to consider this mechanism in the context of vegetative innervation. In view of our analytical consideration we are nevertheless forced to consider this possibility seriously. The pathways which might mediate this adaptation could at least in part be a reflex at the retinal level, for example, in terms of an intramural network of vegetative innervation. Other adaptive mechanisms may emanate from the central nervous system, which could be particularly important during sleep.

In evaluating this thought we should be aware that the regulation of light reception by retinal adaptation and the regulation of light exposure by pupilloconstriction are mutually interdependent and interacting. Both mechanisms seem to secure a generous visual stimulation and yet protect the nervous machinery as much as possible against excessive excitation. Scientists who are not satisfied with a purely descriptive physiology and who dislike considering the organism solely in terms of partial functions are quite prepared to ask the question as to how the organism integrates partial functions, and they must be therefore interested in this problem. For we are seeking to link together two mechanisms in order to identify them as components of one and the same process. The motivation for our search is supported by the fact that the pupillomotor apparatus belongs ontogenetically to the retina [2]. According to Bach and Seefelder [2] the iris is differentiated from the anlage of the retinal cup at the rostral border where internal and external leaflets of the retina are joined. The retinal origin of the dilator muscle is also very likely; its origin are the cells of the external retinal leaflet. Unfortunately, the embryological data which could inform us about the origin of the vegetative innervation of the retina are lacking. However, the idea that the latter may arise from the common matrix is not too farfetched.

Finally, we should draw attention to experimental data in support of the idea that the regulation of retinal sensitivtiy is subject to the influence of the vegetative nervous system.

A few years ago C. von Hess [23] reported remarkable effects of pilocarpine and physostigmine on retinal function: visual acuity was disturbed and this disturbance was shown to be of retinal origin. In addition, certain entoptic phenomena were observed. Obviously, C. von Hess excluded the possibility that the miotic effect and accommodation changes could be the real cause of these disturbances, and he concluded that the latter were due to the direct toxic effect of the miotic drugs upon the retina. Hess was also puzzled by the fact that such minimal amounts which may reach the back of the eye after instillation of the substances into the conjunctival fold were effective and that the retinal effect of drugs could be overlooked for such a long time.

According to C. von Hess's observations it is obvious that the retinal function is extremely sensitive to parasympathomimetic drugs. The nature of the observed changes of visual functions does not permit us, however, to go beyond this conclusion. It is not possible to establish a direct relationship between von Hess's observation and the mechanism of retinal adaptation. We have therefore carried out our own experiments with parasympathetic excitants, and the results are unambiguous [29]. Under the influence of pilocarpine and physostigmine the sensitivity of the retina in comparison to the control side is considerably lowered. If equated with respect to the pupillary size, the attenuation is nearly one-half to one-third. In terms of retinal sensitivity this change does not mean too much. Yet, it is not negligible when we consider that only minimal quantities of excitant drugs reaching the retina by a long diffusion pathway through the sclera and vitreous body can elicit this effect. The massive amount of work on retinal adaptation which has already been carried out, and the fundamental significance of the problem force us to cast our conclusions in very cautious terms. We shall not hesitate, therefore, to undertake additional experimentation to clarify the relationship between the vegetative innervation and the regulation of retinal sensitivity. We are all the more motivated to do this research because the retina is one of the most important animal organs and because our knowledge of the retina may represent a clue to cerebral organization, since this organ represents the first stage in the stimulus-processing mechanism of the brain.

If we ignore for the moment all the unsolved problems, we can summarize the following facts: The mechanism which manifests itself in the sleep function encompasses even the port of entry of visual stimuli and, by imposing a hindrance upon sensory reception, it reduces the general readiness of animal functions. The corresponding change is mediated by the vegetative innervation.

Reduced Reflex Activity During Sleep. Verworn [54], Trömner [53], Pietrusky [46], and MacWilliam [40] agree that sleep consists in a reduction of reflex excitability, especially with respect to certain spinal reflexes. This reduction does not seem to be of equal and consistent intensity, which may explain certain disagreements between the data of various authors. Thus, Rosenbach found an extinction of patellar, cremasteric and abdominal reflexes. The observations of Kuttner seem to contradict these findings at least partially. The reduction of reflexes, however, has been confirmed by Trömner [53]. The results of Kleitman [32] and Lee and Kleitman [38] are unambiguous and especially important because they were obtained by quantitative methods, and the data are documented by means of graphical registration. These authors approached the problem by examining the effect of experimental sleep deprivation upon various reflexes. Besides other phenomena they examined the patellar reflex, which was elicited by a standardized mechanical stimulus. The response was recorded graphically. During artificially prolonged wakefulness the subjects fell asleep while sitting on the experimental chair. When this happened the patellar reflex remained refractory even though the same stimulus intensity was applied as in the waking state. If an arousal stimulus preceded the reflex stimulation, the patellar reflex returned to normal strength. This was repeated several times during the same session.

In order to evaluate the significance of this diminished reflex activity we first have to know which physiological functions are accomplished by these reflexes. Let us take the patellar reflex as an example.

This reflex is physiologically elicited in the same way as all the so-called stretch reflexes (Eigenreflexe of Hoffmann [30]), i.e. stretching of the corresponding musculature. This is the case when during rest or motion the ankle is moved by a flexor impulse which is not counterbalanced voluntarily. In clinical examination this effect is achieved in an uncommon way by a sudden blow to the tendon. Normally, the ankle during an active movement is suddenly pushed against an unexpected object. The reflex is then integrated into the on-going movement and smoothens the discontinuity by overriding the resistance. In this function, the patellar reflex, as other stretch reflexes, represents a dynamic element in the hierarchy of physically induced, goal-directed actions. The primitive nature of this dynamic element is documented by the fact that its response occurs only in the stretched muscles [30] and perhaps in their antagonists [55]. Relatively low segments of the central nervous system are responsible for the relay between afferent and efferent limbs of this reflex arch. In hierarchically organized motor acts the patellar reflex represents merely a peripheral link of effector functions. This statement is of importance to our problem. It means that patellar and pupillary reflexes – with respect to the main chain of animal function – operate at opposite ends. The pupillary reflex is found at the entry-side of excitations leading to conscious behavior. The patellar reflex, in contrast, belongs to the hierarchy of motor actions at a stage where the last touches are applied to the dynamic pattern formed by psychic functions and higher reflex activity, a stage which consists of a smoothing and ironing out of the motor response just before it is finally released. The pupillary reflex stands at the beginning of the receptor phase, while the patellar reflex is the final link in the effector phase of the same function. Hence, we have found a basis for understanding the relationship which exists between the attenuation of animal reflexes and sleep-induced pupillary constriction. In both instances we are dealing with the same fundamental mechanism, namely, with the reduction of the "functional readiness" *(Leistungsbereitschaft)* of the animal system. We can conclude that sleep inhibition may reach into the most distal elements of the effector side of animal functions. The fact, however, that cutaneous reflexes *(Fremdreflexe)*, which are of higher rank order than stretch reflexes, are also affected during sleep, proves that sleep is not limited to the most distal elements.

In view of this situation we find it impossible to follow Trömner [53], who postulates a sleep organ which is basically confined to prohibiting sensory stimuli from entering conscious perception. We disagree especially with Trömner's claim that the sleep center is localized in the thalamus opticus. Sleep function is not to be identified with a blockade within a relatively circumscribed area. The inhibiting factor, whatever its nature may be, reaches far into the periphery where it affects animal functions *(animale Teilleistungen)*. In this respect, i.e., in the assumption of a broad spectrum of sleep functions, we are in agreement with Küppers [36, 37], even though we deviate from the rest of his concept of the nature of sleep.

If we now attempt to explain the *mechanism* which attenuates spinal reflex activity we have to confess that here the situation for research is not as favorable as with the pupil. Nevertheless, a few facts which may elucidate these events can be mentioned.

In the first place, the removal of cortical functions in accordance with Braunstein's hypothesis concerning the sleep constriction of the pupil does not play a role. If the removal of cortical influences were a primary cause, it would result in an increase of spinal reflexes.

In the second place, the prompt reappearance of patellar reflexes upon arousal excludes the possibility that the extinction of reflex activity during sleep could be due to fatigue of the reflex-mediating pathways. A chemically induced disturbance of the excitatory process would make it hard to understand that the patellar reflex is extinguished when sleep occurs after a 16-h waking period, while the same reflex can be normally active after 80 h of waking. Finally, it seems particularly noteworthy that reflexes occasionally can be sustained during sleep, and even give way to an exaggeration of reflex activity.

MacWilliam [40] has mentioned this phenomenon with respect to the patellar reflex in sleeping dogs whose behavior indicated that they were dreaming and that their dreams involved motor activity. The same has been observed by Lombard in dreaming human subjects. This behavior clearly shows that individual functions may be spared from the sleep-inhibition mechanism and thus contradicts the assumption that the factor which inhibits the reflex pathways during sleep could be of hormonal or toxic nature. In contrast, the sparing of individual functions from general sleep inhibition is better understood if we assume that the inhibiting influences are conveyed to the reflex arc by the differentiating action of nervous transmission.

The nervous nature of inhibition offers nothing new. It is encountered in the animal as well as in the vegetative nervous system.

The question mainly arises as to which of the two systems is involved. The unmistakable vegetative nature of sleep function suggests that the vegetative nervous system is responsible, and this notion is supported by the fact that sleep inhibition lacks a specific property which is encountered in inhibitory functions of the animal system. An example is the reciprocal antagonistic innervation in which the inhibition participates actively in the organization of an animal, i.e., goal-directed response. The relaxation of the antagonist means functional support of the agonist. Inhibition is an auxiliary response which enhances the perfection and the economy of animal goal-directed actions.

Even when the suppression of an entire reflex is at stake, the subordination of this inhibitory mechanism on behalf of an animal function is not to be ignored. Such is the case if two stimuli, each of which is capable of eliciting a reflex, collide for mechanical reasons. The collision may be prevented, however, if one of the two reflexes is suppressed. The more vitally important one prevails. The inhibition under these conditions is not to be considered a negative phenomenon. In contrast, it is achieved in the interest of optimizing an animal performance. In this respect the undisturbed reflex is more useful than the interference of two colliding reflexes. If we contrast this positive aspect of animal inhibitory functions with the passive role and the helplessness of a sleeping subject, we can all the more appreciate the contrast between inhibition during sleep and inhibition in the realm of animal functions.

The Extinction of Psychic Functions During Sleep. The problem of psychic activity during sleep concerns the central issue of our main topic. It seems fully justified, therefore, that we introduce this problem only now, after having dealt with the receptor and the effector phase of animal functions. The suspension of consciousness is the most prominent sign of sleep. It will be remembered that the sharp difference between psychic functions during sleep versus wakefulness was originally at the root of our interest in the sleep problem.

According to Trömner [53], the inhibition which prevents sensory stimuli from entering consciousness has an important role in physiological sleep induction. After having dealt with the behavior of the eye during sleep we should add that the "denial" of sensory reception by the "sleeping" sense organs should also be considered as playing a role in the suppression of cortical excitation and extinction of psychic functions. Against this notion we argue that even the strictest sensory deprivation fails to lead to the loss of consciousness as we find it in sleep. The engrams of previous stimuli allow psychic functions to continue without the inflow of new excitations from the senses. Aside from daily life experiences, the observations of Kreidl and Herz [35] in mentally deficient patients prove that external stimuli lack the decisive significance for sleep induction for which they have been credited on the basis of an observation reported by Struempell. Sensory rest is but one condition for the development of central sleep inhibition. It does not, however, represent sleep in itself. Thus, sleep is not consistently interrupted if excitations reach the cerebral cortex. People who are accustomed to living near a railroad track are no longer aroused by a passing train that makes a loud noise and shakes the house. Others who are not used to it are thrown out of sleep. Sensory excitations, if attenuated by the sleep mechanism, nevertheless arrive at the cortical level. They will even be processed and stored during sleep and may be remembered on the basis of engrams. The incoming excitations will vanish after a short interlude which is connected with stimulus identification. The fact that they vanish proves that the phase of stimulus processing is also affected by inhibitions during sleep. Only intracentrally acting ones can explain the interruption of mental activity during sleep. However, there may be a relationship with the inhibition of sensory functions as well, because the effect of central inhibition is supported by the coordinated suppression of peripheral sense organs.

As concerns the nature and the mechanism of these central inhibitory functions, we can only repeat what we said about pupillary and patellar reflexes: Functional insufficiency caused by exhaustion is out of the question. Herz [22] has studied reaction time and memory functions in the state of sleep deprivation in volunteers. This deprivation in one case lasted 84 h and in another, 90 h. The mental functions remained unchanged aside from spells of sudden intensive sleepiness. This finding agrees completely with observations by Lee and Kleitman [38] which we have mentioned before. During waking periods of 40–115 h the responses to visual and auditory stimuli were not notably impaired. Associative functions remained likewise intact. Only the tasks which demanded a finely differentiated response required more time, and more errors were made than during normal waking periods.

The preservation of psychic functions during long hours of sleep deprivation proves that, at a time when we normally sleep, the conditions within the animal nervous system are not such that we would be unable to react. If at the end of a 16-h waking period the psychic functions are nevertheless suspended when we fall asleep, we have to conclude that somehow resistances have been introduced which inhibit the exchange of excitations.

The findings of Straub [52] and co-workers concerning the increase of CO_2 in the blood and in the tissues of sleeping persons may suggest that the interruption of psychic activity could be due to the CO_2 effect. This mechanism can be excluded for the same reason it was excluded as an explanation for pupilloconstriction during sleep. The time course of this factor differs considerably from that of sleep inhibition. The difference is especially evident when we are woken up by an arousing stimulus. The return to normal blood gas concentration is bound to an extremely sluggish mechanism due to the buffering of blood and tissues. In contrast, the resistances responsible for the suppression of psychic functions during sleep are extremely flexible. If we wake a sleeping person, we can see that the normal play of excitatory mechanisms returns within just a few seconds without leaving a trace of preexisting conditions except for beneficial feelings. This point reveals a principal difference between loss of consciousness during sleep and during anesthesia. In the anesthetized person the psychic functions are also suppressed or eliminated. The chemical factor of inhibition, however, has a much more lasting effect on the substrate of psychic functions, and this is why its elimination takes far more time.

We may insist all the more on the rejection of the CO_2 theory because the diminution of gas exchange during sleep is perfectly integrated into the general mechanism of sleep functions. Respiratory activity is most intimately related to the function of skeletal musculature. To a certain degree it represents an auxiliary function within the animal system and is closely coupled with it. Under such circumstances it is conceivable that fluctuations in excitability which characterize the difference between sleep and wakefulness involve the respiratory center as well. Obviously the decrease of systolic blood pressure during sleep is likewise a manifestation of sleep inhibitory actions. The increase of ventilation and the rise of blood pressure upon awakening are indicators of arousal occurring within the vegetative system, and which is in close connection with the animal apparatus. This interpretation is consistent with evidence presented by Gollwitzer-Meier and Kroetz [19] showing that changes in blood and tissue during sleep and wakefulness are due to the differential excitability of the respiratory center. The depression of this center parallels other changes which we have already discussed.

Instead of CO_2 one might postulate another chemical factor which may be responsible for inhibition of psychic functions, for example, a hormone or a hypnotoxin as proposed by Piéron [45]. The flexibility of the inhibitory factor which we have used as an argument against the CO_2 theory also speaks against a hormone or hypnotoxin, at least in the role of a decisive factor. The possible contribution to a slowly developing change in psychic activity by a chemical (hormonal) factor is not entirely excluded.

The fact that individual functions may be spared by sleep inhibition also speaks against a decisive role of an inhibitory principle mediated by blood and cerebrospinal fluid. The best-known example is the mother who pays attention to her child even during sleep. A minor noise caused by the child immediately arouses her. At the same time, the most intensive acoustic stimuli coming from the street remain ineffective.

Dream processes likewise seem to prove that sleep inhibition spreads selectively. Motor activity may be experienced intensively during dreaming in the absence of actual motor manifestations. This may indicate that specific regions of the central nervous system are excited while others are asleep. Somnambulism represents another example, and loud articulation during sleep still another.

This selective spread of sleep inhibition and its rapid reversibility correspond with other experiences of regulation mechanisms, provided that sleep inhibition of psychic functions represents a neural mechanism. This confirms the conclusion which has been reached in the discussion of reflex excitability. Here too the inhibition can only be mediated by elements of the vegetative nervous system. The loss of consciousness during sleep is an obvious manifestation of vegetative rather than animal regulations, and even if we still refuse to accept the possibility that the cerebral cortex is the target organ of vegetative innervation, we cannot deny that the cortical function is dominated by vegetative regulation during sleep. The latter, however, is the main requirement for the former to occur.

Only recently, Pavlov [44] has expressed the opinion that "internal inhibition" of conditioned reflexes and sleep are one and the same process. He supports this conclusion by his observations on conditioned reflexes where the experimental animal easily falls asleep if the interval between conditioned and unconditioned stimuli exceeds the period which was allowed during the training, for example, 30–60 s instead of 10 s. The conditioned reactions disappear during the delay period. Together with the inhibition of vegetative functions one frequently observes inhibition of psychic functions, i.e., the animals fall asleep. This observation provides positive evidence in favor of the neurodynamic nature of sleep inhibition. Pavlov seems to base his analysis of these phenomena on a corticocentric interpretation of sleep and to limit them to the animal pathways. Additional experiments are needed for further clarification of this problem.

An extended analysis is provided by observations derived from human pathology. Mauthner [41] has pointed to the relationship between brain regions in the vicinity of the oculomotor nuclei and sleep function. This conclusion was based on the frequent coexistence of insomnia and abnormal eye movements. A similar topography has been found as a condition of the somnolence which is frequently associated with encephalitis lethargica. In two cases of embolism, Lucksch [39] limited the area which is to be associated with pathological somnolence to the central gray matter surrounding the caudal portion of the third ventricle extending into the aqueduct. Under the impact of such clinical evidence the localization of a sleep center was accepted by several authors, although other investigators, for example, Lucksch himself, seem to recommend a more cautious attitude. It should be pointed out that a wide gap exists between the notion of a sleep center in the mesencephalon and a clear concept concerning the mechanism by which sleeplike activity may be induced. An explanation of cortical inhibition can be rarely obtained on the basis of the commonly accepted hierarchical organization of the brain, particularly as concerns the animal layers. In contrast, we should accept that the rank order within the animal functions may not at the same time determine the priorities of interaction with the vegetative regulations. The cerebral cortex, which seems to control the animal functions, constitutes but one element among others within the range of vegetative functions whose responsibility remains with the control of the internal conditions of the organism. It is the main task of vegetative regulations to protect the organism against wear and tear. The inactivation of the animal system during sleep is one important mechanism in this respect. When individual functions of the animal system are inhibited in order to induce a resting state, these animal elements become target organs of vegetative innervation. An important confirmation of this conclusion consists in the fact that areas of the central nervous system which, on the basis of histopathological findings, are topically related to sleep function, according to Greving [20] are morphologically comparable to the vegetative centers in the medulla oblongata and in the spinal cord.

With this reversal of rank order regarding the susceptibility to vegetative regulations, it is possible to bridge the gap between the postulated sleep center and the cortex. If cortical excitability is the result of competing sympathetic and parasympathetic influences, it is conceivable that a lesion in the corresponding regulation center may influence the balance. Parasympathetic predominance leads to diminished excitability and eventually to sleep. Sympathethic domination of the cortex prohibits sleep and predisposes to states of excitation. This would also explain the lability of symptoms, which is characteristic of encephalitis lethargica, for instance, with regard to oculomotor deficits. The association of purely vegetative disturbances, e.g., salivary secretion or sweating, with mental functions is easily understood when the cerebral cortex (with respect to the regulation of its excitability) and the vegetative organs are conceived to be equivalent in the pathogenesis of symptoms.

To these observations, which were obtained from physiology and pathology, we shall now add pharmacological data which confirm the impact of vegetative innervation on psychic functions in no less certain terms.

III. "Central" Effects of Vegetative Excitant Drugs: The Vegetative Nervous System and the Pathogenesis of Psychic Disorders

A number of drugs are known to influence the functions of the target organs of the vegetative nervous system. Since the excitability of the animal system may be subject to vegetative regulation, we have to consider the possibility that the same drugs can also influence animal functions, although not directly, but rather by changing their responsiveness to animal stimuli. The paradigm for such effects is the behavior of the pupil. Vegetative excitant drugs can influence visual functions by way of pupillary innervation. While they are unable to substitute the visual stimulus, they may nevertheless – and quite effectively – influence the response to visual stimulation by affecting the vegetative function which is set up in parallel. A

Table 1. Pilocarpine and retinal sensitivity in a human subject (values represent the mean of three determinations)

	Retinal adaptation	Size of the pupil in mm^2
Dark adaptation	11,960	41.6
Pilocarpine (40 min after instillation)	351	3.14

sympathomimetic drug, by acting on the pupil, may increase the responses of the retina by admitting more light to it. In contrast, a parasympathomimetic drug may diminish the responsiveness of the eye, since the constriction of the pupil introduces a barrier to the animal stimulus.

In order to demonstrate such influences more concretely, we can quote some quantitative data from a series of experiments in which changes of the sensitivity of the eye were studied under the influence of pilocarpine. With the aid of Nagel's adaptometer we determined in the classical manner dark adaptation of the eye. The adaptation curve showed the time that was necessary for the retina to reach maximum sensitivity. At this moment pilocarpine was applied to one eye of a human subject and the adaptation curve was further pursued (see Table 1 for results).

The threshold of retinal sensitivity increased 34 times under the influence of pilocarpine, while the pupil size diminished 13 times.

In other experiments [29] ratios of retinal sensitivity were found to be 57:1. By measuring the pupillary size with the aid of flashlight photographs it was possible to refer the cause of these changes partially to the pupil and partially to the retina. We have discussed retinal adaptation in Sect. II of this chapter. The example of the eye served the purpose of elucidating the subsequent experiments and of reminding the reader that the vegetative nervous system may well influence the excitability of animal organs.

Inhibition of Psychic Activity by Ergotamine. A number of excitant drugs are available which may change the balance between sympathetic and parasympathetic innervation; for example, adrenaline, cocaine, and atropine may act in favor of the sympathetic component while acetylcholine, pilocarpine, physostigmine, and ergotamine support the effects of the parasympathetic system. It should be added that the precision by which selective effects can be induced may vary from drug to drug and is dose dependent. The mechanism can be mediated by exciting one component as well as by blocking the antagonist. In the latter case the physiologically effective nerve impulse to the target organ is rendered ineffective.

It is from the above-mentioned substances that we chose our tools for influencing the central nervous system pharmacologically. These may be applied subcutaneously or intravenously. It is to be expected that psychic changes may accompany vegetative effects. However, the general symptoms are not merely due to a summation of changes in the primary target systems. Any imbalance in the function of an individual organ may cause secondary repercussions in the remaining organs of the body. This is easily demonstrated in the circulatory system. If the heart activity is inhibited by an artificially introduced stimulus, the organism fails to obtain the normal blood supply necessary to maintain oxygenation and nutrition of the tissues. Failure of internal respiration develops and in extreme cases the danger of suffocation ensues. Under such conditions we may see excitatory or paralytic states which are only indirectly due to the effect of the drug. It is likely that compensatory mechanisms are put into action which tend to restore the normal balance. For instance, adrenaline release may occur and by its action contaminate the primary drug effect.

The same parasympathomimetic drugs which inhibit the heart may enhance gastrointestinal motility, i.e., peristalsis. Spasms of the intestinal musculature may occur, which in turn may cause pain and lead to altered behavior. Thus we cannot expect to obtain a true picture concerning the primary undisturbed and uncontaminated effect of a given drug on psychic activity. Unrest may prevail where sedation could otherwise be expected.

In order to avoid or minimize contamination with secondary effects, we have to apply the drug in such a manner that it acts in isolation on a given organ. There are two ways in which to reach this goal. Firstly, we may resort to biologically active substances which have a specific affinity to certain organs or organ systems. The second consists in the local application of the drug which heightens the concentration in a given organ. In the case of the brain this method consists of intrathecal application. The excitant drug, by its distribution in the cerebral spinal fluid, can affect a wide target area.

In a first experimental series we applied pilocarpine intraventricularly because of its conspicuous parasympathetic effect on the eye. We hoped to see the effect of a parasympathetic predominance in the realm of psychic activity. However, the result was complicated by the extremely rapid resorption and diffusion throughout the entire organism. Evidently the high solubility which is characteristic for pilocarpine enhanced this unintended effect (the situation in the brain is less advantageous than in the eye because the choroid plexus provide an enormous resorptive surface). Aside from central effects we saw so many general symptoms that we felt no conclusions should be reached for the time being.

The experience with pilocarpine prompted us to search for parasympathomimetic substances which are less soluble and could be applied to the ventricle in the form of fine suspensions. A prolonged resorption and higher concentrations might be expected in the immediate area of application as compared with the rest of the body, and under these improved conditions a more or less pure central effect attained. Neither physostigmine nor muscarine seem to have such properties. Low solubility, however, is found with ergotamine. For this reason we switched over to this alkaloid, whose central effects, as far as we know, have not yet been studied.

Ergotamine was purified from *Secale cornutum* by Stoll [51], and its biological effects are nearly identical with those of ergotoxine [3, 33, 34]. According to Dale and Spiro [14], ergotoxine acts only on those sympathetic components in which the sympathetic system exerts an enhancing (positive) influence. Ergotamine, in contrast, may also act on organs normally inhibited by sympathetic innervation [48]. Rothlin [48] also observed that ergotamine affects the parasympathetic system and concluded that an increased responsiveness to electrical and chemical stimulation of the vagus occurs under the influence of this drug. This indicates a sensitization in the realm of the parasympathetic system.

Observations such as these suggest that the biological effects of ergotamine are in the direction of our experimental intentions. In order to verify these claims we added further observations. It seemed especially important to define the relationship of this drug to the parasympathetic system in more detail. In confirmation of earlier results by Zunz [62, 63], we could make the following observations on the eye: instillation of ergotamine into the cat's eye causes pupilloconstriction within a few hours. This might be the result of sympathetic inhibition (blocking the dilator of the pupil) or of parasympathetic excitation (by way of the sphincter muscle); both effects might, however, also be involved. In order to decide between these possibilities we extirpated the superior cervical ganglion bilaterally, which meant the elimination of the sympathetic innervation of the iris. The effect of this surgical removal could then be compared with the effect of ergotamine. Evidently the ergotamine miosis was far more pronounced, and this suggested that the elimination of the sympathetic innervation cannot be the sole cause. More clarity can be added if ganglionectomy is combined with ergotamine application. The pupil is now narrowed within a few hours to such a degree that the lumen can hardly be recognized. This extreme effect can only occur when a parasympathetic stimulus is added to the surgical removal of the sympathetic innervation. In confirmation of Rothlin's sensibilization experiments we can conclude that ergotamine induces a dual response by acting on differential sites, yet has the end effect in the same direction: the balance between sympathetic and parasympathetic influences is changed in favor of the parasympathetic.

We now raise the question as to whether a drug with these properties may influence psychic activity. We chose the cat as an experimental animal because this species reacts clearly to ergotamine and also because alterations in demeanor are easily recognized in cats – especially in kittens. Finally, we chose cats because they have a finely regulated pupillary control whose behavior may clearly reflect changes in psychic activity.

Injections in most cases were made into the lateral and occasionally into the third ventricles (Mr. F. Brumann, medical student, aided in these experiments.) The localization of injection sites and the extent of diffusion were tested in preliminary experiments with injections of dye solutions and subsequently examined on the basis of autopsy sections of the brain. In order to reach the lateral ventricles we prepared fine holes by means of a dental bur drill 2 mm behind the coronal suture and 3.5 mm lateral to the sagittal suture. By vertical insertion of the needle one can reach the lateral ventricles at a depth of 14 mm below the external surface of the skull. The third ventricle can be found with a hole situated 5 mm behind the coronal suture and 2 mm lateral to the sagittal suture. When the needle is tilted slightly toward the median plane, the third ventricle is reached at 20 mm depth. These figures are typical for a cat of medium size. However, there is relatively little difference in these measurements between cats of different size. The ventricles are large enough to offer a certain tolerance.

The injection must be performed under deep anesthesia, because any movement of the animal may cause injury to the brain and this in turn may lead to the rupture of a choroidal vessel and subsequent intraventricular hemorrhage. Excitation of the animal should also be avoided because adrenaline may be released under such circumstances, as was demonstrated by Cannon [10] and his collaborators. In view of the sympathomimetic properties of adrenaline this effect might critically antagonize that of ergotamine.

The injected volume is to be kept as low as possible in order to avoid the increase of intracranial pressure. This point is the more important because we discovered that it is rather difficult to remove the equivalent volume of cerebral spinal fluid through the injection needle. Our procedure was as follows: we injected 1 cm^3 of 1‰ solution of ergotamine tartrate precipitated by adding sodium bicarbonate. Controls with neutral red showed that this precipitate occurs precisely at the turning point from neutral to alkaline, i.e., in the range of biological indifference with respect to pH. The fine precipitate was concentrated by sedimentation to a volume of 0.3–0.4 cm^3 – containing approximately 1 mg ergotamine – and distributed equally to the two lateral ventricles. Injections into the third ventricle were made in a single dose of 0.3 cm^3. Injections were made under sterile conditions and at a low speed, constantly checking the resistance. Fine platin-iridium cannulas whose length corresponded to the depth of the injection site were applied. The injection fluid was warmed up to 37 °C. All these precautions were taken in order to avoid side-effects. Control experiments with physiological sodium chloride and with 1‰ sodium tartrate solutions were also undertaken.

The animals recover from chloroform anesthesia within 5–15 min. They begin to walk around and seem initially somewhat dizzy. Within 10–30 min after the injection they usually vomit and occasionally defecate. Otherwise nothing unusual is observed except for a certain unrest which manifests itself by constant locomotion. After 50–100 min the animal is notably quiet. It searches for a dark and warm corner and stretches out as if tired. Soon it lowers the head, closes its eyes, and falls asleep. Moderate intense acoustic or tactile stimuli awake the animal which then opens its eyes, looks around, and soon falls asleep again. It is noteworthy that the animals assume a typical sleep posture before falling asleep. As in physiological sleep they respond to minimal tickling of the ears with a typical pinna reflex, i.e., by shaking the ears, without awakening. The whisker reflex is abolished, but it reappears as soon as more intense arousing stimuli are applied. Now the response to discrete forward movement of individual whiskers consists, as in normal animals, in the retraction of the ipsilateral upper lip.

Shaking or displacing the sleeping cat can completely arouse it. It tends to return to the previously held dark corner of its own accord, but only to resume its sleeping position. However, it can be distracted by being offered milk, to which it responds by sniffing without usually drinking. In contrast, if a mouse is presented at some distance, the cat immediately pays attention to it and catches it with skill. If the mouse is removed by the experimenter the cat will continue to search for a while, then forgets and begins to sleep again.

This condition might last for 2–6 h. The drowsiness gradually subsides, but even after 10 h one may recognize symptoms of sleepiness.

From this description, which is based on numerous experiments, it can be concluded that intraventricular application of ergotamine induces a response which may be characterized as inhibition of psychic activity and behavior. This condition is strikingly different from anesthesia or narcosis and is characterized rather by a form of pathological somnolence with undeniable similarity with encephalitis lethargica. It is notably accompanied by profuse and long-lasting salivation. We underline the duration because cats kept in captivity seem to have a tendency to respond rather frequently with salivation. The behavior of the pupils is of special interest. If an ergotamine-treated cat assumes the sleep position one can note an increased pupilloconstriction. During sleep the pupils are extremely narrow. This constriction is maintained even in rooms with dimmed light, in which case we used a flashlight for the observation of the pupil.

The behavior of the pupil, however, is quite different from that under peripheral ergotamine application, e.g., ergotamine instillation into the conjunctival fornix. Peripherally induced pupilloconstriction is then accompanied by a complete inability of the iris to respond to psychic excitations. In contrast, the central ergotamine effect fails to abolish pupillary reactions. For example, acoustic or tactile stimuli are instantly followed by pupillodilatation. This effect subsides gradually as the stimulus-induced arousal gives way to somnolence. The functional character of this type of pupilloconstriction is particularly well recognized if the cat is confronted with an object of vital interest, for example, a dog or a mouse. The degree of pupillodilatation which ensues is in no way smaller than that of a normal animal. In contrast, peripherally applied ergotamine induces a pupillary constriction which is influenced by psychic excitation only when the alkaloid effect is either initial or terminal. And even then the response is less marked.

When observing the pupil one cannot infrequently see an additional phenomenon: the pupil may react to light in a *paradoxical* manner. In response to a flashlight stimulus the pupil may initially dilate before the normal light reflex occurs. This paradoxical behavior is particularly evident if light exposure occurs in the initial stage of sleep in which the pupil is not yet fully constricted.

This paradoxical behavior is known to occur in physiological sleep. Thus, the light stimulus acts primarily like a tactile or acoustic stimulus, i.e., it produces an arousal reaction during which the pupilloconstriction is overcome. Only after reaching full vigilance can the regular light reflex occur.

When attempting to interpret the central effect of ergotamine we have to consider the fact that spontaneity of movements and actions is abolished and general reacitivity to external stimuli strongly diminished. This is sufficiently obvious since a normal cat in a strange environment observes unfamiliar objects attentively and is rarely seen asleep.

Furthermore, one should keep in mind that the cessation of psychic functions may not be due to an incapacitation. After all, the state in which we find the animal after ergotamine application is by no means similar to general anesthesia. On the contrary, we are confronted with a change in the central excitatory state associated with potentially intact psychic functions. This explains why the inhibitions can be overcome by an external stimulus of high intensity; the responses then reappear, appropriate and in full conformity with the stimulus.

In examining the mechanisms mediating the action of the drug it should first be considered that the neurological substrate of psychic activity could itself be the target of the drug.

If this is the case, we should expect the spread of the alkaloid within the cerebral gray matter to correspond to the sequential changes in terms of functional deficits or excitations; in other words, the sequence must correspond to the site and direction of the concentration gradient.

Since the structure of the central nervous system in its detail is not directly correlated with the functional organization, one may predict that individual functional components of behavior may not necessarily be abolished in an orderly manner. At certain stages, residual functions may remain in the absence of a harmoniously organized behavior; i.e., a complex texture of partially intact and partially abnormal functions may prevail.

A different syndrome may be expected if the target of the alkaloid regulates the responsiveness of the animal system. In such a case, we have to consider that the substrates representing the functionally active elements may consist of widely separated regions of the central nervous system. The physiological consequence of this arrangement is a cooperative action of those elements which are characterized by a common function. An experimentally applied drug may affect the animal regulation in a similar way, and the experimentally induced disturbance may elicit a harmoniously organized behavior which may resemble the normal pattern. The only abnormality consists in the fact that this behavior occurs inappropriately, i.e., not in conformity with the actual environmental situation. The indirect action of the drug may involve additional peculiarities: since the drug is acting on the central regulation and not on the peripheral organ itself whose changes are being studied, one can consider the nature of these changes as "functional." Even a slight influence on the regulation may induce quite conspicuous effects. Conversely, a relatively minor impulse in the opposite direction may prevent the effect or reverse it. Such responses may be remarkable indeed, but not necessarily consistent.

All these considerations lead to the conclusion that the influence of ergotamine on the animal system is an indirect one. For it is not the function of the animal elements per se which is impaired but rather their responsiveness *(Funktionsbereitschaft)*. The fact that this change is due to a vegetative excitant drug confirms our general notion about the role of the vegetative nervous system with respect to the regulation of psychic activity. The changes which can be observed show the same direction as those of vegetative functions, namely a shift in the balance between sympathetic and parasympathetic innervation in favor of the parasympathetic. This interpretation seems to be relevant also for the effects that have been observed by Barger and Dale [4, 5] following subcutaneous application of ergotoxine: lethargy, early fatigue, and numbness.

Critics of our interpretation of the ergotamine effect may assert that there is no reason to search for specific mechanisms since the analogy with hypnogenic drugs is close at hand. However, this argument provides no explanation, at least not one that clarifies the physiological context. In fact, we believe that our observations and conclusions are useful for the elucidation of the pharmacodynamics of various hypnogenic drugs.

Thus, even for these drugs the question must be raised whether vegetative mechanisms are responsible for the changes in behavior. We are inclined to accept this assumption all the more since artificially induced sleep differs notably from narcosis and is remarkably similar to physiological sleep.

Another argument directed against our conclusion may be as follows: the sleeplike effect might be due to cerebral injury, possibly edema or nonspecific alterations of the cerebrospinal fluid, all of which can elicit shock effects. This interpretation gives us the welcome opportunity to discuss certain functions of the vegetative nervous system which are directly related to the problem at hand.

We have already mentioned that a number of control experiments have been carried out with Ringer's or sodium tartrate solution. It is true that even in these experiments one can notice certain psychic alterations, especially signs of inhibition. However, these changes have never reached the intensity of ergotamine effects. Also, the maximal pupilloconstriction was missing. Nevertheless, it should be kept in mind that even surgical interventions on the brain may to a limited degree bring about inhibitions.

In this context we should remind the reader that inhibition of function in connection with an organic lesion is a well-known phenomenon even outside the psychological realm. This is well known, for example, for the digestive system. If the gastrointestinal tract is seriously injured there is a depression of digestive activity and its auxiliary functions, which is in no way limited to the site of the lesion and its immediate vicinity, but reaches far beyond. This expansion prevents the contamination of the lesioned area with feces. Occasionally, the same effect may be achieved by retrograde expulsion of the gastrointestinal content, i.e., by vomiting. In other words, we consider these mechanisms as protective reflexes which are elicited from the lesion area and projected into all parts of the system which maintain functional relationship with the latter. The protective reflexes may lead to inhibition (cessation of gastrointestinal activity), as well as excitation (vomiting), depending on the prevailing situation. The functional relief is certainly helping to promote recovery, and the physiological significance of these protective reflexes corresponds entirely to the regulatory mechanisms which are active under normal situations by inducing excitations and inhibitions. Quantitatively, however, these reflexes are characterized by a greater intensity. In accordance with their urgency they have a compulsory character in order to provide optimal conditions for repair. Also, they may well outlast the duration of those seen under physiological conditions.

The reflex-induced immobilization of organs is found also in the animal system, for example, in the case of injured legs. This is seen quite strikingly in animals with serious leg injuries. A pertinent case report has been published earlier on this subject [26]. Since mechanisms subserving tissue restitution may represent a function of the vegetative system which is responsible for the nutrition and oxygen supply of the tissues as well as for the elimination of metabolites, we are convinced that it is the vegetative nervous system which primarily influences the excitability of the animal system.

This interpretation of the nature of protective reflexes corresponds to our concept that reflex-induced immobilization of an injured extremity occurs not only in those regions where the musculature serves animal functions (e.g., locomotion), but also where the muscles are involuntarily activated, for example, in the case of shivering, which is a thermogenic mechanism. If the vegetative nervous system dominates the excitability of animal organs under physiological conditions, it seems only natural that it may also be called into action in the case of protective reflexes. By imposing inhibition on the pathways of animal excitation, as is the case in fatigue and sleep, it promotes restitution processes by immobilization.

Reflexes whose aim is immobilization by inhibition or diminution of general excitability parallel those shifts in excitability by which the sensitivity of animal organs is increased. A typical example is represented by the so-called Head zones of the skin. Here we encounter as increase of sensory excitability in the domain of animal functions which is due to vegetative innervation. It is quite typical that the effect of this vegetative reflex upon the animal system does not correspond to that of an animal stimulus but rather to a change of the response threshold.

Let us return to the original theme and consider the changes in psychic activity in relation to vegetative protective reflexes. Especially the diaschisis concept of von Monakow includes certain aspects which are closely related to such mechanisms and should be discussed here. We should like to point out that psychic alterations which are due to the mechanical injury of the brain cannot be used as arguments against our interpretation of ergotamine effects. On the contrary, it seems to us that lesion symptoms point in the same direction as those caused by ergotamine. By including these examples, we have shown how the principle of vegetative regulation of animal function is important in various systems. In fact, we believe that this may be the physiological basis for the understanding of many pathological symptoms which were heretofore difficult to explain.

Increase of Psychic Activity Under the Influence of Cocaine. Since we know that a drug with parasympathomimetic action may also induce an inhibition of psychic activity, we may expect that a substance with antagonistic action may induce the contrary effect, i.e., an increased psychic activity. Naturally, such influences might be related to the side-effects of peripheral drug action which we have mentioned before. We have cited two possibilities by which to limit the action of the drug. One is the local application which we have utilized in our ergotamine experiments. This time, however, we may have to use the second possibility: the selective sensitivity of the brain with respect to certain drugs. Cocaine is an alkaloid with sympathomimetic properties and marked psychic effects. In this respect, a great many observations are available because cocaine is a well-known addictive stimulant, and this body of knowledge is all the more important because it is concerned with human beings. We hope that our discussion will provide some insight into questions which may be relevant with regard to the specific problem of cocainism.

It is known that dose thresholds of cocaine vary substantially between individuals. Sufficient quantities may initially elicit a facilitation of activity. This is expressed in the acceleration of thinking and psychomotor activity, but also in verbal expression and general motor activity. Investigations of various authors, for example von Anrep [1] and Mosso [43; also 7, 13, 31, 61] have shown beyond doubt that the increased general acitivity is paralleled by a synergistic increase of excitability in the animal system, i.e., concerning the efficiency of muscle contraction which is particularly evident if the muscles were fatigued prior to cocaine application. The effect of this improvement is also seen when peripheral nerves are electrically stimulated. This means that the effect cannot be entirely due to the central nervous system. Direct myographic registration and dynamometric measurements provide objective evidence in support of this interpretation.

Finally, there are analogous phenomena concerning the afferent limb of somatic reflex activity, and we owe the relevant observations to Mosso [43]. By using an induction coil, he demonstrated that under the influence of cocaine, weaker skin stimuli than normal may reach the sensory threshold. Cocaine addicts also report the phenomenon of hyperacousia. The mechanism of raising sensory excitability can be studied in the eye, in particular the question of pupillary size and its effect on retinal responses. We mentioned this phenomenon previously. By dilating the pupil, cocaine enhances the mechanism by which the threshold for adequate stimulation of the eye may be decreased.

The various effects of cocaine can easily be regarded in a common context. The unifying formula consists in an increase of excitability in all aspects of animal activity. The multiple ways by which the coordinating factor may reach the target organs are already known from the analysis of sleep function. The fact that this principle prevails in the range of pharmacological action is truly remarkable and yet understandable. Target sites with pharmacologically equivalent reactivity are distributed throughout the animal system. These are the same sites which are open to the regulatory influences of vegetative innervation. The multiplicity of target sites ensures the success of this regulation and strengthens the innervatory effects.

Nothing new is added if we learn that cocaine also influences vegetative organs, since the functions of vegetative organs are so tightly bound to the animal system that they can be considered to be auxiliary to the latter. Consider circulation and respiration, which are activated when physical stress is imposed! And in the same context the constriction of the splanchnic vessels with synergistic inhibition of gastrointestinal functions should be mentioned as mechanisms which help to augment the efficiency of the animal system. It is not surprising, therefore, that increased pulse rate and blood pressure, dilatation of the bronchial musculature, increase of the respiratory volume, and inhibition of food intake belong to the spectrum of cocaine effects. Pronounced changes in the activity of vegetative functions are essential features of this syndrome. This conclusion is particularly compelling if one considers individual cocaine effects from a more general point of view.

We know that the branch of the vegetative nervous system which enhances animal functions is the sympathetic system and that cocaine acts as a stimulus on the latter. Therefore, cocaine-induced increase of psychic activity is a further indication that the vegetative nervous system is involved in the regulation of psychic activity.

The consequences of cocaine addiction, i.e., deterioration of the entire organism, clearly reveal the significance of this regulation. If the parasympathetic system is artificially suppressed in favor of the sympathetic, the tissues of the body are deprived of those elements which should normally prevail in the struggle between animal and vegetative activities; i.e., they function without protection against exhaustion and therefore are bound to be exhausted!

In order to prevent misunderstanding we would like to point out, however, that, among the various symptoms, certain effects of cocaine have been described which seem, at first sight, incompatible with the concept that has been proposed. In view of the notorious complications which occur even when only one factor disturbs a physiological balance, we prefer to refrain from presenting arguments which might possibly resolve these inconsistencies. Rather, we would like to admit that many points of the concept need further discussion. The relationship between cocaine and adrenaline effects which have recently been disclosed by Froelich and Loewi [16] may offer an important clue.

The Vegetative Nervous System and the Pathogenesis of Psychic Disturbances. In the introduction we proposed to devote this essay to the physiological foundations of psychopathology. We have therefore committed ourselves to present evidence in support of such relationships. It is not difficult to find an approach to the domain of abnormal psychic reactions, for example – as suggested in our account – by following up the problems of sleep and those of pharmacological actions. In our interpretation of the central effects of ergotamine we have concluded that the experimentally induced disturbance of psychic activity can be attributed to two essentially different mechanisms. One concerns the substrate of psychic processes of the animal system itself, and this presumably by chemical factors interfering with the basic mechanism of excitation. The other is mediated by the vegetative nervous system which controls the excitability of the animal system. Exactly the same can be said about mechanisms responsible for psychopathological alterations. Of special interest is the mechanism by which the substrate regulating the central excitatory state is affected. Here we can expect the abnormal reactions to be due to an exaggerated increase or decrease of excitability. The ensuing pathological conditions can be of varying and rapidly changing intensity.

Manifestations of this kind are clearly the result of functional changes in the substrate of psychic activity. The animal components forming the neural basis of these functions are not actively but passively involved. Thus, the abnormal behavior of the essentially intact animal system is conditioned by the influence of the vegetative nervous system.

This interpretation of psychological dysfunction provides a new approach for the understanding of the relationship between psychic disturbances and the impaired function of vegetative organs. Psychopathological defects appear in direct parallelism to the vegetative syndrome. Both seem to be influenced by the same causal factor, but the vegetatively innervated organs respond in their own language – the vegetatively innervated cortex being no exception. This approach, in addition, offers an explanation for "psychophysical parallelism" which has been discussed in detail by Weinberg [56–58].

We would like to point out that this way of thinking also touches upon the question of heredity of mental diseases. Insofar as the causal element of psychic disturbances is identical with the causal element of vegetative disturbances, we may expect that this parallelism is likewise expressed in the heredity of certain mental diseases. Connections between hereditary diseases may thus be uncovered which may be manifested in purely psychic or purely somatic symptoms.

Finally, we should like to emphasize that our concept concerning the pathogenesis of mental diseases does not preclude other explanations concerning the relationship between vegetative and psychic symptoms.

Summary and Conclusion

The fundamental distinction between vegetative and animal functions lies in the difference between the functional goals. Vegetative functions subserve the functions of the tissues. They are responsible for the maintenance of functional conditions in the cellular environment. The vegetative nervous system, in combination with the system of hormones, represents the regulator of these functions. It exerts this control by facilitating or inhibiting the functions of various and organ-specific tissue elements and by integrating these individual functions into a concerted action of the organism.

Whenever this organism as a unity of higher biological order rises to the level of a perceiving and acting entity we speak of animal functions. The aim of animal functions is the regulation of the reciprocal interactions between the living being and its environment. Psychic functions are part of these animal regulations.

There is a double reciprocity between visceral and animal functions. On the one hand the vegetative functions create the conditions required by the animal system for an optimal performance. On the other hand, the elements of the animal system subserve vegetative functions by obtaining food or by providing conditions in the external environment of the individual which are necessary for the undisturbed functioning of the vegetative organs. The two systems move in a circle whereby each operates for the other. The relation between the two does not consist in one-sided subordination but rather in mutual interdependence. It is for this reason that the regulation systems which control the two are likewise not in one-sided dependency. The harmonious interaction of the two systems is only possible if their regulations are likewise mutually interdependent. The influence of the animal system upon the vegetative is obvious and uncontested. It is manifested in the dependency of heart and lung activity upon bodily exercise and furthermore in the way in which psychic processes can influence the internal organs. In this context we refer to a monograph by von Wyss which will appear shortly from this institute [60].

Much less clarified is the problem of those reciprocal relationships between vegetative and animal systems in which the vegetative overrules, and this problem is discussed here. In the phenomenon of sleep we can recognize the consequences of this relationship in their most accentuated form. The physiological role of sleep, in our opinion, leaves little doubt that this state is the result of vegetative regulation. In accordance with this notion, we have to consider sleep as a functional manifestation of the vegetative nervous system. If the somatic system is deprived of its control during sleep, it is not because of exhaustion or because it is capable of sacrificing its autonomy, but because functional rest is imposed by the vegetative nervous system by inhibiting animal innervation pathways. These inhibitions also affect psychic functions. Therefore, sleep exemplifies the fact that the animal system as a whole, including the substrate subserving psychic activity, is subject to vegetative innervation. A peculiarity of this influence concerns the mechanism of its action. The latter does not consist in the conduction of excitations from one system to the other, but rather in the change of responsiveness of the animal system toward animal stimuli. In this way the readiness *(Leistungsbereitschaft)* of the animal system can be adjusted to the conditions prevailing in the bodily tissues. The excitability is diminished if rest is required at the cellular level and, depending on the degree of inhibition, fatigue or sleep will ensue. Responsiveness, on the other hand, is increased if the conditions in the external environment require an active response of the individual. Man at rest or asleep will be aroused and this arousal – as an indirect consequence of animal stimuli – is also mediated by the vegetative nervous system. Sleep is a particularly striking example of functional bias provided by the parasympathetic system. The sympathetic on the other hand favors a bias which is directed toward the facilitation of animal functions. It is for this reason that the signal for the disinhibition of the animal system is mediated by the sympathetic branch.

The influence of antagonistic vegetative innervations upon psychic processes is not only manifested in the sleep-wakefulness cycle. Each of these two states may be quantitatively and qualitatively differentiated. Maximal vigilance and excitement are reached in the affective state and a specifically directed behavior is seen. The opposite extreme within the range of animal functions regulated by the vegetative nervous system is represented by deep dreamless sleep. Dreams demonstrate the differentiating potencies of this regulation.

It is of importance that the regulation of animal activity by the vegetative nervous system can be directly observed, for instance, in the case of the pupil. Pupilloconstriction reduces retinal responses. It is induced by parasympathetic innervation and is thus a hindrance to the light beam which represents an animal stimulus. Pupillodilatation increases responsiveness of the retina. The sympathetic stimulus opens the gate for the light beam. If the pupil is narrowed during sleep it means that the eye denies its services in order to support the resting condition of the body. In its behavior the pupil is a reflection of the sleep orientation of the entire animal system, including psychic activity. If the pupil opens during arousal it means that visual operations return to normal, and this is the visible indication that the normalization of animal acitvity is controlled by the vegetative nervous system. The pupillary reflex demonstrates that parasympathetic innervation – consistent with a general law – assumes the role of a protective device which is ready to release inhibitions even during the waking state if the organism needs to be protected from taxing situations and the ensuing damage. An analogous interpretation is given in the case of functional inhibition of animal and vegetative organs which have already been injured and which require functional rest for optimal rehabilitation and restitution.

To the data taken from general and special physiology, we have added clinical and pharmacological observations which support the idea of a regulatory dependence of cortical functions on the vegetative nervous system. It is appropriate here to state that abnormal psychic activity, e.g., that associated with encephalitis lethargica, may be due to pathological alterations in brain regions known to be in close proximity to the areas regulating the vegetative system, not to mention the changes in psychic activity which can be induced by specific drugs known to act on selectively sensitive regions of the vegetative nervous system. Resonance to vegetative drugs of the substrates of psychic activity – because of their relationship to the vegetative nervous system – is clearly demonstrated by the fact that a so-called central effect can be observed. Analogous conclusions can be reached with respect to the pathogenesis of mental diseases (psychoses). Here we recognize the possibility that the same principle which harmonizes relations between vegetative and animal requirements in the psychic domain may turn out as the mediator of a disturbing factor. In this way the otherwise healthy organ of psychic activity may become affected even though the cause of dysfunction may lie primarily within the vegetative nervous system.

References

1. Anrep, B.v.: Über die physiologische Wirkung des Cocain. Arch. Ges. Physiol. Menschen Tiere *21*, 38–80 (1880)
2. Bach, L., Seefelder, R.: Atlas zur Entwicklungsgeschichte des menschlichen Auges. Leipzig, Berlin: Engelmann 1914
3. Barger, G., Carr, F.H.: Notiz über Mutterkornalkaloide. Chem. News *94*, 89 (1906)
4. Barger, G., Dale, H.H.: Ergotoxine and some other constituents of ergot. Biochem. J. *1*, 240–299 (1906)
5. Barger, G., Dale, H.H.: Über Mutterkorn. Arch. Exp. Pathol. Pharmakol. *61*, 113–131 (1909)
6. Bass, E., Herr, K.: Untersuchungen über die Erregbarkeit des Atemzentrums im Schlaf (gemessen an der Alveolarspannung der Kohlensäure). Z. Biol. *75*, 279–288 (1922)
7. Bieberfeld, J.: Ergebnisse der experimentellen Toxikologie. II. Organische Substanzen. A. Alkaloide. Ergeb. Physiol. *17*, 2–362 (1919)
8. Braunstein, E.P.: Zur Lehre der Innervation der Pupillenbewegung. Wiesbaden: Bergmann 1894
9. Brown-Séquard, M.: Du rôle physiologique et thérapeutique d'un suc extrait de testicules d'animaux d'après nombre de faits observés chez l'homme. Arch. Physiol. (Paris) *21*, 739–746 (1889)
10. Cannon, W.B.: Bodily changes in pain, hunger, fear and rage. An account of recent researches into the function of emotional excitement. New York: Appleton 1922
11. Claparède, E.: Esquisse d'une théorie biologique du sommeil. Arch. Psychol. (Genève) *4*, 246–349 (1905)
12. Claparède, E.: La question du sommeil. Année Psychol. (Paris) *18*, 419–459 (1912)
13. Cloëtta, M.: Die Vergiftungen durch Alkaloide und andere Pflanzenstoffe. In: Handbuch der inneren Medizin, Vol. VI. Mohr, L., Staehelin, R. (eds.), pp. 685–725. Berlin: Springer 1919
14. Dale, H.H., Spiro, K.: Die wirksamen Alkaloide des Mutterkorns. Arch. Exp. Pathol. Pharmakol. *95*, 335–350 (1922)
15. Endres, G.: Atmungsregulation und Blutreaktion im Schlaf. Biochem. Z. *142*, 53–67 (1923)
16. Froelich, A., Loewi, O.: Über eine Steigerung der Adrenalinempfindlichkeit durch Cocain. Arch. Exp. Pathol. Pharmakol. *62*, 159–169 (1910)

17. Gaupp, R.: Über Wesen und Behandlung der Schlaflosigkeit. Verh. Dtsch. Kongr. Inn. Med. *31*, 9–44 (1914)
18. Geiger, E., Loewi, O.: Über Änderung des Cholingehaltes der Froschmuskulatur durch elektrische Reizung. Biochem. Z. *127*, 174–180 (1922)
19. Gollwitzer-Meyer, K., Kroetz, Ch.: Über den Blutchemismus im Schlaf. Biochem. Z. *154*, 82–88 (1924)
20. Greving, R.: Zur Anatomie, Physiologie und Pathologie der vegetativen Zentren im Zwischenhirn. Z. Ges. Anat. Abt. III: Ergeb. Anat. Entwicklungsgesch. *24*, 348–413 (1923)
21. Hartmann, F.A., Waite, R.H., McCordock, H.A.: The liberation of epinephrin during muscular exercise. Am. J. Physiol. *62*, 225–241 (1922)
22. Herz, F.: Selbstbeobachtung über freiwillige Schlafentziehung. Pflügers Arch. Ges. Physiol. Menschen Tiere *200*, 429–442 (1923)
23. Hess, C.v.: Eine merkwürdige Schädigung der normalen Fovea durch Miotika. Arch. Augenheilkd. *86*, 89–98 (1920)
24. Hess, W.R.: Die Regulierung des peripheren Blutkreislaufes. Ergeb. Inn. Med. Kinderheilkd. *23*, 1–65 (1923)
25. Hess, W.R.: Über die Wirkung von Acethylcholin auf den Skelettmuskel. Q. J. Exp. Physiol. Suppl. Vol.: Proc. XIth Int. Physiol. Congress, Edinburgh 144–146 (1923)
26. Hess, W.R.: Die reflektorische Ruhigstellung schmerzender Körperteile. Pflügers Arch. Ges. Physiol. Menschen Tiere *203*, 539–541 (1924)
27. Hess, W.R.: Zur Physiologie der Vasomotoren. Schweiz. Arch. Neurol. Psychiatr. *14*, 20–29 (1924)
28. Hess, W.R.: Diskussion zu Michaud, L.: Le rôle du système nerveux autonome dans la pathogénèse des maladies internes. Schweiz. Arch. Neurol. Psychiatr. *14*, 308–309 (1924)
29. Hess, W.R., Lehmann, F.: Der Einfluß vegetativer Reizstoffe auf den Adaptierungszustand der Retina. Pflügers Arch. Ges. Physiol. Menschen Tiere *211*, 603–611 (1926)
30. Hoffmann, P.: Untersuchungen über die Eigenreflexe (Sehnenreflexe) menschlicher Muskeln. Berlin: Springer 1922
31. Joel, E., Fränkel, F.: Der Cocainismus. Ein Beitrag zur Geschichte und Psychopathologie der Rauschgifte. Berlin: Springer 1924
32. Kleitman, N.: Studies on the physiology of sleep I. Am. J. Physiol. *66*, 67–92 (1923)
33. Kraft, F.: Über das Mutterkorn. Arch. Pharmazie *244*, 336–359 (1906)
34. Kraft, F.: Kristallisiertes Hydroergotininsulfat. Nachtrag über Mutterkornalkaloide. Arch. Pharmazie *245*, 644–645 (1907)
35. Kreidl, A., Herz, F.: Der Schlaf des Menschen bei Fernbleiben von Gesichts- und Gehörseindrücken. Pflügers Arch. Ges. Physiol. Menschen Tiere *203*, 459–471 (1924)
36. Küppers, E.: Der Grundplan des Nervensystems und die Lokalisation des Psychischen. Z. Ges. Neurol. Psychiatr. *75*, 1–46 (1922)
37. Küppers, E.: Weiteres zur Lokalisation des Psychischen. Z. Ges. Neurol. Psychiatr. *83*, 247–276 (1923)
38. Lee, M.A.M., Kleitman, N.: Studies on the physiology of sleep II. Am. J. Physiol. *67*, 141–152 (1923/24)
39. Lucksch, F.R.: „Schlafzentrum". Z. Ges. Neurol. Psychiatr. *93*, 83–94 (1924)
40. MacWilliam, J.A.: Some applications of physiology to medicine. III. Blood pressure and heart action in sleep and dreams. Br. Med. J. *II*, 1196–1200 (1923)
41. Mauthner, L.: Pathologie und Physiologie des Schlafes (Verh. ärztlicher Gesellschaften und Vereine. Protokoll der K.K. Gesellschaft der Aerzte in Wien). Wien. Klin. Wochenschr. *3*, 445–446 (1890)
42. Meyer, H.H., Gottlieb, R.: Experimentelle Pharmakologie als Grundlage der Arzneibehandlung. Berlin: Urban & Schwarzenberg 1914

43. Mosso, U.: Über die physiologische Wirkung des Cocains. Arch. Ges. Physiol. Menschen Tiere *47*, 553–601 (1890)
44. Pavlov, J.: „Innere Hemmung" der bedingten Reflexe und Schlaf – ein und derselbe Prozeß. Scand. Arch. Physiol. *44*, 42–58 (1923)
45. Piéron, H.: Le problème physiologique du sommeil. Paris: Masson 1913
46. Pietrusky, F.: Das Verhalten der Augen im Schlaf. Klin. Monatsbl. Augenheilkd. *48*, 355–360 (1922)
47. Reinhart, A.: Die epidemische Encephalitis. Ergeb. Inn. Med. Kinderheilkd. *22*, 245–359 (1922)
48. Rothlin, E.: Über die pharmakologische und therapeutische Wirkung des Ergotamins auf den Sympathikus. Klin. Wochenschr. *4*, 1437–1443 (1925)
49. Specht, G.: Vegetatives Nervensystem und Psychopathologie. In: Die Lebensnerven; ihr Aufbau, ihre Leistungen, ihre Erkrankungen, 2nd ed. Müller, L.R. (ed.), pp. 542–554. Berlin: Springer 1924
50. Stern, F.: Epidemische Encephalitis. Berlin: Springer 1922
51. Stoll, A.: Zur Kenntnis der Mutterkornalkaloide. Verh. Schweiz. Naturforsch. Ges. *101*, 190–191 (1921)
52. Straub, H.: Über Schwankungen in der Tätigkeit des Atemzentrums speziell im Schlaf. Arch. Klin. Med. *117*, 397–418 (1915)
53. Trömner, E.: Das Problem des Schlafes. Wiesbaden: Bergmann 1912
54. Verworn, M.: Schlaf. In: Handwörterbuch der Naturwissenschaften, Vol. VIII. Korschelt, E. et al. (eds.), pp. 906–918. Jena: Fischer 1913
55. Wachholder, K., Altenburger, H.: Über die Wechselbeziehungen zwischen den Sehnenreflexen und der antagonistischen Innervation unserer Muskeln. Arch. Ges. Physiol. Menschen Tiere *203*, 620–631 (1924)
56. Weinberg, A.A.: Psyche und unwillkürliches Nervensystem. Ein Versuch zur Darstellung einer psychophysiologischen Theorie I. Z. Ges. Neurol. Psychiatr. *85*, 543–565 (1923)
57. Weinberg, A.A.: Psyche und unwillkürliches Nervensystem. Ein Versuch zur Darstellung einer psychophysiologischen Theorie II. Z. Ges. Neurol. Psychiatr. *86*, 375–390 (1923)
58. Weinberg, A.A.: Psyche und unwillkürliches Nervensystem. Ein Versuch zur Darstellung einer psychophysiologischen Theorie III. Z. Ges. Neurol. Psychiatr. *93*, 421–445 (1924)
59. Wieland, H., Schoen, R.: Die Beziehungen zwischen Pupillenweite und Kohlensäurenspannung des Blutes. Arch. Exp. Pathol. Pharmakol. *100*, 190–216 (1923)
60. Wyss, W.H. v.: Über den Einfluß psychischer Vorgänge auf die Innervation von Herz und Gefäßen. Schweiz. Arch. Neurol. Psychiatr. *14*, 30–33 (1924)
61. Zangger, H.: Vergiftungen. Leipzig: Thieme 1924
62. Zunz, E.: Action de l'ergotamine sur la pupille chez le chat et chez le chien. C.R. Soc. biol. (Paris) *90*, 379–381 (1924)
63. Zunz, E.: Observations complémentaires à propos de l'action de l'ergotamine sur la pupille chez le chien. C.R. Soc. Biol. (Paris) *91*, 392–393 (1924)

Chapter 8

Sleep*

W.R. Hess

Contents

Sleep with its companion, the dream, is certainly one of the biological phenomena with which the sense of scientific inquiry of man busied itself from earliest times. With the establishment of a scholarly physiology, many facts that characterize the state of sleep ever more profoundly have become known; thus, at least for the natural scientist, sleep and dream have lost to a last residue the nimbus of the supernatural.

For many of the clarifying insights we are thankful to the observations of *clinicians* and *pathologists* who, for lack of opportunities to carry out physiological experiments, were limited to the evaluation of the symptoms that were observed in certain forms of disease as the result of experiments presented by nature. I wish to recall only the work of von Economo [6] and to mention the valuable contribution of the psychologists, particularly the interesting work of Claparède [3]. Finally, a number of zoologists and physiologists have concerned themselves with the phenomenon of sleep, although perhaps not in full correspondence with the fact that it is a physiological problem of the first order.

Whoever expects that a clear and convincing idea of the essential substrate and mechanism of sleep could be developed through the collation of the data that have been presented from different sides will be disappointed. I believe therefore that I may dispense with a repetition of theories which have been sufficiently presented in handbooks and monographs on sleep and utilize the available space better to pursue the query of *why* the mech-

* Hess, W.R.: Der Schlaf. Klin. Wochenschr. *12*, 129–134 (1933)

anism of sleep has remained essentially unclarified until most recently, and to show farther along what path we shall approach a solution.

What directs me in this is the notion that the problem of sleep may be studied and perceived *not isolatedly* but only with reference to the functions of the *entire organism.*

General Commentary on the Functional Organization of the Organism

Whatever physiological process we may extract for closer study from the vital manifestations of an organism, we must always dissociate it from the context of an ordered system. Because we have nothing substantial in sight in the domain of the functional it is not easy to see through this order (which integrates the individual functions into organic unity). In investigating the functional organization of the organism the feeling may arise that one is concerned with a theory more than with a description of factuality. However, this is not the case, since order is something factual whether it concerns arraying a structure or a functional system. We should like to emphasize this and so obviate any misunderstandable interpretations of the following presentation.

What we can express now about the functional organization of the organism is the following: an individual – observed in its entirety – is exposed to the influences of the environment, both harmful and useful. To utilize the one and to evade the other is the task of a special system of function that is *oriented toward the environment.*

A second system of regulation has an entirely different orientation. We show this best with a concrete case, for example, that of respiration. Here, too, it is a matter of mutual relationships between organism and environment, namely in the form of the respiratory gas exchange, but there is a decisive difference in that the consequence of this functional system attains its validity within the body, i.e. in the *milieu of the tissues.* The contrast of the two systems may be formulated in yet another way. The one system fulfills its function in the service of *the individual as a whole;* the second system fulfills conditions in the vital range of the *cells* of which the individual is composed. Thus the two systems of regulations are to be attributed to *two different levels of organization,* that of individual cells and that of the cell collective. In seeking support from the usual classification we recognize that the proposed division corresponds most readily to the one between animal and vegetative functions. I do not wish to enter here into the question of whether the congruence of the concepts is complete. Suffice it to know what we must observe in the further application of these terms, namely the distinction that ensues from the criterion of *functional achievement (Funktionserfolg).*

In this context we are interested primarily in the *relationship between the two systems*. This is best illustrated with the help of a concrete situation. With this intention let us observe an individual when he fulfills any purposive act connected with bodily work, for example, locomotion! It is common knowledge that in physical exercise pulmonary function as well as circulation react with increased activity. This implies a central organization which provides for the *coupling of animal and vegetative functions*. We do not want to examine the mechanism of this linkage here, but rather its significance. It becomes clearly apparent when the respiratory and the circulatory apparatuses are hindered by any disturbance or are adapting themselves to the function of the animal system. The consequence is a far-reaching impairment of the *efficacy* in the execution of the intended purposive act. Thus the circulatory and respiratory apparatuses, which belong to the vegetative system, regulate the efficacy of the other, namely the animal system.

Another example is digestion. Here, at first glance the relationship seems to be different, namely in so far as the failure of digestion has no such momentary effects on the action of the animal organs. At best, an inhibition of digestive function can have a passing beneficial effect (in terms of the redistribution of the blood). The interdependence is shown from another vantage when the closer and the more distant future are taken into view. The consumption of material that is connected with every release of energy leads to a limitation of the efficacy unless it is followed by a quantitatively appropriate restitution of the material. Here too, we meet fundamentally the same situation: the *effective operation* of the vegetative apparatus is the prerequisite for the efficacy now in the animal counterpart. Still, with both the cited examples we have noted a difference, notably with respect to the temporal relation: the functions of circulation and respiration are the bases for the *short-term improvement* of efficacy, a digestion is for *prolonged subsistence*. With the former there is the interest of the present, with the latter, the interest of the future.

With this differentiation we have found an important guide for a subgrouping of the vegetativ functions, namely the separation of two realms in which for the one part the sympathetic and, for the other part, the parasympathetic system take responsibility. Here too we want to put in abeyance the question of whether the subgrouping has been developed in conformity with the usual subdivision. I merely want to indicate that deviations which show themselves in detail do not in any way lead to a controversy as to which of the two partitions, that based upon the anatomical or that *based upon the physiological argument*, is the right one. From the extant deviations from a complete conformity we can extract only the motivation to carry out further investigations on the relationship between morphological and functional order and so we stride new ways to further knowledge.

We continue to speak of the relations between the vegetative and the animal system when we finally enter into the examination of a third example of the relationships that we meet in the *eye*. Visual orientation within the external environment belongs without doubt to the functions of the animal system. Yet the latter is unambiguously dependent on the vegetative innervation which regulates the width of the pupil. However, the vegetative system does not participate directly in the visual act. Its role is an indirect one in that it either diminishes the incident light stimulus by narrowing the pupil or grants it a free path by widening. The significance of this mechanism becomes clear when we think of the conditions in which this mechanism of the pupils becomes effective. We think, for example, of the wide pupil in dim light. With this we recognize again the relation of the vegetatively regulated process to the animal visual function. Whoever wants to become critically aware of this fact needs only to try in sparse light to find his way correctly when the pupil is narrowed by a miotic drug; the eye denies its service. The preponderance of the miotic influence in the counterplay of the sympathetic and parasympathetic components of the iris also confirms the rule mentioned earlier concerning the differential functions of the two divisions: the narrow pupil hinders an excessive stimulus and thereby helps to maintain the functional integrity of the neural apparatus. With this example we recognize that this securing has at once an *alleviatory* and a *protective function*.

These considerations derived from the eye are especially informative for us, because they show that besides that influence of vegetative innervation that comes via the regulation of vegetative organs (circulation, respiration, nutrition), an effect can occur in the animal organ directly and thereby achieve what is necessary for its effective and sustained operation. And in another aspect, the clear arrangements of the eye offer us a valuable insight into the functional interactions. We recognize namely that the narrowing of the pupil is a *positive accomplishment* in spite of its *inhibitory nature*. By increasing the efficiency of the photic stimulus the pupil economizes and protects at the same time. Inhibition is then not itself a goal of function but a means of purpose. This confirms a concept derived from other general physiological observations that the goal of biological order is never to be sought in a negative function, but rather in a positive one.

Perhaps it is not superfluous to use a *schema* to briefly elucidate the relationships we have discussed. In considering this schema it should become clear that functional relationships are represented in an abstract manner, i.e., are entirely separated from the anatomical substrates. In this sense we represent the total function of the body as a closed circular area (Fig. 8.1). From the total function we separate an outer ring which corresponds to the system that regulates the relation to the environment (*A*). In this ring we differentiate three segments as equivalents of the sensory (*S*), the motor (*M*), and the central (*C*) functions.

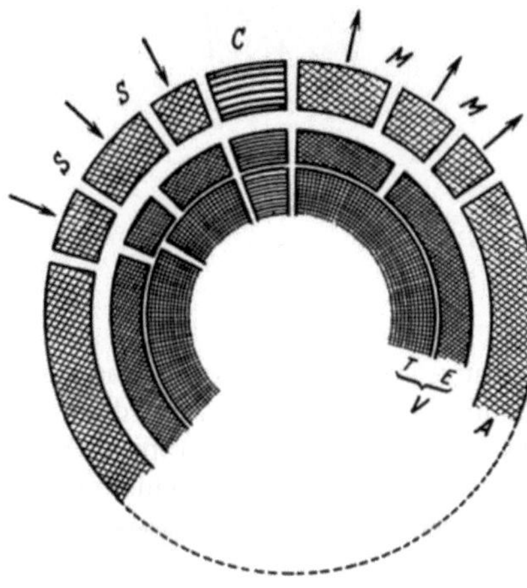

Fig. 8.1. Functional organization of the organism. The outer ring *A* represents the animal system which is responsible for the system of functions through which the organism controls its interactions with the external environment. This ring is divided radially into various segments attributed to different subsystems: *S*, sensory; *M*, motor; *C*, central. The two subjacent concentric rings together represent the vegetative system *V* as a whole. Its two components correspond to the ergotropic *E* and trophotropic *T* divisions, respectively. The former increases the momentary readiness for action of the animal system; the latter tends to promote protection, restitution, and economy. The radial division of both rings into different segments symbolizes the functional subsystem

The subjacent ring is divided into two concentric compartments (*E*, *T*) which are deliberately left incomplete. The significance of these compartments is easily recognized: they correspond to the functions that regulate the internal milieu of the body and thus represent the vegetative system (*V*). The ring that lies nearest to the animal system corresponds to the compartment most closely connected with the latter; it receives the activating influences of the sympathetic nervous system and we would like to designate it as the *ergotropic* division (*E*) because it facilitates the release of energy. The innermost compartment represents the other division, which is oriented toward economy, restitution, and protection. We identify this division with the necessary reservations, with the target of the parasympathetic division, and we would like to denominate it as *"trophotropic"* or *"histotropic"* (*T*).

After referring the individual spheres in our scheme to their functional meaning we shall examine their interactions. From the total system we select a single sector for this discussion.

We are interested at first in the functional connection between the vegetative and the animal system (Figs. 8.2, 8.3). For evaluating the effect of the sympathetic division we must keep to those organs upon which the sympathetic stimulus exercises a facilitatory effect, i.e., circulation and respiration. For the parasympathetic system the given example is the digestive ap-

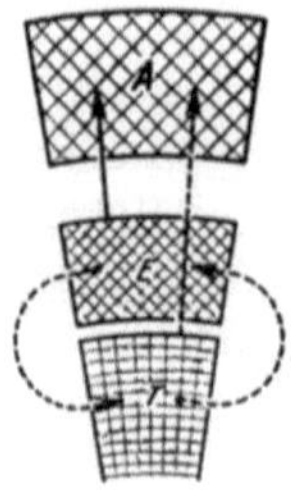

Fig. 8.2. Actions of the vegetative system upon the responsiveness *(Funktionsbereitschaft)* of the animal system. *A*, animal system; *E*, ergotropic division of the vegetative system, which enhances the responsiveness of *A*; *T*, trophotropic division of the vegetative system responsible for restitutive processes, e.g., by inhibition of the animal system

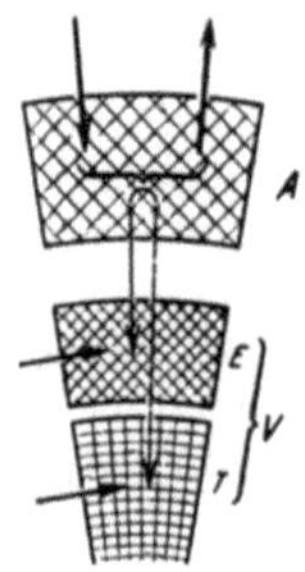

Fig. 8.3. Actions of the vegetative system. Afferent and efferent excitations in *A* (animal system) activate the ergotropic *E* and inhibit the trophotropic *T* divisions, respectively. Both divisions are under the influence of the internal milieu of the bodily tissues which they regulate

paratus, in which parasympathetic impulses heighten the activity. Here we must establish a fact that we have ignored hitherto. The enhanced activity of the digestive apparatus is connected with an increase of blood supply to the gastrointestinal tract. From this results a repartition of the blood supply in that the increased need on the one side corresponds to a diminished supply on the other. This regulatory effect, which ensues throughout in the sense of a *positive* accomplishment (that of secretion and resorption), is escorted by a negative component which impedes other functions, especially within the animal system. When we then direct our attention to the influence of the parasympathetic division upon the animal system we establish an *instantaneous* inhibitory effect which we express with a dashed connecting arrow. An additional fact is to be added which has been established by manifold observations. One segment of the vegetative system never acts in isolation but is always combined with another segment whereby the increased activity is coupled with the decrease in the other division. This is a consequence of reciprocal action of the two divisions in the sense that the increasing state of excitation in the one exercises a damping effect upon the other. It is expressed by the dashed arrows that connect the rings (*E*) and (*T*). Figure 8.3 mainly expresses the influences acting upon the equilibrium dominating the vegetative system, as well as the way in which the increase of activity in the ergotropic and the inhibition of the trophotropic division may be accomplished in interdependency with incoming and outgoing impulses of the animal system. The final modification that arises from the milieu of the tissues is represented in Fig. 8.2 by two horizontal arrows.

As pointed out before, we have dealt with the problem of functional organization by limiting ourselves to only one sector, and here we selected a number of examples that illustrate the situation as it exists in reality. Whether these conditions correspond to a generally valid law is another question. Is what we found in the eye also true of other sensory organs? The question has already been attacked experimentally. I recall the investigations of Foerster et al. [7], Achelis [1], and also of our own.

Suffice it to point out here that the questions regarding the functional organization of the organism, which we have discussed in a more general context, are really in urgent need of and ready for an experimental clarification. Besides the sensory functions one should also take the motor system into consideration. I mention, for example, the problem of the vegetative innervation of the skeletal muscles. But here too we must pass over the many aspects connected with this controversial topic.

The question is, rather, to what extent *the vegetative nervous system* influences the *central functions*, including the psychic activity. There exists no doubt that the influence that is mediated by the vegetative organs and which creates the essential conditions in the tissues not only serves the momentary readiness for action (*Leistungsbereitschaft*), but may also promote the restitutive function. To be discussed is the question of an influence such as that which we know from the eye where the *relation between stimulus and response* is regulated by the vegetative system. Such a direct influence is to be sought in two directions, namely in the sense of an increase of the momentary readiness for action and second in a constraint of it. This constraint in itself is *not* to be understood as the true goal of this regulation, but rather as the accompanying phenomenon of the restitutive processes, i.e., as a means to an end. In formulating this problem we have reached the essential point of our theme, for sleep belongs to that complex of phenomena which, by inhibiting the readiness of the animal system, facilitates the restitution of the organism. We want to emphasize here that the solution of this problem in its consequences reaches beyond the special interest into the domain of psychology.

The Symptoms of Sleep

The physician is accustomed to speaking of a syndrome when he has a group of causally related, physiological symptoms in mind. In an analogous way one may well speak of a physiological syndrome when one views a group of *normal* phenomena of life that are intrinsically interconnected. In an attempt to depict the manifestations of sleep we should emphasize first that which is most prominent: the inability of the sleeping person to apprehend external stimuli and to respond to them. This lack of responsiveness of the senses is only a relative one and can be overcome with sufficient intensity of the stimuli, in which case the organism enters into contact at once with its environment. A similar deficiency in the domain of the motor system is also noted. Lastly, that function in which the animal system exerts its highest potency, i.e., mental activity, is also extinguished or at least diminished.

With regard to the motor system we should like to direct the reader's attention to an essential point: not only do we no longer see those manifestations that occur, in the state of wakefulness, in direct interdependence with processes of consciousness, but a number of reflex activities may even become extinguished. Motor rest is then not simply a consequence of the failing consciousness but rather a specific symptom of sleep.

But not in the domain of animal function alone does sleep develop its effects; in the realm of *vegetative organs*, too, a number of functions are altered. This comes to its most conspicuous expression in that group of organs whose activity stands in closest dependence upon the state of function of the animal system. The operation of the circulation is notably lowered. One recognizes this in the slowing of the heartbeat and in the lessening of the blood pressure, together with a shallowing and slowing of respiration, which in turn causes a reduction of the respiratory gas exchange as reflected in blood gas concentrations. The reduced ventilation entails an accumulation of CO_2 which in turn leads to a pH increase. Thus we find blood conditions which, under physiological circumstances, are compensated for by an intensive increase of respiration. That this does not occur during sleep is again a fact that helps to explain its essence. It demonstrates a change within a system that regulates the relation between stimulus and response in the direction of a decreased excitability. Further alterations in the blood have to do with the equilibrium of calcium and potassium, according to the findings of Cloëtta and Thomann [4].

Differences in the digestive functions are less marked. The ingested food is processed properly during sleep. However, one deviation shows itself in that after completion of digestion no feeling of hunger arises, indicating that consciousness does not react to the conditions in the digestive system during sleep as it does during wakefulness. When we are truly tired we can even "oversleep" the regular mealtime. Excretion likewise continues during sleep. No retention of substances excretable with the urine is noted, yet the condition of sleep is not wholly without influence upon excretory functions. Diuresis is clearly reduced so that the bladder becomes filled less rapidly, a phenomenon which is analogous to the limitation of the secretion of tears and saliva and is certainly connected with the avoidance of disturbing stimuli during sleep.

To put it in one word, there is a *diminution* if not a complete extinction of the functions within the animal system, while the functions of the vegetative system continue to some degree. But this characterization is not fully apt and even may lead us astray, and although functional deficits do come into the foreground, there are *exceptions*. A known example is that the orbicularis oculi muscle as well as the sphincter muscle of the pupil, undergoe an increased tonus. The sphincter of the bladder and of the rectum are also spared from the general slackening. When we look around in the animal kingdom we find still more examples where the maintenance or even a strengthening of activity in certain muscles can be observed. In mentioning the muscle groups which are responsible for the sleep position or those regulating postural reflexes, it seems important to emphasize these exceptions because from them we can learn that the sleep state is in no way characterized by the dominance of a *negative principle* but rather that sleep inhibition is *actively selective*, even in adapting to momentary conditions. From this fact we conclude that although functional defect stands in the foreground of the sleep syndrome, if viewed from outside, a *specific organization* can nevertheless be recognized even during the sleep state. We underline this because it is this insight which provides the motivation to seek some ordering factor.

An evaluation of sleep based on entirely *negative* criteria is particularly inadequate when we now examine a different phenomenon which concerns processes that occur in concealment, yet whose existence can be proved indirectly by their functional result. By this we mean the impressive alteration which the entire organism experiences after completion of sleep. The missing enthusiasm for work and the actions of the person wearied by the day's work are transformed by an invigorating sleep into a sense of dynamism and a will to enterprise. With this observation we have placed the finger on that which we would like to designate as *the positive and therewith the decisive aspect of sleep*. We have already emphasized on earlier occasions that the physiological function is directed toward a positive effect, i.e., is based upon the interests of the individual. Thus the functional pause of the animal system as it is seen in sleep represents merely an inevitable corollary phenomenon. For the resting state of the animal system obviously facilitates the restitutive processes of its functional elements, the cells. Of what kind these restitutive processes are cannot be said at this time. It is conjectured that they are not similar to those that release energy by means of oxidation. The fact that the release of energy and its utilization in purposive action is abolished during sleep leads to the thought that the restitutive processes require conditions that remain undisturbed by oxidative cell metabolism, and that they also most likely concern the structural substitution of cell elements which are worn out during the active phase of wakefulness. But as we pointed out, we do not yet have any exact information about the nature of restitution, and it should be rewarding to pursue further the few beginnings which have been undertaken in this direction.

To add more contrast to the picture of sleep we should like to mention other states which in a certain way show similarities. One of these is *narcosis*. Indeed the observable behavior of the sleeping and of the narcotized person is nearly identical. It consists of a lack of responsiveness toward sensory stimuli, the abolishment of consciousness, and motor inactivation. A decisive difference between sleep and narcosis is observed when stronger stimuli are applied. The sleeping person, within the briefest time, will be in full possession of his faculties. The state of narcosis, on the other hand, can be eased somewhat by external stimuli, but never so that the individual regains complete control of himself and his actions. The state of narcosis, in contrast to sleep, is a disturbance in energy requiring metabolism which is not controlled by the organism. The difference between the two states in respect to restitutive processes is also a principal one: after physiological sleep one is refreshed, but after narcosis, exhausted.

Another state reminiscent in some respect of sleep is encountered in *coma*. The vanished consciousness and lack of responsiveness to stimuli are based, however, as in narcosis, not on controlled inhibition but rather on true defective function caused by a disturbing factor of autotoxic nature. But here, the deficit does not show the gradual preference for psychic functions that is typical of genuine narcotic drugs.

Similarities with sleep can also be recognized in somnolence and *stuporous states*. However, with the latter, too, the chief distinction of sleep, the restitutive function, is lacking.

More likely to be related to the state of sleep, according to the judgment of experts, is *hypnosis*. Here I must confess that I have been unable to form a final opinion. It is true that the state of hypnosis can pass easily into sleep, but perhaps only upon the basis of a constellation that is favorable for sleep. It is conceivable and even likely that the primary mechanisms concerned with these two states differ in essential points.

The Mechanism of Sleep and the "Sleep Center"

In the preceding discussion we have given an account of the *symptoms of sleep*, and we should now like to turn to the question of what changes within the body are responsible for inducing this condition. This problem is all the more important because sleep, as we pointed out in the preceding section, is not simply a blackout, but represents a positive function within the context of *purposive order*. This concept leads inevitably to the important consequence that the order is established by an active mechanism. In other words, we are now approaching the much discussed question of the sleep center, and we shall enter it at first only in the sense that it is *in no way conceded by all authors*. As a matter of fact this question is at present in a critical phase, because the concept of centers has been attacked only recently, to mention Bethe and Fischer [2] and Goldstein [8] and to allude to the terms *gleitende Innervation* (dynamic innervation) and *Funktionswandel* (plasticity). I shall soon have the opportunity on the basis of new experimental observations concerning the physiology of the brain stem [9, 10] to participate in this discussion, which is of considerable interest not only to physiologists but to clinicians as well. Suffice it to circumscribe the concept of a "center" as we use it here. We think of a neural apparatus which regulates a physiological function, in our case the function of sleep. I can think of this regulating apparatus only in the form of a *structural order*. Whether this arrangement can be precisely localized is another question which we will put in abeyance temporarily.

To prove the existence of a neural apparatus that is defined in this fashion we must have recourse to the experiment. Nature itself has performed such experiments. I refer to the findings of pathological anatomists concerning cases of somnolence which led von Economo [6] to conclusions which strike not only upon the existence but also upon the localization of the "sleep center." Interpretations of pathological alterations are, however, not as unequivocal as the physiological experiment because we do not know with certainty what to refer to pathological irritation and what symptom may be due to paralysis.

Two ways are open for experimental investigations. One departs from the concept that the sleep-inducing center is inhibited in the waking state. If this inhibition is abolished, sleep must intervene. The second method consists in artificially exciting the sleep-inducing mechanism. No matter which way we take, it is important not to produce disturbances such as narcotic effects or diaschisis. In our own experiments we have followed both ways, the first in the form of intraventricular injections of ergotamine, which is known to inhibit the sympathetic system without narcotic side-effects. The results were entirely positive, i.e., the experimental animals fell asleep after 20–30 min. Nevertheless, I discontinued these experiments because the interpretation of injection experiments meets with great difficulties since too many possibilities with regard to the mode of action must be taken into account, which should in no way detract from the great interest that is being invested in a special aspect of this approach. I am referring especially to the intracerebral injections of potassium and calcium carried out by Demole [5] and also by Marinesco et al. [12]. More clear-cut results are obtained in stimulation experiments, especially when the *electrical* impulses are exactly controlled with respect to intensity, duration, and waveform. Such impulses can be delivered through fine electrodes which are lowered into the diencephalon. In a number of cases a pronounced effect of sleep can be observed.

From the entire behavior of the animal one can conclude that this effect is not comparable to stupor or coma caused, for example, by an unspecific lesion of the brain, but rather that we are confronted with a change of state corresponding to physiological sleep. This impression is corroborated by the fact that stimulation of other sites produces differential effects. Occasionally, changes in mood or complex symptoms of the extrapyramidal motor systems can be seen under such circumstances.

From our experimental observations one can draw the conclusion that excitation of circumscribed areas in the brain is answered by sleep with all its characteristic signs of inhibition; the sites of stimulation are in the brain stem, and its target is the cerebral cortex with the corresponding receptor and effector systems.

By adducing insights which were gained from the study of the functional organization of the entire organism we can extend somewhat our conclusions on the sleep-inducing neural mechanism. We have seen that the regulation of the milieu of the tissues is the decisive task of the vegetative functions, whereas the relations of the individuum to the environment belong to the realm of the animal system. There can be no doubt as to which of the two systems assumes the responsibility for the sleep-inducing mechanism: *it can be only a functional manifestation of the vegetative nervous system*. And it is likewise beyond question for us that the hypnogenic center belongs to the parasympathetic (trophotropic) division, whose functions stand in the service of economization, organ protection, and restitution. In no case, however, do we imply that all organs whose functions are facilitated by the parasympathetic innervation should be activated increasingly during sleep. Sleep is a highly organized state that is accomplished *in the frame* of trophotropic (parasympathetic) functions.

Concerning the loci from which stimuli engender sleep we must still refrain from a definitive answer since the analysis of the extant experimental material is not yet completed. It is not enough to simply determine the sites of stimulation with a positive sleep effect. Additional criteria should be considered, for example, the fact that the sleep effect in single cases appears in direct connection with brain stimulation, while in other cases it may occur as an aftereffect, i.e., not until stimulation has been discontinued. Too, different forms exist in the sleep manifestations and demand an exact analysis. We see, for example, cases with particularly strong inhibition of receptor functions, where the sleeping animal can be aroused only with maximal intensity of external stimulation and only momentarily; yet the animal is sleeping in the sitting position. Other animals, conversely, are recumbent with deeply relaxed skeletal musculature, although they react relatively easily to external stimuli. We cannot decide with certainty whether the different forms of sleep are correlated with different loci of stimulation. What can be said positively is that the hypnogenic effect is also obtained in regions that lie significantly rostral to the area delimited by von Economo [6] for the sleep center. A preference for the deeper brain stem levels is obvious. What appears to be especially notable with respect to pathological somnolence is the close relationship between stimulation effects representing the extrapyramidal motor system and sleepiness. The in extenso presentation of the rather large material acquired during some years will follow in the already-mentioned monographs [9, 10] in preparation.

The clarification of the sleep mechanism should also include the problem of how it is initiated and in what way it is effectuated.

It seems certain that the primary stimulus of normal sleep is to be sought in the body itself. The compelling need for sleep can never be evoked alone by certain external conditions. We have to think in the first place, therefore, of products of stimulation, for instance, according to the theory of Piéron [14], the so-called hypnotoxins. Nothing is known about the nature of such substances, but we can foresee that they scarcely stand in relationship to the oxidative metabolism but rather with the products of cellular degradation. Yet, we do not mean to say that the external conditions are without significance for the triggering of the sleep mechanism. Pavlov [13] observed that a dog habituated to a conditioned reflex will succumb to sleepiness with certain changes in procedure, and beyond that daily experience speaks against it. With this we are inclined to give the observations of Pavlov an interpretation different from his own.

It is important to note that induction of sleep in no way depends only upon the removal of external stimuli, but rather on positive signals from the surroundings. We surely value these influences correctly by bringing them into relationship with the mechanism of the conditioned reflex and use the analogy of the increased feelings of appetite and even salivation that are induced by auditory and olfactory stimuli involved in the initial phase of feeding, especially at eating time. A number of authors, including von Economo [6], have expressed the opinion that a certain autonomic and inborn periodicity is intrinsic to the sleep-regulating apparatus. We ourselves should like to set a question mark here and admit a periodicity only in the sense that the influences of the diurnal rhythm – to which belongs also the trickling away of the hours – become factors in the sense of *the conditioned reflex*. The need for sleep orders itself according to the rules of the conduct of life, which, like the time of eating, are in part a matter of habit, and are also rooted, in part, in stimuli that remain unconscious.

With the second question, namely the means and ways through which the sleep apparatus exerts its influence, we must, in the "era of internal secretion", think of the possibility that a genuine sleep hormone is involved. The investigations of Zondek and Bier [15, 16] seem to provide some concrete evidence for this even though we must await further confirmation and an extension of their findings. Another way in which the sleep apparatus can reach the target organs is by direct neural connections which, for example, influence and possibly inhibit the exchange of excitation in the animal system by alteration of its excitability. The mechanism of humoral mediation discovered by Loewi [11] perhaps fits itself into this chain of events, for instance, in that the nervous impulse releases a humoral principle. These are thoughts and possibilities which indicate the way in which sleep research should advance if a systematic approach is desired.

Opposite to the question of how sleep is initiated is the problem of how awakening or, as may be, the artificial interruption of sleep comes to exist. Insofar as spontaneous awakening is concerned, one should first of all consider the disappearance of the hypnogenic factor that is effective from within the organism and which determines (with others) the depth of sleep. Besides this, however, the shifting balance of the vegetative system affected by neural excitation should be taken into account, for it is beyond question that the vegetative system not only influences the state of the animal system, but that it also depends upon it (see Fig. 8.3). Thus, based upon manifold observations, we may conclude that every stimulus, be it a photic or an auditory or a pain stimulus, is transmitted to the ergotropic division which in turn intensifies the sympathetic and inhibits the parasympathetic nervous system. In this way the arousing stimulus intervenes in the mechanism of sleep. Whether a *specific waking center* acts as an antagonist of the sleep center is another problem which we do not want to examine further here. For reasons of "symmetry" one could think of it. Various facts, however, seem to permit a different and less schematic arrangement.

References

1. Achelis, J.D.: Über Umstimmung der Sensibilität. Pflügers Arch. Ges. Physiol. Menschen Tiere *226*, 212–238 (1931)
2. Bethe, A., Fischer, E.: Die Anpassungsfähigkeit (Plastizität) des Nervensystems. Einführung und experimentelles Material. In: Handbuch der normalen und pathologischen Physiologie, Vol. 15: Correlationen 1/2. Bethe, A. (ed.), pp. 1045–1130. Berlin: Springer 1931
3. Claparède, E.: Opinions et travaux divers relatifs à la théorie biologique du sommeil et de l'hystérie. Arch. Psychol. (Genève) *21*,, 113–174 (1928)
4. Cloëtta, M., Thomann, H.: Chemisch-physikalische Untersuchungen zur Theorie der Narkose. Arch. Exp. Pathol. Pharmakol. *103*, 260–281 (1924)

5. Demole, V.: Pharmakologisch-anatomische Untersuchungen zum Problem des Schlafes. Arch. Exp. Pathol. Pharmakol. *120*, 229–258 (1927)
6. Economo, C. v.: Schlaftheorie. Ergeb. Physiol. *28*, 312–329 (1929)
7. Foerster, O., Altenburger, H., Kroll, F.W.: Über die Beziehungen des vegetativen Nervensystems zur Sensibilität. Z. Ges. Neurol. Psychiatr. *121*, 139–185 (1929)
8. Goldstein, K.: Über die Plastizität des Organismus auf Grund von Erfahrungen am nervenkranken Menschen. In: Handbuch der normalen und pathologischen Physiologie, Vol. 15: Correlationen 1/2. Bethe, A. (ed.), pp. 1131–1174. Berlin: Springer 1931
9. Hess, W.R.: Beiträge zur Physiologie des Hirnstammes. I. Die Methodik der lokalisierten Reizung und Ausschaltung subkortikaler Hirnabschnitte. Leipzig: Thieme 1932
10. Hess, W.R.: Beiträge zur Physiologie des Hirnstammes. II. Das Zwischenhirn und die Regulation von Kreislauf und Atmung. Leipzig: Thieme 1938
11. Loewi, O.: Über humorale Übertragbarkeit der Herznervenwirkung I. Pflügers Arch. Ges. Physiol. Menschen Tiere *189*, 239–242 (1921)
12. Marinesco, G., Sager, O., Kreindler, A.: Experimentelle Untersuchungen zum Problem des Schlafmechanismus. Z. Ges. Neurol. Psychiatr. *119*, 277–306 (1929)
13. Pavlov, I.: „Innere Hemmung" der bedingten Reflexe und der Schlaf – ein und derselbe Prozeß. Skand. Arch. Physiol. *44*, 42–58 (1923)
14. Piéron, H.: Le problème physiologique du sommeil. Paris: Masson 1913
15. Zondek, H., Bier, A.: Brom im Blute bei manisch-depressivem Irresein. Klin. Wochenschr. *11*, 633–636 (1932)
16. Zondek, H., Bier, A.: Der Bromgehalt der Hypophyse und seine Beziehungen zum Lebensalter. Klin. Wochenschr. *11*, 759–760 (1932)

Chapter 9

The Sleep Syndrome as Elicited by Diencephalic Stimulation*

W.R. HESS

Contents

The experimental investigation of central functions which are in control of *comprehensive behavior* demands a special method. Precautions must be taken that allow the experimental animal to be observed under conditions that leave it with a *minimum of disturbances* and provide a *maximum of freedom of expression.* The fulfillment of these requirements is especially important for the investigation of sleep, because in view of the lability of sleep even subtle influences can have very disturbing effects. This is consistent with the fact that the *responsiveness* of the central nervous system is the resultant of a fine play of excitations and inhibitions whereby competing external and internal factors come to the fore. The extremes, on the one hand, are states of highest tension with corresponding readiness for action and, on the other hand, an extreme decrement of sensory reception, of central processing, and of motor performance.

Among the data of more than 300 experiments performed in this laboratory with a method described earlier [26], there are a considerable number

* Hess, W.R.: Das Schlafsyndrom als Folge dienzephaler Reizung. Helv. Physiol. Acta *2*, 305–344 (1944)

in which the animal's readiness for action and reaction *(Aktions- und Reaktionsbereitschaft)* underwent a notable decrease. The symptoms which appear in this situation raise the question as to whether they are not correlated in any way with a physiological function or whether a nervous apparatus has been activated which fulfills a regulatory task in terms of the inhibition that occurs. If the latter is the case, we would be dealing with experimental interventions in the mechanism which regulates sleep and wakefulness. In order to decide this important issue, it is necessary to compare the experimentally induced symptoms with the phenomenology of normal sleep and also with abnormal syndromes of the central nervous system. For this purpose we present the following observations which were made in our stimulation experiments.

Experimental Observations and Localization of Stimulation Effects

The following report deals with the *behavior* of the cats that serve as experimental animals, as well as with the *conditions* under which this behavior became apparent. As far as clues concerning the disposition of the animal are available, these too are noted as something of importance. It should be emphasized that the experimental program was not initially planned as a study of sleep, but rather as a systematic point-by-point investigation of the diencephalon and the adjacent areas of the brain stem. We wished to analyze the effect of liminal and slightly supraliminal stimulation for every site that could be reached by our electrodes. Thus, the experiments discussed here were carried out at irregular intervals. Interspersed among them were cases with symptoms of another kind which in large part have been analyzed and discussed in other communications. With respect to the *method*, which has been published elsewhere [26, 28], only those points that are particularly important for the interpretation and critical evaluation are touched upon briefly here.

Stimulation was done with pulsating direct current; measures were taken to delay the rise and fall of the stimulus pulse and to prevent the polarization of the steel electrodes in the intervals between impulses. The first measure avoids as much as possible the disproportionate response of elements with short chronaxia, i.e., of the animal system, to the disfavor of the more slowly reacting elements of the vegetative nervous system. Usually the work was done with a stimulation frequency of 8 Hz. For a control, lower frequencies (to 2 Hz) and higher frequencies (12–15 Hz) were also used occasionally.

Fixation of the electrodes upon the skull took place under nitrous oxide anesthesia combined with local anesthetics (novocaine). Half an hour after the operation the animals recovered and were ready for the electrical stimulation experiment. Bipolar and unipolar stimulation was applied. In the first case, every single electrode was used once as an anode and next time as a cathode. In unipolar stimulation the socket that had been screwed to the skull served as the indifferent electrode. Stainless "double electrodes" with a diameter of 0.25 mm were used.

Behavioral responses were recorded in the protocols during the experiments. The written notes were supplemented by cinematographic recordings that illustrate the behavioral changes of the animal in the critical phases of the experiment. For the comprehension of the procedures and the designations used in the protocols the following is to be noted: the electrodes bilaterally carried in one holder were designated from front to back, by *A*, *B*, and *C*. In the earliest cases, that is, until experiment 49, four electrodes *(A, B, C, D)* were used unilaterally. As a rule, the left side (l) was stimulated first, then the right (r). From experiment 135 on, every group of electrodes – after the first series of stimulation – was lowered 2 mm for a second series of experiments so that the responses were obtained from an upper and from a lower level. Thus, 12 sites of stimulation were available in each experiment and these were separated by distances of 1.5 mm (rostrocaudally) and 2 mm (dorsoventrally). By placing electrodes of different length (18–28 mm) more laterally or medially, more rostrally or caudally, a systematic mapping could be carried out. This study was further extended by using different stimulus intensities, that is, at least three levels, applied at each site. The intensity of the stimulus was given in volts (V), the duration of the stimulus train in seconds. From experiment 20 on, the position of the electrodes was always verified in serial histological sections. For every individual case, the sites of stimulation were entered in photographs of a normal histological brain series, and these plots were then added to the protocol and films of the corresponding experiment.

In the following presentation, we shall relate the stimulation sites in a rostrocaudal, mediolateral, and dorsoventral direction to conspicuous landmarks such as the fornix (columna fornicis descendens), mammillothalamic tract (Vicq d'Azyr), the habenulo-interpeduncular tract (Meynert), and the cerebral peduncle (pedunculus cerebri). In the dorsoventral direction the localizational parameters correspond to the horizontal projection of the stimulation sites upon the median plane. For the hypothalamus an anterior, a middle, and a posterior, a medial as well as a lateral section, have been differentiated.

In experiments in which the left and the right groups of electrodes do not lie symmetrically, the identification of the stimulation sites is restricted to the group of electrodes from which a pronounced decrease of responsiveness was elicited. The numbers of the sections refer to the histological atlas series, as published earlier for the topographical identification ([26], see especially Tables IV-XVIII, pp. 88–117).

Based upon the symptomatology, we shall present the results in three groups.

Group I: "Elite" Cases with Sleep Syndrome

Because of the complex nature of central stimulation effects, very small differences in the localization may have important consequences with respect to behavioral changes. Conversely, surprisingly similar effects may be elicited from sites of stimulation that lie relatively far apart. Under such circumstances it is impossible to support the conclusions by a brief summary. In fact, every case offers its singularity when one confronts symptoms with localization. Indeed, this is the chief task of our investigation. Nevertheless, in the following account we have to limit ourselves, because of lack of space, to that which is most important for the conclusions.

Experiment 43 (27–28 February 1928). Lively cat. After inserting the first pair of electrodes (AB), she begins to purr and to fondle. The nictitating membranes are slightly prolapsed. Insertion of a second (posterior) pair of electrodes (*CD*) alters nothing in the behavior. Upon stimulation of *CD* with 0.5 V an increase of the prolapse of the nictitating membranes occurs. Now the cat lies down; she gives the impression of bodily weariness. When the table is knocked upon, she raises herself and attempts to escape. The stimulation is repeated (*CD*, 0.5 V). First the cat walks about on the experimental table. After 20 s she lies down in a corner, then she curls up. At first, the head is still slightly raised; afterwards it is laid upon the front paws. The eyes are now completely closed, and the pupils (visible upon artificial opening of the eyes) are narrowed to 2 mm transverse diameter. When the table is struck with a hammer, the cat opens its eyes somewhat, but remains recumbent. During the stimulation of *CD* with 1 V the described effect becomes more profound. Tickling of the pinna is – as in normal sleep – answered with a defensive reflex (quick retraction of the pinna). After termination of the stimulation the condition remains unaltered. The reaction to strong acoustic stimuli (shaking a bunch of keys beside the cat and knocking the table with a hammer), and to tickling the nose, is minimal and lasts only a short time. Conversely, the cat remains standing after she has been lifted and has been shaken somewhat; however, she soon lies down again. The stimulation is now discontinued, but the animal is kept under further observation. It persists 5 h in the curled-up position. The cat scarcely reacts to acoustic stimuli such as thumping or clapping, though somewhat more to sounds caused by quiet walking in the room. The behavior when one comes near the animal is noteworthy: it lifts its head and hisses slightly. Obviously the environment is perceived to a certain degree despite closed eyes and a position of rest concordant with that of sleep; a threatening disturbance is answered with the defense reaction that is typical of the cat. From the 6th hour on, the described state disappears. After 7 h the cat is lively again, very nearly like before the experiment. The animal is connected again to the stimulation device and a new stimulation period follows (1.5 V). Even shortly after the onset of stimulation, the cat lies down again and rolls up, with the nictitating membranes prolapsed and the eyes closed. To arousing stimuli, e.g., shaking a bunch of keys, the animal scarcely reacts, as little as formerly. The pinna reflex is also not impaired. The animal is disconnected from the stimulation apparatus and stays in the cage overnight.

The next morning it again shows spontaneous interest and normal activity. The animal is reconnected to the stimulus device, and repeated stimulation shows anew the described sleep effect. It is noteworthy that the cat brought onto a damp mat shows signs of weariness (blinking, closing the eyelids, lowered head), but does not lie down. It lies down only after the damp mat has been replaced by a dry one. Thus, despite central induction, the behavior is still influenced by the *external situation*. Stimulation with faradic current at increasing intensity has a clear arousal effect, but only in passing. A few minutes after cessation of this stimulation, the cat lapses again into a condition of strongly diminished responsiveness.

Localization of electrodes: medial to the line connecting the mammillo-thalamic and habenulo-interpeduncular tracts, 1.5 mm lateral to the midsaggital plane, at the level of the lower half of massa intermedia, corresponding to sections 349–387 of the horizontal atlas series [26].

Experiment 180 (15 March 1935). Friendly animal. Insertion of the electrodes causes no symptoms. Stimulation (1 V, 30 s) also has no effect. In contrast, a pronounced damping ensues immediately following stimulation (unipolar) with 2 V for 30–45 s. Following repetition the cat assumes the *typical position of sleep* and this occurs even while the stimulation is in progress (with about 20-s latency).

Increase of the stimulus intensity has a clear arousal effect, yet after cessation of the intensified stimulation, the animal falls again into a strongly damped condition. The pupils are found to be maximally narrowed, and the pinna reflex is preserved. The damping effect *first influences the general motility*. The cat regains its full activity as soon as the electrodes have been lowered. Stimulation of the new site produces a syndrome that corresponds in every respect to that of a sleeping animal. External stimuli (knocking on the table, lifting the animal) are responded to slowly. Higher stimulus intensities (3–4 V) elicit pronounced awaking effects with motor symptoms. After the end of stimulation, a strong damping of the reactivity occurs again.

Localization of electrodes: between mammillo-thalamic and habenulo-interpeduncular tracts (near the latter), 1.5 mm lateral to the midsagittal plane, at the level of the lower half of massa intermedia (sections 349–374). The most caudal electrodes seem to be those responsible for the sleep effect.

We met with similar behavior in other cases which will be mentioned only briefly:

9.1

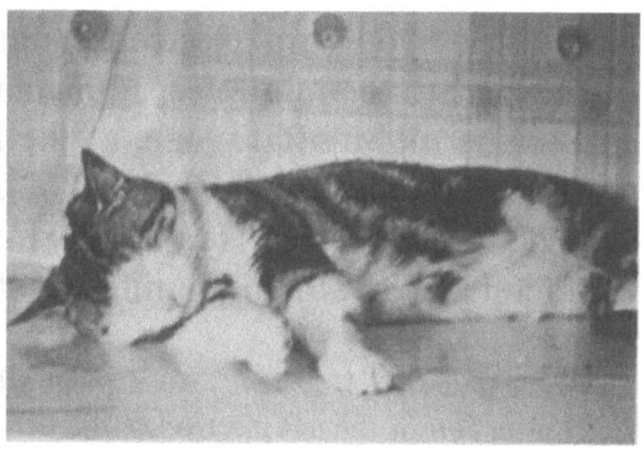

9.2

Fig. 9.1. Following electrical stimulation the previously agile cat lies in the typical position of sleep. The pinna reflex is well retained; to acoustic stimuli the animal reacts only little (experiment 211)

Fig. 9.2. Same cat as in *Fig. 9.1*, but the cat is lying on one side (experiment 181)

Experiment 181 (19 March 1935). Shy animal; after insertion of the electrodes, the cat closes its eyes and lies down. After stimulation (unipolar, 2–3 V, 30 s) the sleeplike effect becomes more profound. From time to time, the cat is curled up; at other times it lies on its side with the extremities stretched out (Fig. 9.2). The breathing is quiet and slow; the pinna reflex is readily obtained. If meat is held under its nose, the animal wakes up. After a while, it begins to stretch and groom itself (Fig. 9.3). Renewed stimulation again elicits the previously described condition. Clapping of hands and small cuffs awaken the animal, but only for a few seconds.

Localization of electrodes: between the fornix and the habenulo-interpeduncular tract (closer to the latter), 1.5–2 mm lateral to the midsagittal plane, at the level of the upper half of massa intermedia (sections 312–349).

Experiment 211 (15 February 1939). Mobile animal; insertion of the electrode has no effect. After stimulation (unipolar, 1.5 V, 60 s) the responsiveness decreases; the cat lies down and curls up (Fig. 9.1). Meat held under its nose awakens the cat and is eaten (Fig. 9.5). Awakened, the cat is somewhat slowed down. Renewed stimulation reinduces a change in the sense of deepened somnolence.

Localization of electrodes: axis through the three electrodes lies left and right, parallel to the connecting line between the fornix and the tract of Meynert (habenulo-interpeduncular), 2.5 mm lateral to the midsagittal plane, at the level corresponding to the upper half of massa intermedia (sections 299–349).

Experiment 303 (29 April 1943). Hostile animal, very shy. Upon being approached, it responds with aggression. After insertion of the electrodes, one observes a slight prolapse of the nictitating membrane, narrowing of the pupils, and closing of the right eye. Marked change of mood by stimulation (unipolar, 1.0–2.5 V, 30 s). The cat reclines, turns in its paws, but not with maximal relaxation. From time to time the eyes are held closed. Pupils at the beginning are labile, then narrow; breathing is quiet and shallow. On both sides prolapse of the nictitating membrane is seen and after this, continuous lid closure.

Localization of electrodes is of interest because the most caudal of the three electrodes (*C*), even at the low intensity of 1 V, elicits a clear waking effect. In contrast, the substrate responsible for somnolence lies in the region of the two rostral electrodes, namely between the fornix and the tract of Meynert (habenulo-interpeduncular), nearer to the latter, 1.5–2 mm lateral to the midsagittal plane, at the level corresponding to the middle of massa intermedia (sections 324–362).

The following case is especially interesting because here a change in the character of the animal manifested itself.

Experiment 304 (6 May 1943). Conspicuously shy cat, always ready for flight or defense. After insertion of the left electrodes, the behavior is changed to that of clear somnolence. This effect is intensified by the following (unipolar) stimulation series (1 V, 4 min) of intermittent stimulation. In the course of the stimulation light motor effects (head, nose) in synchrony with the stimulation frequency become visible. The movements rapidly *decrease in amplitude, thereby indicating a diminished excitability*. With higher intensity (2.0 V) the motor effects reappear. Now they *increase* in amplitude with increased duration of stimulation, and the reactivity increases at the same time. The stronger stimulation thus changes the

a

b

c

Fig. 9.3a–c. Same cat as in *Fig. 9.2.* In the interim it has been awakened by electrical stimulation at higher intensities. While it is grooming itself *a*, a new stimulus is applied. The effects show up within 30 s, as seen in *b* and *c*

reactivity, i.e., it causes an arousal. After cessation of stimulation, a pronounced somnolence again develops. Attempts to disturb the animal by lifting it up and shaking cause a transitory arousal which soon gives way to the typical position of sleep. The cat that was extremely defensive before stimulation no longer hisses upon approach, not even when touched – a sign that the tendency to manifest affective discharge has experienced a change.

Localization of electrodes: between the mammillo-thalamic and habenulo-interpeduncular tracts, 2.0 mm lateral to the midsagittal plane, at the level of the lower half of massa intermedia (sections 349–387).

Experiment 306 (10 January 1944). Friendly and agile animal, stimulation with 0.75 V induces a very pronounced, nearly maximal change of mood, in the sense of damping of the general activity and reactivity. At 2 V, motor effects appear which diminish successively in the course of stimulation, thus indicating the lowering of excitability. One stimulation site seems to be particularly active. The condition is nearly but not completely reversible; a slight somnolence remains even after strong waking stimuli and even with application of higher intensities.

Localization of electrodes: ventromedial thalamus 3.0–3.5 mm lateral to the midsagittal plane; level corresponding to the upper half of massa intermedia (sections 299–349).

Experiment 308 (29 March 1944). Good-natured, trusting, and agile cat. Insertion of the left electrodes induces a transient motor weakness of the hindlegs and a slight prolapse of the nictitating membranes. Soon after onset of stimulation (0.5–1.0 V) with the rostral electrode (A), a clear inclination to sleep appears. It becomes significantly stronger with continuation of the stimulation at the more caudal electrode (B) with intensities up to 3 V and leads to a completely sleeplike condition. Raising the stimulation intensity to 4.0 V at electrode B causes a very impressive gradual awakening; the cat opens its eyes, raises itself, and arches its back like a naturally awakening cat.

Localization of electrodes: ventromedial thalamus, 3.0 mm lateral to the midsagittal plane, at the level corresponding to the lower half of massa intermedia (sections 349–399).

Experiment 110 (22–23 September 1932). Agile, trusting animal. Insertion of the right electrodes causes no change in behavior. Electrical stimulation periods with 1.0–4.0 V and up to 60-s duration induce primarily motor effects which are in synchrony with the stimulation frequency. As an aftereffect a condition develops that is in every aspect similar to sleep. Upon mechanical and acoustic as well as electrical stimulation (especially with higher intensities), the animal becomes completely awake again. Upon stimulation with 1 V by means of the left electrodes an unequivocal sleeplike state develops. Motor effects appear with 2 V. After cessation of stimulation a deepening of the sleeplike state is noted. The hypnogenic effect is reproducible on the following day; it is not, however, as intense as on the 1st day.

Localization of electrodes: region of the thalamic radiations, lateral to the mammillo-thalamic tract (Vicq d'Azyr) and tract of Meynert, 3.5–4.0 mm lateral to the midsagittal plane, level corresponding to the lower half of massa intermedia (sections 337–387).

In summarizing the reported observations we may reach the following conclusions:

The responsiveness of the central nervous system is decreased in a series of cases even with the *insertion of the stimulation electrodes. This effect is enhanced by electrical stimulation at low intensities*, i.e., at the level of 0.75–1.5 V, and in exceptional cases, up to 3.0 V. In another series of cases the insertion of the electrodes remains without effect; however, a change in the

Fig. 9.4. Arousal effect elicited by strong sound stimulus (knocking a bunch of keys on the table) (experiment 205)

sense of damped responsiveness to stimulation appears at low-intensity stimulation. In animals that react rapidly it is at least 20–30 s before clear-cut sleep effects are seen; otherwise it requires several minutes (including the stimulation intervals of up to 60 s). After each stimulation, there is *further deepening* of the sleep state whereby the animal assumes a posture corresponding to that of a naturally sleeping cat: after lying down, the paws are turned in, the head is down, or the whole body curled up (Fig. 9.1). More rarely the cat lies on its side (Figs. 9.2, 9.3 c). The eyes close, the nictitating membranes prolapse, and the pupils are narrow. While this change takes place one can occasionally observe a distinct *fluctuation* – a waxing and waning of individual symptoms, e.g., ocular responses. An important criterion of responsiveness is fulfilled when diencephalic stimulation excites nearby elements related to the extrapyramidal motor system. The saccade responses become gradually less marked and finally disappear. This phenomenon is not to be confused with the well-known effect of summation whereby rhythmic movements become tonically smooth.

The animal responds to external stimuli according to the depth of the sleep state in a more or less accentuated way (Fig. 9.4). Conspicuous is the relatively strong response to the olfactory stimulus of meat (Fig. 9.5). When the cat is lifted and shaken, its attention and readiness return in part or completely. Thus, the somnolent state is *reversible*. Awakened, the animal moves with perfect coordination. It does happen that, after a massive arousal, the cat simply stays awake. Usually, after a short time it reclines again and continues to sleep. It can happen in deep somnolence that, despite strong acoustic stimuli and also lifting and shaking of the animal, a low-level reactivity remains. Then, repeated strong waking stimuli are needed to relieve the dampening gradually. The greater resistance of shy or aggressive animals to the change caused by stimulation has been observed

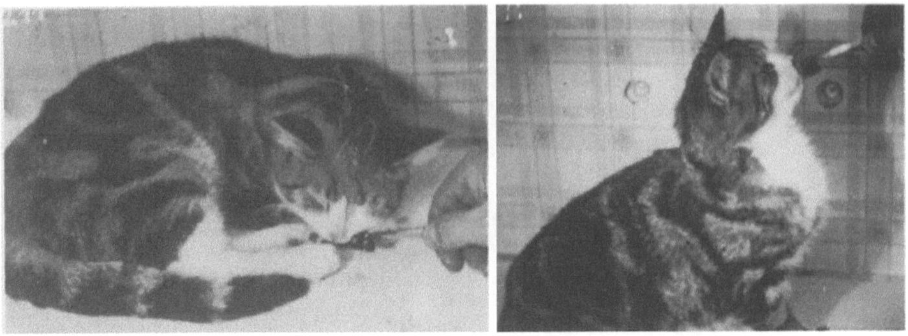

a b

Fig. 9.5a, b. Sleep induced by diencephalic stimulation. Meat is held under the nose of the sleeping cat *a*. The odor exerts a strong arousal effect *b* (experiment 211)

repeatedly. Occasionally an animal which showed vicious behavior before stimulation becomes friendly and begins to purr in the course of the experiment. When the stimulation intensity is raised successively to 3.0 or 4.0 V one can always observe an *impressive reversal in behavior;* the eyes open; the pupils become wide; the animal stands up, walks about, and increasingly shows sympathetic (ergotropic) symptoms. This often occurs at 3 V or somewhat lower; it regularly occurs at 4 V or somewhat higher.

The *localization findings* are shown diagrammatically in Figs. 9.8 and 9.9, together with the cases which will be discussed in the following section.

Group II: Cases with Sleeplike States

This group contains observations concerning diencephalic stimulation experiments in which sleep symptoms have been elicited but complete sleep was not achieved. The series comprises 22 experiments. Only the best cases will be briefly reported here.

Experiment 90 (15–16 July 1932). Shy animal, though not aggressive. Immediately after insertion of the electrodes somnolence occurs. A pronounced lability is noteworthy in that despite constant daylight, the width of the pupils varies relatively fast between very narrow and rather wide. With bipolar stimulation the damping effect deepens only very gradually, i.e., in the course of about 20 min. The cat reclines; the eyes are closed and the front paws turned in, the head still raised. After further stimulation the head is laid down upon the forepaws. Strong acoustic stimuli frighten the cat. On the next day the pupils are abnormally wide. Following renewed stimulation the pupils begin to dilate and contract, periodically and without any obvious reason. The palpebral fissures change continually and then close. The breathing is quiet.

Localization of electrodes: between the fornix and the habenulo-interpeduncular tract (Meynert), nearer the latter, 2.0 mm lateral to the midsagittal plane, corresponding to the middle level of massa intermediate (sections 312–362).

Experiment 109 (21–22 September 1932). Somewhat lazy cat. After insertion of the electrodes, prolapse of the nictitating membrane occurs. In the course of the electrical stimulation (bipolar, 1–2 V, multiple trains) the cat lies down after about 8 min and turns in its forepaws. Attentiveness is diminished only slightly, however. The damping effect shows up mainly by a decrease of motor reactivity. The eyes are closed. A fright-like reaction is induced by acoustic stimulation.

Localization of electrodes: ventral level of massa intermedia; infundibulum and entrance of aqueduct (sections 387–424).

Experiment 128 (14 November 1932). The cat, which behaves normally after the insertion of the electrodes, lies down in the course of electrical stimulation (bipolar, 1–4 V, intermittent stimulation during 25 min). Although passively placed in an upright position she allows her head to droop a number of times; intermittently, however, the head is raised spontaneously, then sinks down immediately again. At times, after lying down, the animal curls up. The whole picture develops so slowly that it is impossible to relate the effect to a single electrode. Stimulation of symmetrical sites causes bodily weariness to appear more conspicuously, but the cat continues to respond to sounds. Spontaneous movements of the tip of the tail indicate a certain inner disquiet.

Localization of electrodes: caudate nucleus, lateral ventricle, fimbria of the fornix, 2–3 mm lateral to the midsagittal plane (sections 211–262).

Experiment 174 (26 February 1935). Shy cat. Soon after the insertion of the electrodes she lies down and rolls up. Lifted up and shaken, her behavior is again approximately normal. Stimulation (unipolar, 1 V with intervals of 3 min) allows the somnolence to develop more strongly. At first the animal remains crouched with closed eyes; then it lies down with its head on the forepaws. An attempt to disturb the animal by pinching the nose is answered solely by slight opening of the lids. After being lifted up, the cat lies down again, closes its eyes, and curls up. Stimulation with two electrodes elicits motor effects in the face in synchrony with the stimulation frequency. As in an earlier case (experiment 304) the intensity of the rhythmic movements subsides slowly. Through this a successive lowering of the responsiveness of the motor elements is expressed.

Localization of electrodes: lateral to the connecting line between fornix and Meynert's tract, 3 mm lateral to the midsagittal plane, at the level of the lower half of massa intermedia (sections 349–387).

Experiment 175 (1 March 1935). Friendly animal, very mobile. Insertion of the electrodes has no effect. During multiple stimulation periods (unipolar, 30 s each) the animal becomes visibly tired; pupils narrow. At first the cat remains in sitting position. Afterwards she reclines with forepaws turned in. The reaction to sound and tactile stimulation of the nose is rather prompt. Meat is accepted and the cat becomes fully awake. Stimulation at 2–3 V has a waking effect. After cessation of the stimulation intensified somnolence ensues as an aftereffect. It appears as though two opposing mechanisms are activated through the stimulation, whereby the initially more prominent exciting component is outlasted by the damping one. At a neighboring electrode during stimulation with 1–3 V, accelerated breathing appears.

Localization of electrodes: rostrolateral to the habenulo-interpeduncular tract (Meynert), 2.5–3 mm lateral to the midsagittal plane, corresponding to the middle level of massa intermedia (sections 337–362).

Experiment 205 (10–11 January 1939). Friendly animal, very mobile. Insertion of the electrodes remains without effect. During electrical stimulation (unipolar, 2 V, 32 s) the cat lies down, initially with raised head; the eyes become closed. Now and again she seeks, without raising herself, to take another position. Soon she lays the head upon the forepaws, which are stretched forward. Upon artificial opening of the palpebral fissures, one finds the pupils narrowed. Through this manipulation they widen! The cat reacts relatively strongly to sounds. A latent vigilance still exists. The breathing is mostly quiet and slow.

Localization of electrodes: almost exactly between the mammillo-thalamic (Vicq d'Azyr) and habenulo-interpeduncular (Meynert) tracts, 2.5–3 mm lateral to the midsagittal plane; corresponding to the lower level of massa intermedia (sections 337–387).

Experiment 215 (6 March 1939). Shy and ill-tempered animal; she hisses and shows her claws when approached. Insertion of left electrodes has a slight but clear damping effect. Motor effects which appear even with 1 V compete with somnolence, which develops in parallel. Diminished responsiveness is noted, especially during the stimulation intervals through narrowed pupils, prolapse of the nictitating membranes, and lowered head. Stimulation on the right side fails to elicit motor effects up to 3–4 V. The somnolence is well pronounced even during stimulation, most strongly at one electrode with 1.5 V. The eyes are closed and the pupil narrowed. Higher intensities of stimulation (3–6 V) evoke rhythmic motor effects with markedly expressed arousal. After cessation of stimulation the pupils narrow again and the lids close. The overall behavior, at the close, is significantly quieter than before the beginning of the experiment, although the animal remains alert.

Localization of electrodes: left side, between the mammillo-thalamic (Vicq d'Azyr) and habenulo-interpeduncular (Meynert) tracts, 1.5–2 mm lateral to the midsagittal plane, at the level of the middle portion of massa intermedia. Right side: massa intermedia, caudal half (sections 299–337).

Experiment 219 (17 April 1939). Shy animal; insertion of the electrodes has no effect. Electrical stimulation (unipolar, 2 V) elicits dilatation of the pupils accompanied by movements of face, head, and forelimbs. After cessation of the stimulation the cat reclines and rolls up. Repeated stimulation has a waking effect, followed again by marked somnolence after stimulus cessation. The motor symptoms clearly diminish during the stimulation. In the competition between damping and waking components, at 2 V, the latter predominates.

Localization of electrodes: lateral to the connecting line between the fornix and Meynert's tract, rather lateral (4.5 mm from the midsagittal plane), in dorsoventral direction corresponding to the upper half of massa intermedia (sections 312–349).

In summarizing the observations that have been made with this group, we see that, in general, the findings confirm those made with the "elite" cases. Differences of more quantitative nature nevertheless exist. Of special interest is the observation that activity or, as the case may be, reactivity in the motor and sensory realms, diverge considerably. With signs of marked bodily relaxation, a *relatively* good response to visual and acoustic stimuli remains. In other cases the animal reacts only slightly to sensory stimulation; it relaxes its musculature only to a certain degree and stays fixed in the sitting, lying, or crouched position (Fig. 9.6).

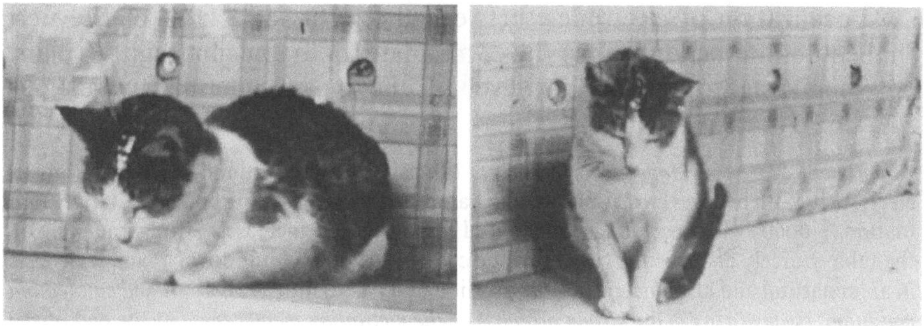

a b

Fig. 9.6a, b. Stuporous state elicited by diencephalic stimulation. In *a* the head is still held up in spite of drowsiness. In *b* the forepart of the body is erect. Reactions to external stimuli are, however, strongly reduced

Group III: Cases with Reduced Responsiveness of Undetermined Nature

In more than 26 cases an impairment of spontaneous behavior with a reduced level of responsiveness was noted; however, the entire picture clearly deviated from that of sleep. The animals remain apathetic; they intermittently succumb to a cataleptoid stupor. External stimuli elicit a slight increase of the responsiveness even though the animals do not become properly awake. There remains rather an important residue of sluggishness and paucity of movement. Postures such as those in physiological sleep may occur, but neither consistently nor in complete form. There is also a partial lack of temporal correlation between the response and the electrical stimulation or insertion of electrodes. It should also be said that since the criteria of sleep are generally negative in nature and subject to wide variability during physiological sleep, it is rather difficult to separate the various stimulation effects from one another. Nevertheless, our observations with respect to this group have a certain interest because neuro- and psychopathology have to do with such conditions and encounter the same difficulties. For orientation, short reports on three cases with somewhat differing symptoms have been added.

Experiment 70 (6–7 November 1929). Stimulation with 1–2 V, bipolar, for 6 min (including stimulation intervals) elicits a clearly decreased reactivity. The cat remains fixed in a crouched position; the eyes are mostly closed. Lifting up and strong acoustic stimuli are responded to only slowly. At times the cat purrs. On the following experimental day the cat is alert again. Renewed stimulation reproduces the same condition as before, now, however, at somewhat higher intensities (3–4 V). Faradic stimulation with intensities that elicit rather strong motor effects has a certain, but rather limited waking effect. After cessation of the faradic stimulation a deepening of the drowsy condition occurs again. The relationship to the symptoms of physiological sleep is obvious, yet the state is only partially reversible.

Localization of electrodes: between the fornix and tract of Meynert (habenulo-interpenduncular) 1.5–2 mm lateral to the midsagittal plane corresponding to the middle level of the massa intermedia (sections 312–362).

Experiment 123 (27 October 1932). The cat is clearly drowsy 2 min after insertion of the electrodes. With her eyes closed, she assumes a posture between sitting and lying. After stimulation (bipolar) with low voltage the reduced responsiveness increases. Strong knocking on the table scarcely disturbs the cat. After the cat is lifted up and put down again, it persists in an unnatural and also uncomfortable position and passes into a *stuporous and cataleptoid condition.* Noteworthy is the *strong prolapse of the nictitating membranes.* At the end of the experiment, milk and meat are accepted, although the cat does not raise itself.

Localization of electrodes: left lateral geniculate body (sections 312–349).

Experiment 232 (13 July 1939). Insertion of the stimulation electrodes is followed by a short twitching spell with sudden but transient pupillodilatation. Palpebral fissures and pupils become gradually narrower; the nictitating membranes prolapse. Two min after lowering the electrodes to the deeper stage, the animal is fully apathetic. It lies down, the front paws stretched out in an abnormal position. Typical sleeping posture is never observed.

Localization of electrodes: lateral part of the hypothalamus, the tips of the electrodes reaching down to the base of the brain (sections 462–500).

Criteria of Sleep

Having described our experimental observations we are faced with the problem of how to evaluate the profound alteration in the mode of behavior of the experimental animal. The answer is frankly not a simple matter. Therefore, it was necessary to report in detail on the different experiments. The principal difficulty lies in the fact that judgment must be essentially based upon *criteria of negative nature.* This becomes particularly apparent when we think of behavior patterns which involve a positive array of symptoms, for example, the affective defense reaction [36; see translation in this volume, p. 183] or defecation and urination, likewise induced by central stimulation [37]. Here the coordinated character of the responses and the concordance with well-known physiological functions and behavior leave little doubt that the stimulation bas brought the corresponding central organization into play. Returning to the question that occupies us here, it is to be emphasized that in the framework of regulatory functions, *facilitation* and *inhibition are equally valuable means to the end.* Correspondingly, they are both involved in the organization of successful performance, here perhaps with predominance of the one, there, of the other mode of function. Sleep entails (seen from without!) *inhibition.* To clarify this situation and to create an exact basis for the interpretation of the stimulation effects, the *picture of normal feline sleep* must be characterized. Despite the familiarity

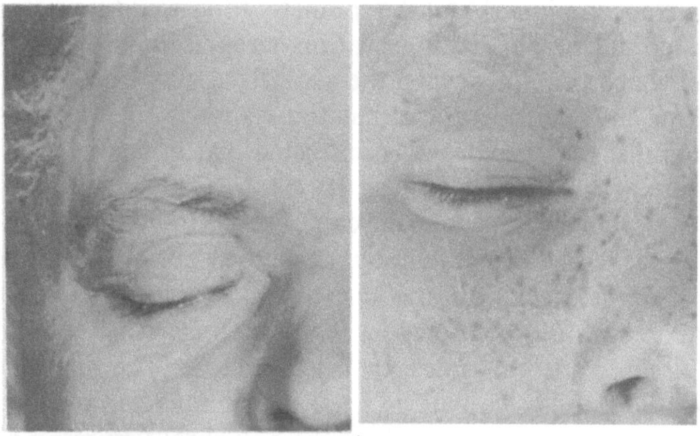

Fig. 9.7. a Passive closure of the lid as a sequel of innervation defect (paresis of the levator palpebrae muscle). *b* Active closure of the palpebral fissure in sleep. The lower lid touches the upper lid; the palpebral fissure is approximately horizontal. An attempt to actively open the eyelid is answered by a reflex increase of muscular tone in the orbicular muscle of the eye

of the phenomenon and of certain symptoms we may not, in view of the eminent significance of the sleep function, allow ourselves to be dissuaded from this, especially since the answer to critical objections to which we shall direct ourselves is already prepared.

One of the most perceptible symptoms of sleep is the closure of the palpebral fissure. This is not at all due to failure of the muscles that open the eyelids, so that the eyes, as one expresses it in lay terms, "shut." Participating positively is an *increase of the tension* of the orbicular muscles. The comparison of the closing of the eyes in sleep to the condition where the levator palpebrae muscle is paralyzed is a very important one indeed. During sleep the eyelids are held closed by an *objectively demonstrable force*. Yet, one can easily ascertain the intactness of the lid closure reflex by attempting to open the lids passively; even during deep sleep the force of closure is clearly heightened. It is to be observed further that the lower lid comes up against the sunken upper lid, so that the line of closure lies higher than with the simple loss of levator tone (Fig. 9.7a, b). Although this may appear as a minor issue this observation reveals a goal-directed coordination, in terms of relaxation here and tension there – one which is of fundamental significance. *It is precisely this mechanically oppositional and, in terms of functional orientation, synergistic combination which documents sleep as an "ordered" condition.*

This insight is confirmed by the behavior of the pupils, because in sleep the constriction is more pronounced than with the pure default of the dilator muscle. Hence an increase of the sphincter tone is additionally involved. The third step of coordinated innervation is represented by the nictitating membrane, which prolapses when the cat falls asleep. Through this mechanism (taking into consideration that the response to external stimuli is strongly diminished), the cornea is protected and optical stimuli that disturb sleep are averted.

Under favorable conditions, one can observe an interesting play of antagonistic tendencies during the initial stage of sleep. This is the case when we have before us a cat with artificially induced sleep, but external conditions that are not fully comfortable. The status then fluctuates in that the palpebral fissures open at times, then again close. The nictitating membranes and the pupils fluctuate in the same manner and direction. One might expect that with the opening of the palpebral fissures, a reflex narrowing of the pupils would occur, because more light falls upon the retina. In reality they dilate! This behavior reveals a centrally induced synergistic reorientation of the eye's reactivity (one which is clearly opposite to the function of the pupillary reflex).

One of the more obvious attributes of sleep is the position of the body which involves *extensive relaxation of the skeletal musculature* and offers the force of gravity a minimal target. Cats and dogs lie down when they feel sleep approaching; the cat often folds its front paws and retains a certain active posture. At a moderate environmental temperature it is curled up, as is also the case with dogs. This observation is of considerable importance because it demonstrates that a specific group of muscles is kept under coordinated tension on behalf of a thermoregulatory mechanism. Thus, the curled-up position of the cat is an effective measure against undercooling during sleep. This vegetative regulation stands in competition with the general relaxation of the sleeping animal, whereby a compromise tends to be reached. The *relativity* of the sleep posture is best documented in the horse, which in unfamiliar surroundings "reserves" itself a certain preparedness (flight or defense). Thus it sleeps standing so that the entire static apparatus, including proprioceptive and vestibular reflexes, retains a certain degree of activity. A further exception from otherwise comprehensive relaxation is represented by the sphincters of the anus and bladder. This exception, however, is consistent with the concept that sleep is dominated by an *order directed toward biological necessity*.

Apart from the symptoms in the motor domain we see sensory responsiveness strongly reduced, especially with respect to acoustic test stimuli which are routinely used for the determination of the *depth of sleep*. The heightened threshold for tactile and proprioceptive stimuli (e.g., by passively moving the body or the extremities) also demonstrates the low level of receptor functions in sleep. The disconnection of the organism during sleep from its environment is, however, neither absolute nor unconditional. If the intensity of the stimulus is great enough, it breaks through the separating barrier. Not only does the threshold of the directly affected sensory system decline rapidly, *but all the other symptoms of sleep vanish concurrently*, in particular in the motor sphere. Thus, *sleep is reversible in its totality*. This effect, despite its intrinsic logic, even for the layman, is of greatest scientific significance! To be noted is the relatively short time span required for such a sweeping and fundamental change in the waking condition. Human beings in general and certain animal species (e.g., dog and horse) manage to regain full contact with the environment and coordinated motor control within a few seconds after an arousing stimulus. Some individuals and certain species of animals, *to which the cat belongs*, seem to have a somewhat longer "running-in time," so that in this regard species-specific and also individual differences obtain. The development of sleep from the first symptom until its full depth requires in every case essentially more time than the changes in the opposite direction. Finally we should like to mention that certain (local) *protective reflexes* can be *retained* in sleep. Thus the cat responds, without obvious alteration of the sensory threshold, to tickling of the pinna with a one-sided rapid shaking of the ear, a reflex which has evidently remained "awake."

The third level of responsiveness concerns associative mental functions which organize the content of consciousness. These functions, and even consciousness itself, are extinguished in fully developed sleep. This *subjective* aspect of sleep, which each person can assess in himself alone, is of particular interest because a *synergistic coordination* is revealed by the fact that the sensory, the motor, and the associative domains are simultaneously suppressed. On the other hand, the experience of the dream proves a certain (qualitative and quantitative) autonomy of particular levels which become even more pronounced in the pathophysiology of sleep.

The individual's failure to take up *contact with the surroundings* and the absence of active interaction with the outside world should not distract us from the fact that the vegetative system which regulates the internal environment of the organism also undergoes certain changes during sleep, particularly those functions which are closely related to the animal system. Thus we find cardiac activity and blood pressure lowered and respiration diminished.

With this, the picture of sleep is completed. The fact that other processes take place within the central nervous system should not be overlooked. It is true that they do not become directly visible to us. All the more impressively do we see them confirmed in the final result, namely, in the recuperation of the energy lost during the waking period. Finally, the functional relations between external and internal aspects remain to be established. Here again the situation seems clear. The outer quiet of sleep is subordinated to the inner necessity, in that the inhibitions represent an indispensable prerequisite to the complete restitution of the centers responsible for the organization of behavior.

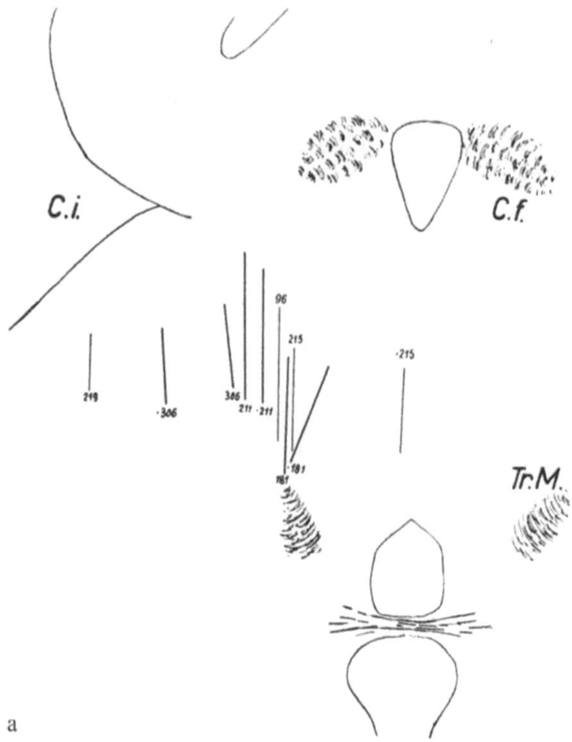

a

Fig. 9.8a, b. Localization of stimulation sites which were responsible for sleep effets. The positions of the electrodes are plotted in two horizontal planes, section 324 (*a*) and section 362 (*b*) of the atlas [26]. The lines connect the three or possibly two electrode positions which refer to a given experiment. *Thick lines* indicate group I ("elite" cases), *thin lines* those of group II. All points are transferred to the left side for comparison. *C. a.*, anterior commissure; *C.f.*, fornix; *C. i.*, internal capsule; *C.p.*, posterior commissure; *Tr. M.*, tract of Meynert (habenulo-interpeduncular); *V.d'A.*, mammillothalamic tract (Vicq d'Azyr)

Symptoms and Localization

Our next task consists in the comparison of natural sleep with the conditions which were observed in our experiments. As concerns group I, the so-called elite cases, we can confirm an earlier statement, namely that there is no demonstrable difference from natural sleep. For group II this conclusion has validity insofar as the natural symptoms of sleep are present to a certain extent, although varying in regard to depth and completeness. In group III we encounter here and there certain similarities with sleep behavior, although deviations have been recorded which indicate a *limited reversibility as is characteristic for stuporlike states.* These cases are nevertheless of interest because they offer certain contrasts to the former two groups and because they are reminiscent of certain *pathological conditions* such as stupor and catalepsy. The difference from the fully developed picture of *adynamia*, as it was described in a preceding publication [35; see translation in this volume, p. 171], is obvious.

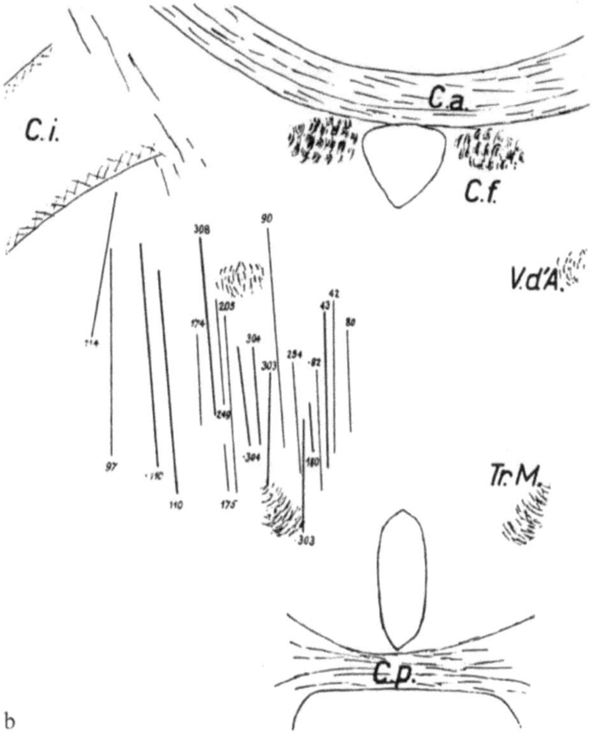

b

In reviewing the localizations of electrodes corresponding to the different groups we can make the following comments:

In nine elite cases of group I we find the bare points of the electrodes *in the lower part of the thalamus*, namely at the level of the horizontal plane which passes through the middle and lower level of massa intermedia. The medial boundary of the effective area lies about 2 mm from the midline. The outer boundary reaches rather far to the side; it is marked less sharply, in part perhaps because we enter a zone in which the electrical stimulus elicits motor effects that disturb the sleep induction. In the rostrocaudal direction the active area is situated between the mammillo-thalamic and the habenulo-interpeduncular tracts (Figs. 9.8 a, b and 9.9). From a topographical point of view it is of interest that fibers of the brachium conjunctivum and proprioceptive afferents radiate into the *paleothalamus*, alongside the trigeminal lemniscus which may carry proprioceptive impulses controlling the facial (mimic) musculature. It is likewise of interest that *in no case where all three electrodes are situated in this region does the sleep syndrome remain absent*. The effect is less striking when only one or two needles are in the described area. We should point out, however, that the experiments concerned were not exclusively carried out in connection with the sleep syndrome, but rather with the control of the *extrapyramidal motor system* [29–31, 32–34] (see translation of [32] in this volume, p. 223). Wherever such effects occurred we have stimulated with short intervals and from various stimulation sites in order to obtain a detailed analysis by establishing similarities and contrasts. With this procedure the electrodes, which lay extrafocally, may have contributed to the disturbances in the induction of sleep. Furthermore the latter was not given adequate time to set in.

A second area from wich sleeplike states belonging to the cases of group II and in part to group III (experiments 21, 124, 127, 128) were elicited is the *caudate nucleus*. On closer examination though there is a difference from the typical behavior of sleep that is easily recognized, namely that the motor functions are predominantly affected by stimulation in the caudate nucleus. The head is held low, so that it lies upon the table; the animal may be curled up, with the front paws turned inward, or it lies flat upon the table with flaccidly outstretched extremities. Despite this it reacts to relatively weak or moderately strong stimuli. Respiration is accelerated in two cases, namely when the electrodes lie in the caudolateral region of the caudate nucleus (experiments 21, 124). The observations described refer to electrodes that had exclusive contact with the cellular mass of the caudate nucleus. This is mentioned because an already slight contact or, as may be, electrical stimulation of neighboring fiber tracts can cause a disturbance.

Attention should also be directed to some cases in which the bare points lay in the *paraventricular* tissue of the hypothalamus, in the direction of the infundibulum and at the entrance of the aqueduct. In four cases the symptoms corresponded to those of group II (experiments 29, 109, 222, 227). The short latency, lying between 6 and 10 s, is remarkable in two of these cases (experiments 29, 222). The pupils are narrowed. Multiple pinching of the eyelids precipitates the lid closure reflex and the prolapse of the nictitating membrane. The animal seeks a relaxed position by rolling up. The breathing is slowed.

In an additional six hypothalamic-paraventricular cases (experiments 38, 103, 111, 243, 245, 251) the effect has rather the character of drowsiness and stupor, in part with signs of adynamia. The classification of these cases on the basis of symptoms is especially difficult here, since transitions exist alongside some completely unambiguous cases. Disturbing influences from neighboring structures involving mechanisms of excitation may also occur. Even a minor coexcitation can lead to a competition with the damping effect. The area responsible for affective defense lies in an adjacent zone and is already known to us. Areas from which general excitation can be elicited will be discussed in one of the next publications. Between the *preoptic area* and the *anterior hypothalamus* we have localized the stimulus sites of four cases (experiments 28, 231, 233, 301) which, according to their symptoms, belong to group III. It should be noted that in three cases one or, occasionally, two electrodes came to lie posteriorly in the hypothalamus, between fornix and mammillothalamic tract, from which affective excitation may be elicited [36, see translation in this volume, p. 183]. It is our experience that stimulation of this zone elicits excitation followed without delay by a *nega-*

tive afterphase with damped activity and reactivity. This effect, which is the opposite of the direct effect of the stimulation, is certainly to be differentiated from the *positive aftereffect* which one obtains especially from the lateral thalamic region. Here the primarily induced negative changes can develop (because of the motor effects) only after cessation of the stimulation (for example, experiment 219). We must also consider the fact that the hypothalamic cases refer to a location in which the typical picture of *adynamia* is induced (we refer to an earlier publication [35]; see also the subsequent section concerning "interrelations").

Similar effects to those seen in group III are encountered in four cases, in each of which one of the three electrodes lies in the vicinity of the intersection between *fornix* and *anterior commissure* (experiments 223, 224, 237, 238). Again, with the beginning of the stimulation, one observes closure of the eyelids, prolapse of the nictitating membrane, and narrowing of the pupils. Slight motor effects which appear during the stimulation periods indicate a decrease in the amplitude and therefore an increase in the stimulus threshold. In some circumstances the animal curls up. There is relatively slight response to external stimuli, indicating that the motor functions (rather than the sensory) are predominantly affected. The same is true for a case in the septum (experiment 22). In view of analogous stimulation effects from the caudate nucleus (see p. 150), we should, however, remember that the caudate nucleus lies directly adjacent to the septum if the anterior horn of the ventricle is nearly collapsed, and so the observed effect is perhaps to be attributed to the caudate nucleus.

Finally, two cases will be described which are noteworthy because of an especially conspicuous location of the electrodes. In one case (experiment 123), the bare point lies directly in the *lateral geniculate body*. Here the behavior of the animal changed immediately upon insertion of the electrodes. By stimulation with 1 V the signs of drowsiness are further accentuated in that the animal reacts only slightly to disturbing external stimuli. After being lifted up and set down again it reclines, but not in a way suitable for sleep, for it persists in an uncomfortable position. The strong prolapse of the nictitating membrane is conspicuous.

In the second special case the electrodes lie in the reticular formation between the fibers of the *brachium conjunctivum* not far from their exit from the cerebellum (experiment 288). After the insertion of the left electrodes, a proprioceptive disturbance in the ipsilateral foreleg is observed. Then the animal wants to lie down. Upon stimulation one observes motor effects, and the animal begins to walk about in a circle. In view of the very low intensity (0.75 V) the motor effects are conspicuous. While the stimulation is still in progress, a *pronounced tiredness* develops. The animal now lies down with front paws turned in, and closes its eyes; the head, however, remains upright, with a slight rotation. The posture of the body is also not entirely

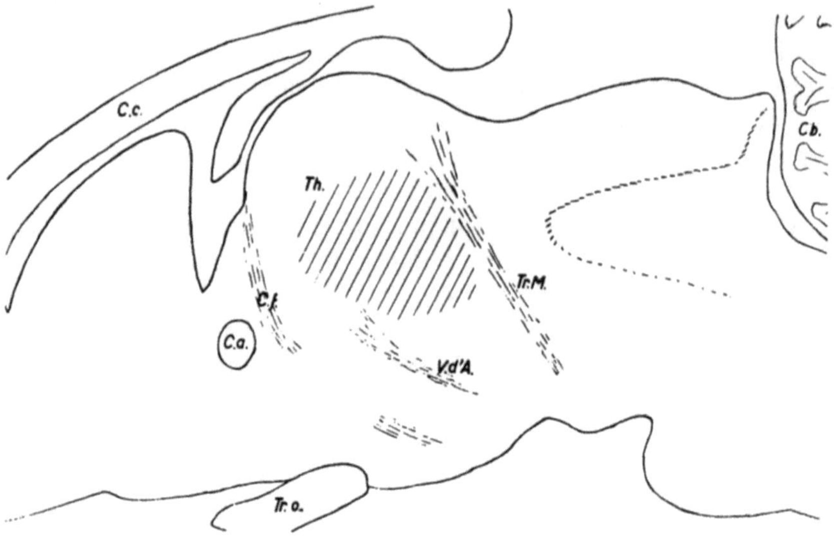

Fig. 9.9. Sagittal section 110 (2 mm lateral to the midsagittal plane) [26]. *Diagonal hatching* shows the region from which, through stimulation, the best sleep effects were obtained. The region extends laterally where motor symptoms compete with the sleep effect. After cessation of stimulation the sleep effect emerges as a "positive afteraffect." *C. c.*, corpus callosum; *C.f.*, fornix; *Cb.*, cerebellum; *Th.*, thalamus; *C.a.*, anterior commissure; *V.d'A.*, mammillothalamic tract (Vicq d'Azyr); *Tr. M.*, tract of Meynert; *Tr. o.*, optic tract

symmetrical in that it inclines somewhat to the left side. It subsequently curls up and scarcely any reaction ensues when a bunch of keys is thrown onto the table. The breathing is somewhat accelerated and not entirely regular.

In summarizing our observations concerning the relationship between symptoms and localization, we encounter the clearest situation in the region of the thalamus with respect to the definition of the localization as well as to the characteristics of the responses. The latter correspond perfectly with the behavior in natural sleep! The results upon stimulation in the caudate nucleus are also not difficult to analyze. Here, the motor system is primarily affected. In other regions in which a pronounced reduction of the reactivity was seen, the results are not so uniform. Besides typically sleep-like states one encounters behavior similar to general drowsiness, accompanied in part by stupor, or signs of weakness in the sense of the adynamia described earlier. Judgment is made especially difficult where one or two electrodes lie within or near substrates whose stimulation, as mentioned above, induces states of excitation. In this case, the unselective electrical stimulus can lead to a competition of opposing effects. The unambiguous correlation between site and effect of stimulation is made difficult, in part impossible, by the fact that a primary excitation can be followed by a "negative afterimage," i.e., by a lowering of the general responsiveness as a secondary effect.

Experimental Conditions

Under this heading we would like to emphasize that the stimulation effect is not determined exclusively by the position of the electrodes. Earlier experimental data concerning centrally induced micturition and defecation have shown very clearly that other factors can play an important role. For instance, in the latter cases the filling of the bladder or of the rectum is decisive. On the whole, the *peripheral conditions* are important because afferents with a physiological control function exercise an influence on the central excitability and with that upon the regulatory response. This influence can have a negative aspect, e.g., in the awake animal micturition is not as easily elicited by electrical stimulation as in the anesthetized preparation [37]. Obviously, in the awake animal interference in terms of the perception of external events is responsible for an inhibitory influence. Such interactions play an important role in the dynamic processes of the central nervous system. They also explain a certain relativity of the results of experimental investigation of those brain areas in which sensory information converge from different receptors.

This state of affairs is especially pertinent in the case of sleep. Hence, under experimental conditions, it turns out that the *character* of the experimental animal is of considerable importance for the success of the experiment. If the animal is trusting and well adapted to the surroundings, the prospects for sleep induction are considerably better than when the cat is vicious and suspicious or employed in an experiment immediately after coming to the laboratory. Further, as has been mentioned, it is not a matter of indifference if one electrode or two come to lie in a region from which sleep-disturbing stimuli can be elicited. Even the mechanical contact between electrodes and substrate can have an effect. Furthermore, the aftereffect of centrally induced somnolence is of considerable consequence and sometimes lasts for a rather long period. Thus, it was observed a number of times that stimulation of the right electrodes, for example, even when they lie beneath the focal region, brings about a pronounced deepening of the somnolence, if a sleep effect has previously been elicited by the left electrode, and with it a state or mood that facilitates sleep. Here we note the interesting fact that the analogue of this is also well known in reflex phenomena of the vegetative nervous system.

Understandable without further explanation is the absence or the delay of the sleep effect when the electrodes lie in a somewhat lateral thalamic region, from which motor effects are elicited through contact with the fiber tracts of the thalamic radiations [29–34]. In competition the latter are preponderant. Conversely, diminished responsiveness begins to appear at the moment when stimulation is discontinued. At this time the motor effects cease momentarily and give way to considerable sluggishness in the sense of a positive aftereffect.

This characteristic slowing is also reflected in the relatively long latency typically seen in sleep experiments. The *time factor* plays an essential role under all circumstances. Of decisive significance, furthermore, is the *intensity of stimulation*. Of particular importance are the cases in which the mere insertion of electrodes was followed by a change and where weariness and sleepiness developed. One may question whether the insertion of the electrodes and the continued mechanical contact really function as a stimulus or perhaps cause an injury to specific nervous elements. Attention was given to this latter possibility in a previous communication [35, see translation in this volume, p. 171] in which we suggested that the insertion of electrodes acts mainly as a subtle mechanical stimulus whose effects may be reinforced by electrical stimulation, as long as the intensity lies within a favorable range.

Another question pertains to the significance of the *frequency of stimulation*. Here we can say that eight impulses per second, which we have set for reasons that were given earlier, have proved to be the most suitable. For comparison, in some experiments the work was done with lower frequencies, i.e., with four or only two impulses per second. A clear-cut difference of effect could not be established, and the problem of an optimal stimulus frequency should be further explored. However, in spite of the positive results obtained, at the moment there is no reason to discuss this point here.

An Interpretation

After having dealt with experimental observations and histological findings, we now face the problem of *explaining* the described effects. In doing so it is important to note that the experimentally induced changes in behavior were elicited from various areas, but all rather well defined. In individual cases, the effect was characterized by a number of typical symptoms and, taking into consideration the extremely slight experimental intervention, the effect is of striking intensity. If we take, for example, experiments 43, 110, 180, 211, 303, 304, 306, and 308 into consideration, the zone that is subjected to the direct influence of the electrodes is limited to the region of the middle to lower levels of massa intermedia (Fig. 9.8–9.10). The distant center controlling the pupil, including those elements of the oculomotor and facial nerves which contribute to the closure of the eyelids, is predominantly involved. This effect cannot be explained on the basis of the interruption of afferent pathways at the level of diencephalic synapses (Trömner), because in that case, the pupils during sleep should be dilated. Of importance is the fact that psychomotor activity is likewise extinguished; this implicates the involvement of cortical centers. Also, statokinetic reflexes which are connected with deeper levels of the brain are no longer active during sleep. Thus, the question arises first as to how these *distant effects* occur. An *injury* causing a functional impairment can be excluded, partly because of the distance and partly because the relaxation of certain muscles is often associated with increased activity in others. For the same reason we have to abandon the notion that the stimulation effects could be due to an impairment of the cerebral blood supply via increased vasoconstriction. The observed sleep effect has nothing in common with the blackout that is due to circulatory failure.

In contrast to all these possibilities one should consider a mechanism by which the change of behavior has the character of an *active nervous inhibition*. In pursuing this notion it is important to separate the sensory, associative, and motor functions from the mechanisms which regulate the central excitatory state *(Leistungsbereitschaft)*. With this concept we return to an interpretation of the mechanism of sleep which has been expressed in a previous review [23–27] (see translations of [23] and [27] in this volume, p. 71 and p. 115). This notion is of current interest because our experiments demand an answer to the question of how the stimulation effect is conveyed to the various centers responsible for the behavioral changes. The fact is that histological techniques support, with very positive findings, the notion that sleep and wakefulness represent mechanisms which regulate the general excitability and responsiveness of the organism. Indeed, with silver staining (for example by means of the Reumont [46] technique modified by Frey [21]) one can reveal fine nonmyelinated nerve plexus which obviously belong to an intracerebral innervation system of their own. The existence of such an intramural apparatus wins highest importance where the recognition exists that *sleep with its restitutive function regulates the internal relationship of the central nervous system;* through this it manifests itself as an element *in the functioning of the vegetative nervous system.* Sleep clearly belongs to a group of functions which protect the single organs from overwork and exhaustion.

Here we should remember that when exceeding a certain stimulus threshold the effect is reversed. The cat, which until now has been lying perfectly quiet curled up on the table, opens its eyes; the nictitating membrane is retracted and the pupils dilate. At the same time the head is raised, the animal becomes perceptive, and stands up. The reversal occurs within a remarkably narrow range of stimulus intensities (2–4 V). This can only mean that a mechanism was released that is the opposite of that of sleep. Its threshold lies somewhat higher, at least in those regions that have been explored (perhaps depending on the density of reactive nerve elements). In the change from restitutive behavior to the active state, a competition between ergotropic and endophylactic-trophotropic[1] orientation takes place. It

---

1 Endophylactic-trophotropic (meaning protection, economy, restitution) is identical with the earlier but occasionally misunderstood term "histotropic"

should be pointed out that the restitutive principle is effective in the quiet of the sleeping person to the extent that it creates the appropriate conditions to attain this goal. Correspondingly, despite its outwardly negative aspect, the inactivation of the organism during sleep possesses a positive character. It is not the expression of inability, of "unconsciousness," but rather *a target-oriented reversible inhibition*. The occasional oscillations of the pupils during the transition from wakefulness to sleep, despite uniform luminosity, represent a direct projection of the centrally induced alternation in the predominance of ergotropic and endophylactic-trophotropic regulation, *mediated by the sympathetic and parasympathetic pathways, respectively*.

Thus, our explanation deals not only with the mechanism by which central stimulation elicits sleep, but also with the relationship of the *sleep-waking cycle to the functional organization of the vegetative nervous system* and beyond that, with the *correlation of physiological and morphological facts*.

There remains the discussion of some observations that deal with the *transition* from the ergotropic to the restitutive state, and vice versa. The physiological waking effect can be easily recognized, since it is typically the *exteroceptive stimulus* that brings about the readiness for action which has been suspended in sleep. The widening of the pupils and, in the cat also, the retraction of the nictitating membrane indicate that the enhanced responsiveness also extends into the *vegetative nervous system* whereby the ergotropic division regains preponderance. Lehmann and Michaelis [43] have recently shown that the arousal is accompanied by the release of adrenaline, and this observation represents a most instructive broadening of our previous knowledge. The fact that such an influence can be as effective when arising from the acoustic, tactile, nociceptive sector as when coming from the optic sector, indicates how *comprehensive the functional contact between the animal and the vegetative system* can be. An important factor is that of time. The brevity of the arousal process, to be counted in seconds alone, excludes a humoral mechanism as the sole decisive one. It is quite conceivable, however, that the mediating action of a humoral agent plays a role.

According to the experimental investigations of Cloëtta [13–16] and his co-workers Demole [17], Fischer [20], and others, there is little doubt that changes in the ionic concentrations may represent additional links in the chain of events leading to the awakening effect. Further clarification of these problems offers a rewarding field of continued research that is still capable of expansion. In the present investigation, however, we should like to emphasize the *organization* aspect of sleep function. This brings up the question of how the central nervous system is led into the sleep state. One should be aware that the physiological condition, i.e., the prolonged period of wakefulness, releases the mechanism only *conditionally*, namely when no disturbing external stimuli or exciting associative processes hinder the otherwise impending change. The long period of wakefulness assumes an increasingly imperative character, so that the tendency to fall asleep is imposed upon the healthy individual by an insurmountable force working from within. In this regard the situation is not dissimilar to the primarily endogenous drive for food intake, where outside factors have mainly a *conditioning*, e.g., a supporting or delaying, role. From the point of view of *organization*, it is noteworthy that both functions (sleep and hunger) have to do with restitutional processes! An additional correlation exists in the fact that food intake can be elicited by central electrical stimulation from specific sites in the brain. We refer to the recent work of Brügger [12], in which the relevant data of diencephalic stimulation experiments have been published.

Nothing can be said from our experiments concerning the *nature of the factor* responsible for the initiation of sleep. Added attention should therefore be given to the fact that electrical stimulation is most effective in a region in which fibers of the *proprioceptive system* and elements of the brachium conjunctivum radiate into the paleothalamus. The *pronounced contrast* to the effect brought about by exteroceptive stimuli is striking. In this context we are reminded that vestibular and proprioceptive stimulation reveal rather close relationships to the endophylactic system (which is represented by the parasympathetic division) in still other respects. I think of the symptoms of motion sickness. Experimental findings which we have made outside of this special theme, namely by the analysis of vestibular reflexes [1–8, 47, 48], also confirm the relationships mentioned. Roosters, for example, whose vestibular functions are tested on a turntable, show after *repeated* vestibular stimulation a prolapse of the nictitating membrane and a loss of tone in the neck muscles. We know of course that with artificial vestibular stimulation there may be also signs of an increased activity of the sympathetic nervous system. This is in no way contradictory to what we said before, if we contrast the role played by a single strong stimulus with the effect of summated weak stimulations. Further experimental support can be found in observations on the circulatory depressor reflex whose activation is to be attributed to a mechanism which in *every respect resembles that of a proprioceptive reflex*. Both mechanisms serve the same principle, the regulatory control of a specific performance which stands in the service of economy of resources and protection against overwork. Although it is easy to perceive how mechanisms that exceed the boundaries of the vegetative and animal systems may develop powerful control functions, we leave it to the critical reader to decide whether these phenomena are due to random processes or reflect the manifestations of *order and vital necessity*. The relationship disclosed by the localization data was important to us because it provides an insight into the whole functional mechanism by which exteroceptive and proprioceptive stimuli are competing, the former by alerting the organism and the latter by protecting it from overstress and exhaustion.

An interesting and important problem is raised by those experiments in which conditions were reproduced which resemble the hypothalamic adynamia [35]. These cases demonstrate that stimulation effects, in reality, are not always consistent and clear cut. For instance, an animal that feels weak in the course of *developing hypothalamic adynamia* follows a normal reaction when it lies down and curls up as is usual with natural weariness and physical fatigue. The possibility that *subjective phenomena and conscious perception accompany behavioral changes* must be taken into account. There are additional experiments which lead to the same conclusion, for instance, the fact that centrally induced micturition can be more easily obtained during anesthesia than in the wakeful state [37]. It seems quite obvious here that electrically induced excitation can be influenced by the conscious perception of external stimuli, in this instance in the form of inhibition. The subjective factor which may well exert direct reciprocal actions with electrically induced excitation may also be related to the sleeplike syndrome following stimulation of the caudate nucleus. This syndrome is characterized by the primary inhibition in the realm of extrapyramidal motor functions, whereby the ensuing feeling of weakness may codetermine subsequent behavioral changes. In this way we can understand how the symptoms of accentuated bodily weariness may be combined with the preservation of sensory vigilance. Stimulation of the brachium conjunctivum may be associated with nausea, possibly as a sequel to unbalanced interplay of motor excitation patterns.

Sleep and sleeplike effects may be obtained by *extrafocal* excitation of the endophylactic-trophotropic regulation. A concrete example concerns the *postural mechanism* which is observed when micturition and defecation are induced from the septal area and from more caudally situated regions [37]. The fact that the *entire repertoire* of the evacuation act comes to play (although with remarkable temporal delay of individual components!) demonstrates that *between the distantly localized elements of the coordinating nervous apparatus there exist intimate functional contacts*. The central organization represents a unitary system, which can be triggered from various sites and which by irradiation leads to a *collective response*. Physiological sleep is an instructive example since the primary preponderance of fatigue in the motor system not only brings about motor rest during sleep but also the attenuation of sensory and associative processes. In an analogous way associative-psychic fatigue (within physiological range) brings about a rest for motor and sensory performances by means of sleep induction. Another telling example is represented by the slightly exaggerated circulatory depressor reflex which irradiates into the somatomotor innervation sphere, and even beyond, and can lead to a perfectly sleeplike state [40, 41]. The varying conditions in the physiological experiments under which sleep effects can be induced may provide an understanding of pathophysiological states which have an influence on the changed responsiveness of the organism. This theme, which is of special interest to clinicians, lies however outside the present discussion.

Critique and Conclusions

Every experimental technique and also the evaluation of data are subject to errors! In the present study one has to consider the hypnogenic aftereffect of anesthesia. Here we should remark that the effect of nitrous oxide anesthesia vanishes completely within 20–30 min and, therefore, cannot be held responsible for the induction of the sleep syndrome. It seems more likely that sleep-disturbing influences could arise indirectly from anesthesia, namely when they are connected with irritation of the mucous membranes, as is the case, for example, with ether and chloroform. Nitrous oxide, when given slightly in excess, can lead to dyspneic breathing and indirectly to the release of adrenaline, and generally to an activation of the ergotropic system which exceeds the duration of anesthesia and represents a disturbing factor for some time. Nitrous oxide anesthesia also entails a considerable fall in temperature. After awakening, the animal shivers and is undercooled; one has to work therefore in a warm room, and the experiment should not begin before the shivering has ceased. It happens occasionally that the setting of burrholes in the cranium may injure the diploic veins, causing a hemorrhage which can be stopped only very slowly. If the blood loss is too great, the animal feels weak and should not be used for the experiment.

The transition between opposing central excitatory states can make the evaluation of the stimulation effects quite difficult. Unfortunately, it is not possible in animal experimentation to use the central *restitutive process* as the *primary criterion* for sleep as it obtains in the normal human when (after sleep) the feeling of invigorating power returns subjectively to conscious feeling. We can only repeat that the *successive* and *coordinated* development of sleep posture and the reversibility of these phenomena are of primary importance for the evaluation of feline sleep. Also, the transition from inactivity to maximal preparedness induced by a waking stimulus should occur *abruptly* and be followed by highly contrasting behavior.

In this context we should point to another important observation: one might expect, as mentioned by Bremer in discussions, that the sleep effect would diminish after cessation of stimulation. Instead it is characterized by a considerable inertia, not only during the initial phase, but also in terms of sleep maintenance. The former is understandable in view of the relatively slow onset during the physiological process of falling asleep. On the other hand, it is not easily understandable that the artificially induced sleep state can outlast the stimulation period considerably, sometimes for several hours if the animal is not disturbed. Strong sleep effects are even characterized by the fact that the animal, after being awakened by an external stimulus, lies down again and falls asleep, even without renewed electrical stimulation. While it is true that the animal will respond adequately to an arousing stimulus, the tendency to fall asleep persists, however, if the stimulus is not applied vigorously enough and repeatedly. Conceivably, this behavior may be due to the influence of a humoral factor. The perseverance of the induced sleep state may also be related to the relatively high sensory threshold. As a consequence, the animal is surrounded by deep silence, and no external stimuli, be they auditory, visual, or tactile, can effectively contribute to a change of mood in the ergotropic direction. Sleep produces sleep! This is the true message of the stimulus deprivation theory, and it makes it understandable that true sleep can be superimposed on anesthesia – and perhaps even hypnosis.

The inverse type of symptoms can be observed during wakefulness, because the exteroceptive stimuli and associative processes can favor the predominance of the ergotropic system. Thus a kind of safety device prevents, in the competition between energy-releasing and economizing systems, a continuous up and down of conditions. Rather, these rhythms alternate physiologically in a periodicity which is adapted to diurnal rhythm.

A further criticism was advanced by von Brücke during a discussion. Although he agreed to the fact that the vegetative nervous system can influence the functions of the animal system, he drew attention to an inconsistency with respect to the role of the parasympathetic preponderance and the symptomatology of sleep. Profuse salivation, as a typical symptom of parasympathetic excitation, should be expected to be present in his opinion. This conclusion is evidently based on the notion that parasympathetic innervation should be in a general state of tonic activity during sleep. This is typically not the case, because the trophotropic-endophylactic system – in contradistinction to the ergotropic – *is characterized by a differential organization according to organs and functional systems, respectively*. Sleep function represents a highly differentiated process which protects the animal functions from being overstressed. Thus, the idea that salivation should be associated with the facilitation of restitutive processes and behavioral mechanisms entailing inhibitory components, is a conclusion based on misleading generalization and is not consistent with the true functional order which characterizes sleep.

In this context, we would like to discuss the criticism of Ranson and Magoun [45] who, in rejecting the idea that sleep is based on active inhibition associated with parasympathetic innervation by arguing that atropine fails to prevent sleep, have thus ignored the fact that the effect of atropine varies according to the target organs. *Skeletal muscle* and its response to acetylcholine is not affected by this alkaloid, or only in extremely high concentrations [38] which cannot be applied intra vitam because of the deleterious effect on the heart. Moreover, we know that the effects of atropine cannot be explained in as simple a scheme as that used by Ranson.

As concerns the *mechanisms*, one may consider the effect of *electronarcosis* described by Leduc [42], because this effect has been produced by pulsating direct current. However, this syndrome, with its abrupt onset, with the animal lying down or falling over, with the loss of consciousness which cannot be overcome by external stimuli, not even by treating it with a random combination of relaxation and spasms, has no similarities or identity with sleep induced by electrical stimulation as described in our experiments. Likewise, the confusion with *galvanonarcosis*, which is due to polarization effects on the cellular level, is not possible because of the different experimental conditions and the totally different symptoms. We refer to the investigations which have been carried out more recently by Scheminzky [49] and Martini et al. [44].

The criticism made by Ranson and Magoun [45] in relating somnolence to the *electrotonic and electrolytic effect of the stimulating current* should be discussed here in more detail. It is surprising, first of all, that the two phenomena, which differ in their biological aspects, should be mentioned by these two authors in the same context. More specifically, however, they relate the sleep-inducing mechanisms primarily to an electrolytic lesion and base this conclusion on experimental data described by Harrison [22], who claimed to have carried out experiments with a stimulation device "similar" to that used in our laboratory and failed to produce sleep effects except in cases where histological controls confirmed destruction of the tissue.

We cannot concur with this statement regarding the similarity of the setup. The electrodes used in this laboratory have a cross section which is only a fraction of that used by Harrison; consequently, inexorably smaller lesions of the tissue were produced by us. We feel justified in emphasizing this point, since we took special care from the very beginning as regards technique, and we made a special attempt to refine our electrodes. Another difference concerns the anesthetic: in contrast to Harrison, we have used nitrous oxide instead of ether. Other conditions cannot be safely compared because of the briefness of Harrison's report. This is especially true with respect to the slow development of sleep symptoms and the prevention of disturbing external influences.

Harrison differentiates between five types of stimulation responses: somnolence, somnolence with lack of motor initiative after awakening, catalepsy, excitation, and no response. As an example of the first group, he

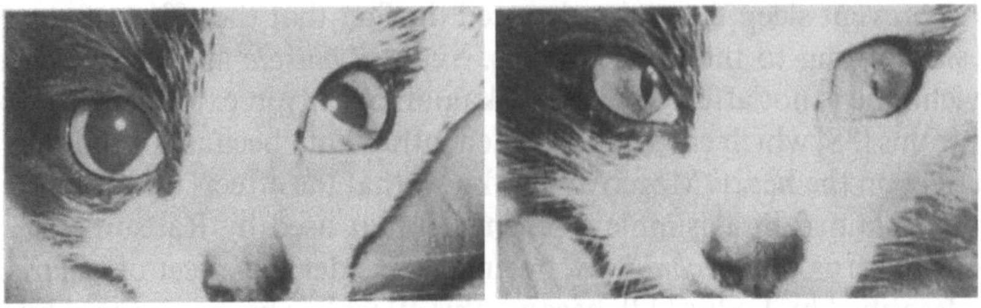

a b

Fig. 9.10. a Sympathetic innervation of the iris has been eliminated by transection of the left sympathetic trunk at the level above the superior cervical ganglion. Flashlight exposure at dimmed light. *b* Same situation, taken during bright daylight. The left eye shows the effect of reflex contraction of the sphincter iridis combined with complete loss of dilator innervation. Note the difference between the two eyes with respect to sleep pupillary size and retraction of the nictitating membranes. An analogous situation is present in the eyes of the sleeping cat. The prolapse of the nictitating membrane is a typical sign of impaired sympathetic innervation

described a case in which stimulation at 0.4 mA and with five impulses per second induced a sleeplike state which could be reversed by arousing stimuli. Microscopic examination revealed not inconsiderable lesions in the hypothalamus, and the somnolent state was interpreted as being a consequence of the destructive action of the electric current. Ranson and Magoun [45] provided an additional explanation by referring the sleep induction to the delayed rising and falling of the current pulses to the point where stimulus effects could no longer be seen. They end up by claiming that "sleep can be easily induced if direct current with corresponding electrolytic effects is sent through the hypothalamus." As concerns electrolysis and the lesions which can be produced with this procedure, we should like to refer to our earlier publication [28]. Our experimental conditions, as well as the histological findings exclude such an interpretation. With due respect to other important experimental results established by Ranson and his school, we would like to propose that a simple calculation could have saved these authors from imprudent statements. At 1 V, the average voltage used in our experiments, and with an estimated resistance of 7,000–10,000 Ω (circuitry and tissue included), the equivalent intensity of the stimulation current is in the range of only 0.1–0.15 mA. And this value (in the case of electrolytic action) should be divided by 5 because of the pulsating mode in which each stimulus pulse is followed by a four times longer interval. In this way, we arrive at a stimulus intensity which, for purely physical reasons, is unable to produce functional impairments which could be due to a focal lesion. Moreover, we have been able to elicit typical sleep effects even in cases (e.g., experiment 205, performed after 1939) in which polarization was compensated for by a four-times weaker countercurrent during the stimulus intervals. In other words we applied an alternating current with phases of unequal duration and reciprocal amplitudes, whose electron flow was equal in both directions. This method strictly prevents electrolysis. The lesion hypothesis, however, may still be sustained by assuming that small hemorrhages could have occurred even at the finest electrode tips. However, such lesions are usually too small to produce an effect. They are in no way comparable to the destructions which are likely to result from the relatively large electrodes used in Ranson's stereotactic apparatus. Thus it can be safely concluded that the behavioral changes with the typical characteristics of sleep that were produced in our experiments cannot be related to lesions but rather to the stimulating effects of pulsating direct current.

The question remains as to how the localized lesions in the hypothalamus as described by Harrison [22] and Ranson and Magoun [45] could have produced somnolence.

Two comments may be relevant: on the one hand we should like to refer to hypothalamically induced adynamia [35] and its relationship with the sleep mechanism as mentioned before; on the other hand, it seems important to point out that the caudal regions of the hypothalamus maintain relationships to the ergotropic system and that the elimination of this region by destruction would automatically entail the preponderance of the trophotropic-endophylactic component. Agreement exists with respect to Ranson and Magoun's observations that a shift in the vegetative balance occurs during sleep, although they assume that this shift may be due to a *diminished sympatheticotonic influence* rather than to an increased activity of the parasympathetic system. To this interpretation we would like to add the following observation: the impairment of sympathetic innervation without antagonistic activation of the parasympathetic would never produce *such a narrow pupil* as can be seen during deep, normal sleep (when the eyelids are opened artificially). During this stage the pupil is narrowed to a slit of less than 0.5 mm and is shortened; in contrast, with the complete elimination of the sympathetic innervation (by transection above the level of the superior cervical ganglion), the pupil is narrowed only to a middle size (Fig. 9.10). In view of Ranson's concept concerning a simple loss of sympathetic tone, it is important to know that both divisions of the vegetative nervous systems are by no means independent of each other. On the contrary, they compete in such a way that loss of excitation in one automatically releases the activity of the other. We are reminded of the instructive example given by von Brücke [11] with respect to the extracardial regulation of heart activity. Thus, the evaluation of Ranson and Magoun's criticism ends up with the conclusion that our concept concerning the antagonistic, i.e., the ergotropic-endophylactic, regulation of animal functions is to a large extent well supported. In considering the organization of the entire vegetative nervous system and its range of regulatory functions one should consider its influence in sleep not merely as accessory. It is indeed basic to the phenomena which characterize the behavior of the sleeping and waking animal.

Finally, we should like to confirm that we are well aware of the effects that may be produced by hypothalamic lesions [9, 10, 39, 50]. In such experiments we observed somnolence which lasted for several days and which the context of focal lesion effects, we should also refer to the exhaustive analysis by von Economo [18,19] in cases of encephalitis lethargica. This analysis by von Economo [18, 19] in cases of encephalitis lethargica. This analysis deserves special attention, because these focal lesions concerned the mesodiencephalic region and produced a functional deactivation (besides activation) of centers connected to conscious perception and associative functions. The fact that the changes in central preparedness last over long periods is consistent with our observations on focal lesions. These long-lasting effects are in conspicuous contrast to the stimulation-induced sleep effects in the acute experiment.

Instructions for Sleep Studies

1. Animals: healthy, adult cats which have been housed and familiarized with their environment for at least 10–14 days before the experiment should be used.

2. The operative fixation of the electrode carrier on the skull should be carried out 3–5 h before the beginning of the study under nitrous oxide anesthesia. Operation and postoperative care should take place in a room with a temperature not under 18 °C.

3. Placement of the electrodes should be 3.0–3.5 mm lateral to the midline and 6 mm (rostral electrode), 7.5 mm (middle electrode), and 9 mm (caudal electrode) behind the coronary suture. Length of the electrodes should be 21 mm (calculated from skull surface) and the bare tip, 2 mm. (These measurements are valid for an adult female cat.) To acquire the necessary practice, it is advantageous to first carry out the procedure several times on dead animals and, after fixation of the brain, to verify the localization of the electrode tips by means of Prussian blue reaction. The electrodes should remain in place for 12–24 h and the formalin-fixed brain then sectioned into 2-mm-thick horizontal slices.

4. Insertion of the electrodes at the beginning of the experiment (usually this remains without immediate reaction). Wait for 10–15 min; transient sleepiness as a result of the mechanical stimulation by the electrode tips may possibly be observed.

5. Maintain glare-free illumination of the experimental table. Motion picture recording requires mercury lamps with Wood filters. The brighter

the light, the less pronounced is the pupillary reflex. Bright light should be avoided because it acts as a waking stimulus.

6. Stimulus parameters: begin with 1.0 V; possibly go down to 0.75–0.5 V if motor symptoms appear. If they are only slight, they are not disturbing and they fade away spontaneously. Maintain frequency at six to ten impulses per second; the duration of a stimulus train is 30–60 s. If a sleep effect fails to occur, pause for 1–2 min; under some circumstances the sleep effect occurs during this pause. Otherwise, repeat the stimulation with 1.5 V, if necessary with 2.0 V. Use intervals between stimulation periods for behavioral observations.

7. Perform arousal tests by increasing stimulation voltage to 3 V or possibly 4 V; under some cirumstances use rectangular impulse forms. Lifting of the animal and shaking may be repeated; also, meat held before the nose may be effective as a rule.

8. Further controls: after spontaneous waking of the animal, renewed stimulation for attaining a sleep effect requires, as a rule, somewhat higher voltage.

9. General technical aspects of the experiments are described under "Methods" [26].

10. After the experiment the cat receives allobarbital (Dial, 2 cc 10%) intraperitoneally and, after the onset of anesthesia, it is killed. The brain is then perfused with warm Ringer's solution followed by a 4‰ formalin solution via the carotid. During the next 24–48 h (in the refrigerator) the brain hardens to such an extent that it remains undamaged after the electrodes are removed from the skull; it can now undergo histological studies. Determine the points of stimulation by use of the Prussian blue reaction; cell (Nissl) and fiber (Heidenhain) stains should also be carried out. When examining the latter preparations it is to be kept in mind that the postmortem diffusion traces of iron around the tips of the electrodes are visible as a dark halo which has nothing to do with the intravital staining of the tissue.

11. After practice with this technique a series of at least six experiments with stimulation in the area to be investigated should be carried out.

Summary

A systematic investigation of the diencephalon and bordering areas of the central nervous system was undertaken with the aid of rhythmic electrical stimulation. Not infrequently there was a striking change in the behavior of the animal in the sense of a sleeplike condition. Preconditions for such an effect are the use of low stimulation voltages, the delayed rise of the pulse amplitude, and not too short a duration of the low frequency

(8 Hz) stimulus train, which may be repeated at longer intervals. The responses are classified into three groups. The symptoms of the first group (I), the "elite" cases, correspond in every respect to those of physiological sleep. In the second group (II), a condition similar to sleep prevails, whereby the inactivation of motor and sensory functions is not uniform and, in general, the effect is comparatively less deep. The third group (III) comprises cases with signs of a stuporous drowsiness. A further group with symptoms of muscular adynamia has already been described in an earlier publication [35].

In addition to cases which can be readily classified in terms of these groups, there are also transition cases.

Based on histological analysis, the following brain areas seem to be related to the syndromes of the three groups: (1) the thalamus between the mammillo-thalamic tract (Vicq d'Azur) and the habenulo-interpeduncular tract (Meynert) at the level of the lower-thirds of massa intermedia, extending a short distance laterally from the midline toward the thalamic radiation; (2) the nucleus caudatus; (3) the periventricular area of the hypothalamus; (4) the preoptic area; and (5) the area of intersection of the fornix and the anterior commissure.

Qualitatively and quantitatively the area of the thalamic stimulation stands in the first place because sleep symptomatology can be elicited from this region in the purest form (group I). From the remaining areas it was possible to elicit conditions of somnolence with bodily fatigue and decreased muscle tonus; such conditions may occasionally develop into complete adynamia in the presence of a relatively clear sensorium; in addition, sleeplike conditions with transition to stuporous drowsiness occur.

Aside from the localization, the effect is also dependent on the conditions of the experiment. Excited and nonfamiliarized animals are less suitable. Nitrous oxide anesthesia has proved to be reliable; external disturbances must be avoided. Working in a warm room and a sufficient pause after discontinuation of the anesthesia are necessary for success.

On the basis of the symptoms of sleep and its physiological function, and taking into consideration the entire organization of the vegetative nervous system, as well as the correlation with histoarchitectonic features of the stimulated areas, waking and sleep functions can be conceived as manifestations of a central regulation of the excitatory state *(regulierte zentrale Leistungsbereitschaft)*. Both functions are characterized by alternating cycles of ergotropic and trophotropic-endophylactic predominance within the vegetative nervous system. One can also recognize an alternation with respect to the readiness of the organism for interaction with the environment and the predominance of a system which protects and secures the coordinating centers at higher levels. The influence of extero- and proprioceptive stimuli takes precedence in the mechanisms which bring about and maintain one or the other of the two conditions.

Finally, certain criticisms as well as the conditions for the successful performance of sleep experiments have been discussed.

This project was supported financially by the Jubilaeumsspende der Universitaet Zuerich; valuable assistance was provided by the Emil Barell-Stiftung in Basel for medical research. I wish to give thanks to both institutions for effective help.

References

1. Bartorelli, C.: Movimenti di correzione per la conservazione dell' equilibrio del corpo visti attraverso immagini cinematografiche a rapida frequenza. Verh. Schweiz. Naturforsch. Ges. 120. Vers., 210–212 (1940)
2. Bartorelli, C.: Esperienze di stimolazione elletrica della regione mesencefalica I. Arch. Fisiol. *42*, 384–415 (1941)
3. Bartorelli, C.: Der heutige Stand der experimentellen Erfahrungen über die Beziehungen von Zwischen- und Mittelhirn zur Motorik. Schweiz. Arch. Neurol. Psychiatr. *48*, 370–396 (1941)
4. Bartorelli, C.: Effets moteurs dus à la stimulation du tegmentum du mésencéphale. Verh. Ver. Schweiz. Physiol. (July), pp. 12–14. Basel: Schwabe 1941
5. Bartorelli, C., Wyss, O.A.M.: Influenza di lesioni encefaliche sui riflessi respiratori vagali. Bol. Soc. Ital. Biol. Sper. *16*, 219–221 (1941)
6. Bartorelli, C., Wyss, O.A.M.: Die Nystagmusphase als Kompensationseffekt des labyrinthären Drehreflexes. Verh. Ver. Schweiz. Physiol. (January), pp. 3–4. Basel: Schwabe 1941
7. Bartorelli, C., Wyss, O.A.M.: Kinematographische Analyse der labyrinthären postrotatorischen Körperdrehreaktionen. Pflügers Arch. Ges. Physiol. Menschen Tiere *245*, 511–523 (1942)
8. Bianchetti, G.: Kinematographische Analyse der vestibulären und optischen Kompensationsmechanismen am postrotatorischen Kopfnystagmus bei Änderung der Körperlage. M.D. Dissertation. Zürich: Orell Füssli 1942
9. Bloch, W.: Beziehungen des Hypothalamus zum respiratorischen Stoffwechsel. Helv. Physiol. Acta *1*, 53–78 (1943)
10. Bloch, W.: Über das Verhalten des Blutzuckers nach herdförmiger Ausschaltung im Hypothalamus. Helv. Physiol. Acta *1*, 177–181 (1943)
11. Brücke, E.Th.: Über die reziproke, reflektorische Erregung der Herznerven bei Reizung des N. depressor. Z. Biol. *67*, 507–519 (1917)
12. Brügger, M.: Freßtrieb als hypothalamisches Symptom. Helv. Physiol. Acta *1*, 183–198 (1943)
13. Cloëtta, M., Fischer, H.: Über die Wirkung der Kationen Ca, Mg, Sr, Ba, K und Na bei intrazerebraler Injektion (Beitrag zur Genese von Schlaf und Erregung). Naunyn Schmiedebergs Arch. Exp. Pathol. Pharmakol. *158*, 254–281 (1930)
14. Cloëtta, M., Thomann, H.: Chemisch-physikalische Untersuchungen zur Theorie der Narkose. Arch. Exp. Pathol. Pharmakol. *103*, 260–281 (1924)
15. Cloëtta, M., Fischer, H., van der Loeff, M.R.: Die Biochemie von Schlaf und Erregung mit besonderer Berücksichtigung der Bedeutung der Kationen. Naunyn Schmiedebergs Arch. Exp. Pathol. Pharmakol. *174*, 589–675 (1934)
16. Cloëtta, M., Fischer, H., van der Loeff, M.R.: Die Verteilung und die Wirkung des Magnesiums im Organismus und deren Beeinflussung durch Calcium. Naunyn Schmiedebergs Arch. Exp. Pathol. Pharmakol. *200*, 6–42 (1942)
17. Demole, V.: Pharmakologisch-anatomische Untersuchungen zum Problem des Schlafes. Naunyn Schmiedebergs Arch. Exp. Pathol. Pharmakol. *120*, 229–258 (1927)

18. Economo, C. v.: Die Pathologie des Schlafes. In: Handbuch der normalen und pathologischen Physiologie, Vol. 17: Correlationen III. Bethe, A. (ed.), pp. 591–610. Berlin: Springer 1926
19. Economo, C. v.: Encephalitis lethargica, ihre Nachkrankheiten und ihre Behandlung. Berlin: Urban & Schwarzenberg 1929
20. Fischer, H.: Die Rolle des Calciums beim Zustandekommen von Narkose und Erregungszuständen am rindenlosen und am völlig dezerebrierten Tier. Naunyn Schmiedebergs Arch. Exp. Pathol. Pharmakol. *138*, 169–189 (1928)
21. Frey, E.: Studien über die Entwicklung der zentralen optischen Bahnen des Menschen. I. Die Entwicklung des Nervus opticus und des Chiasma opticum. (Habilitationsschrift). Schweiz. Arch. Neurol. Psychiatr. *47*, 74–124 (1941)
22. Harrison, F.: An attempt to produce sleep by diencephalic stimulation. J. Neurophysiol. *3*, 156–165 (1940)
23. Hess, W.R.: Über die Wechselbeziehungen zwischen psychischen und vegetativen Funktionen. Schweiz. Arch. Neurol. Psychiatr. *15*, 260–277 (1924); *16*, 36–55, 285–306 (1925)
24. Hess, W.R.: The mechanism of sleep. Am. J. Physiol. *90*, 386–387 (1929)
25. Hess, W.R.: Le sommeil. C.R. Soc. Biol. (Paris) *107*, 1333–1364 (1931)
26. Hess, W.R.: Beiträge zur Physiologie des Hirnstammes. I. Die Methodik der lokalisierten Reizung und Ausschaltung subkortikaler Hirnabschnitte. Leipzig: Thieme 1932
27. Hess, W.R.: Der Schlaf. Klin. Wochenschr. *12*, 129–134 (1933)
28. Hess, W.R.: Beitrag zur Technik des zentralen Reizversuches. Pflügers Arch. Ges. Physiol. Menschen Tiere *243*, 431–438 (1940)
29. Hess, W.R.: Zwischenhirn und Motorik. Pflügers Arch. Ges. Physiol. Menschen Tiere *243*, 634–650 (1940)
30. Hess, W.R.: Motorische Symptome im Gesicht bei Reizung im Zwischenhirn. Pflügers Arch. Ges. Physiol. Menschen Tiere *243*, 678–685 (1940)
31. Hess, W.R.: Diencephale Reizsymptome am Körperstamm; Beziehungen derselben zu Bewegungseffekten im Gesicht und an den Extremitäten. Pflügers Arch. Ges. Physiol. Menschen Tiere *243*, 741–747 (1940)
32. Hess, W.R.: Charakter der im Zwischenhirn ausgelösten Bewegungseffekte. Ein Beitrag zur extrapyramidal gesteuerten Motorik. Pflügers Arch. Ges. Physiol. Menschen Tiere *244*, 767–786 (1941)
33. Hess, W.R.: Experimenteller Beitrag zur Frage der extrapyramidalen Motorik. Z. Ges. Neurol. Psychiatr. *172*, 639–642 (1941)
34. Hess, W.R.: Physiologische Aspekte der extrapyramidalen Motorik. Nervenarzt *15*, 457–466 (1942)
35. Hess, W.R.: Hypothalamische Adynamie. Helv. Physiol. Acta *2*, 137–147 (1944)
36. Hess, W.R., Brügger, M.: Das subkortikale Zentrum der affektiven Abwehrreaktion. Helv. Physiol. Acta *1*, 33–52 (1943)
37. Hess, W.R., Brügger, M.: Der Miktions- und der Defäkationsakt als Erfolg zentraler Reizung. Helv. Physiol. Acta *1*, 511–532 (1943)
38. Hess, W.R., Büsch, J.: Der Einfluß von Atropin auf die Reaktion des Skelettmuskels bei direkter Reizung. Pflügers Arch. Ges. Physiol. Menschen Tiere *216*, 644–650 (1927)
39. Hess, W.R., Stoll, W.A.: Experimenteller Beitrag betreffend die Regulierung der Körpertemperatur. Helv. Physiol. Acta *2*, 461–480 (1944)
40. Koch, E.: Die Irradiation der pressorezeptorischen Kreislaufreflexe. Klin. Wochenschr. *11*, 225–227 (1932)
41. Koch, E.: Weshalb sind die Aorten- und Karotissinusnerven als pressorezeptorische Kreislaufnerven bezeichnet worden? Z. Kreislaufforsch. *24*, 613–616 (1932)
42. Leduc, S.: Production du sommeil et de l'anesthésie générale et locale par les courants électriques. C.R. Acad. Sci. (Paris) *135*, 199–200 (1902)

43. Lehmann, G., Michaelis, H.F.: Adrenalin und Arbeit. VI. Der ergotrope Reflex. Arbeitsphysiologie *12*, 440–448 (1943)
44. Martini, E., Gualtierotti, T., Marzorati, A.: Die Rückenmarkelektronarkose (R.E.N.). Pflügers Arch. Ges. Physiol. Menschen Tiere *246*, 585–596 (1943)
45. Ranson, S.W., Magoun, H.W.: The Hypothalamus. Ergeb. Physiol. *41*, 56–163 (1939)
46. Reumont, M.: Une nouvelle technique d'imprégnation argentique pour la mise en évidence de l'appareil neuro-fibrillaire et cylindraxile. Rev. Neurol. *2*, 53–58 (1931)
47. Sanchez-Calvo, R., Monnier, M.: L'influence des excitations vestibulaires rotatoires sur la mimique faciale et les oreilles du lapin. Helv. Physiol. Acta *1*, 381–387 (1943)
48. Schatzmann, M.: Kinematographische Analyse vestibulärer Kompensationsmechanismen am ruhenden Auge und beim postrotatorischen Nystagmus. M.D. Dissertation, University of Zurich. Zürich: Orell Füssli, 1943
49. Scheminzky, F.: Stromwirkung und Feinstruktur im Zentralnervensystem. Naturwissenschaften *31*, 288–298 (1943)
50. Stoll, W.A.: Beziehungen des Hypothalamus zur Temperaturregulierung. Helv. Physiol. Acta *1*, 329–357 (1943)

Chapter 10

Hypothalamic Adynamia*

W.R. Hess

The phenomenon which is reported here can be elicited by electrical stimulation of the diencephalon and the neighboring areas of the brain. It involves maximal relaxation of the skeletal musculature as well as the absence of spontaneous activity and reactivity to external stimuli. The most striking finding is the contrast between the extremely weak central influence and the strong overall effect on the motor system. These findings are worth examining in more detail because they seem to be relevant to the anatomical pathology of certain nervous diseases and provide important clues for the characterization of differences between various clinical conditions with motor deficit. The following case reports provide information on this phenomenon.

Experiment 118. During lowering of the electrodes (right) turning of the head toward the left develops at the last moment. The position of the head slowly returns to normal again spontaneously. The nictitating membrane prolapses halfway. If the cat is laid on its side, it remains (Fig. 10.1). The cat hardly reacts to banging on the table. This adynamic condition becomes so pronounced in the course of some minutes that the forelimbs retain their abnormal position. Only in the posture of the head can a small amount of motor innervation still be noted. The increased excitability of the reflexes is striking: the extremities are withdrawn as soon as they are touched. Despite the unilaterality of the central origin, the increase of the reflex excitability is bilateral. With weak stimulation (1 V during 15 s) the anterior part of the body is raised. As with lowering of the electrode, there is turning of the head to the left. After some time a certain spontaneity returns; meanwhile, for short periods, the animal behaves as if depressed. In continuing the study by multiple short stimulations (intervals of 20 s) with somewhat higher voltage (2–4 V), the *behavior changes completely*. The animal is now attentive and looks around spontaneously. Later it accepts milk and, after removal of the bowl, seeks it by snuffling. The adynamic condition with hyperreflexia lasts about 20 min.

Localization of the electrode tips. In the lateral part of the hypothalamus and the mammillary bodies (sections 437–481 of the horizontal atlas series [6]).

---

* Hess, W.R.: Hypothalmische Adynamie. Helv. Physiol. Acta *2*, 137–147 (1944)

10.1 10.2 10.3

Fig. 10.1. Complete atonia in the extremities. Only the head is under some degree of postural control. The eyes are open, thus there is no relaxation of the levator palpebrae muscles (experiment 118)

Fig. 10.2. The animal is in an abnormal position; the nose presses against the table. The awkward position of the extremities remains unchanged (experiment 178)

Fig. 10.3. Atonia is expressed especially in the position of the head. The eyelids are closed (experiment 113)

Experiment 178. After stimulation with 1 V (left), then with 2 V over 6 min (with pauses between), frequent closing of the eyes occurs; the cat assumes a sleeplike position with relaxed nictitating membrane on the left. Upon external stimulation the eyes are opened relatively rapidly. There is still a latent attentiveness. While the electrode is advanced 2 mm deeper the animal shows a certain activity, but it later lies down again with folded paws. The position is similar to that of sleep, however, and to some extent unnatural in so far as the forelimbs are somewhat spread out. The head is dropped down so far that the nose hits against the table. When the table is banged the head moves only slightly. After the cat is lifted up, it rapidly collapses again. While the electrode is lowered on the right the animal reacts with a single short twitch followed immediately by a prolapse of the nictitating membrane on the right. Finally motor activity ceases little by little, i.e., within 3 min. Passive positioning of the extremities fails to elicit any attempt to carry out corrective movements. The condition is that of complete adynamia which remains stationary during the course of the subsequent electrical stimulations with 2–4 V (Fig. 10.2). After the electrode is advanced about 2 mm deeper there is essentially no change. When the cat is lifted up and then placed down again, it sinks slowly. There is no response to strong acoustic stimuli. With electrical stimulation at 4 V the cat shrinks back, snarls, and hisses, and the adynamia now completely disappears. On the other hand stimulation with weaker voltage (electrode *A*) reinduces adynamia.

Localization. Left lateral hypothalamus, the anterior electrode just touching the optic tract (sections 424–525 of the horizontal atlas series [6].

Experiment 113. Fully 7 min after onset of the first stimulation with moderate intensity (1–4 V over 70 s) which results in the appearance of some motor symptoms in the head and face, the animal begins to sink down, thereby assuming abnormal positions, primarily of the head (Fig. 10.3) and the forelimbs. These effects often reveal a marked disturbance of proprioception. During strong banging only a sluggish reaction occurs in that the animal lifts the head somewhat, as if it were in slow motion, and opens the eyes. The adynamia alternates with phases of spontaneous activity, e.g. licking, at irregular time intervals. Suddenly, however, an arrest intervenes and, for several seconds, the animal maintains a randomly taken posture. After more than 1 h the cat returns to almost normal behavior. Milk presented, for example, is promptly licked up. Then a brief episode of atonia follows abruptly, and the head hangs down until it touches the chest. A substantial improvement later occurs, but the reactivity to external stimulation remains diminished.

Localization. Subthalamus and mainly lateral hypothalamus (section 387–437 of the horizontal atlas series [6].

Experiment 243. Immediately as a consequence of lowering the electrode (i.e., advancing it to the 2-mm deeper-lying "lower level"), an adynamic-stuporous condition gradually develops. Electrical stimulation can bring about a certain reactivity, after which the animal falls back into the earlier condition.

Localization. Anterior hypothalamus.

Similar observations have also been made in other cases in which the adynamia is not so strongly pronounced, or is limited only to one part of the body.

Experiment 177. Upon electrical stimulation at the right higher level a nausea-like picture develops in which the cat staggers when walking or standing and vomits. Advancing the electrode deeper shows no direct reaction; soon afterwards, i.e. in the course of 1–2 min, the cat lies on its side and the extremities become completely atonic. The eyes remain open. Electrical stimuli at 4 V have no effect. Independent of the stimulation the cat makes attempts to lift itself; it does not succeed, however. Stronger electrical stimulation triggers rhythmic motor effects in the face, extremities, and body. At the same time the atonia disappears transiently. Notable in this case is a fright reaction of the animal to noise, e.g. of the movie camera.

Localization of the electrodes. Lateral hypothalamus; the optic tract is also touched by the bare tip of the electrode.

Experiment 122. After lowering the electrode (left), a weakness of the hind part of the body appears, especially of the *posterior extremities*. Thus there is a regional adynamia which deepens after advancing the right electrode about 2 mm in 1 min. The adynamia gradually subsides. On presentation of food the animal moves toward it, and the hind part of the body shows a clear weakness. The animal, in the adynamic condition, shows a conspicuous fright reaction to touch and auditory stimuli.

Localization. Mesencephalon (red nucleus and reticular formation).

Discussion

In summarizing the characteristics of this behavior we note the following symptoms: sinking *into a random position* with pronounced to maximal relaxation of the skeletal musculature. At the same time spontaneity is nearly or completely extinguished. This abnormal condition is, in most cases, triggered merely by the simple lowering of the very fine electrodes (cross section of a single electrode: about $^1/_{20}$ mm^2). In one case (experiment 113) adynamia occurs only after electrical stimulation and indeed with *low voltage*. In another case (experiment 178) adynamia caused by insertion of the electrode is potentiated by weak electrical stimulation.

With respect to the symptomatology in four cases (experiment 113, 118, 178 and 243) there is complete agreement in all essential aspects. A quantitative difference is shown in cases 118 and 178 only in that the adynamia is complete, while in cases 113 and 243 there is a clear *fluctuation* in which phases of generalized adynamia alternate with phases of activity. Tendon reflexes and responses to tactile stimulation are found to be increased in one case in which this was tested (experiment 118). The four cases cited are consistent in that the head and (except in experiment 118) also the eyes are involved, and the animal scarcely reacts to acoustic stimuli. Two cases (experiments 122 and 177) are peculiar as the eyes normally remain open. Here there is a *fright* reaction in response to acoustic stimuli.

The reversibility of the adynamic condition is typical in that it disappears again either spontaneously after some delay, or as a result of stimulation with a higher voltage. (The location of the electrodes is shown in Fig. 10.4.) The similarity in location of the electrodes in the cases in which the symptomatology agrees (experiment 113, 118, 178, 243) is noteworthy. The electrodes of case 122 (the animal in which the adynamia is limited to the hind part of the body) lie eccentrically, i. e., posteriorly. It is assumed that substrates (fiber tracts?) are involved here which are related to the caudally lying parts of the extrapyramidal innervation.

A possible explanation for the syndrome described here is the assumption that the stimulated area normally exercises a tonic influence on differentiated parts of the motor innervation apparatus. As a result of damage by the electrode, despite the extremely fine needle track, a transient deficit phase might take place, that is, a decrease of tonus; in that case, however, increased motor activation must be expected when stimulation intensity rises. Another explanation might come from the fact that the penetration

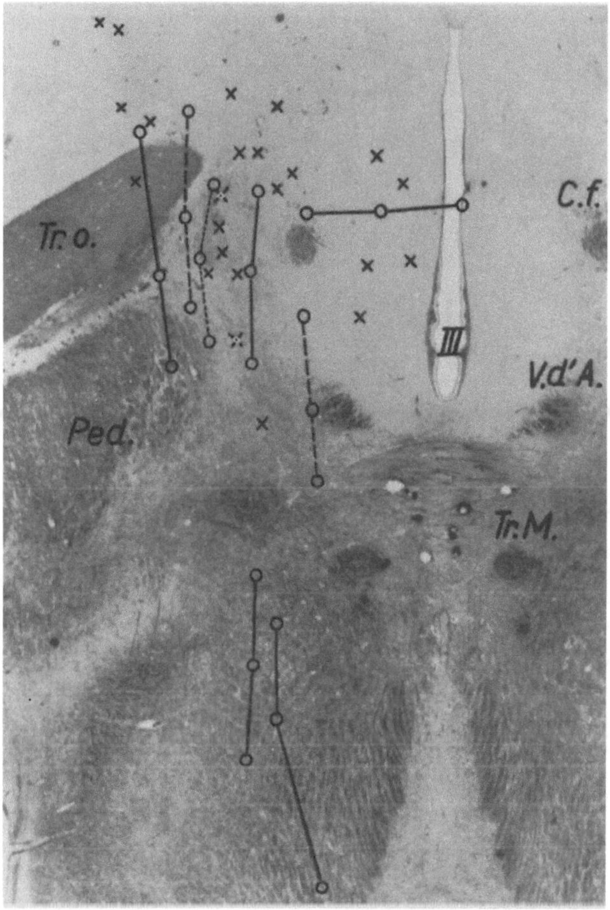

Fig. 10.4. Position of the electrodes in the hypothalamus: Section 437 of the horizontal atlas series (see [7], Table II, Fig. 94a). The position of the bare points is indicated by *small circles*. The electrodes combined in one electrode carrier are shown connected by a *heavy line* where the center of the bare tip lies in the plane of the section; *dotted line* where the center is 1–2 mm higher; *dashed line* where it is 1–2 mm deeper. Rostral positions from left to right refer to cases 178, 177, 113, 178, 118, and 243. Positions caudal to the tract of Meynert refer to cases 122 left and right. All electrodes are projected onto the left side for the sake of comparison. The places noted by an *open cross* (⨯) indicate fall of blood pressure (see Figs. 10.5, 10.6), rostrally case 172, caudally case 135. Further points with blood pressure decreases are noted by a *closed cross* (×). *C.f.*, fornix; *Ped.*, cerebral peduncle; *Tr.M.*, tract of Meynert (habenulo-interpeduncular); *Tr. o.*, optic tract; *V.d'A.*, mammillothalamic tract (Vicq d'Azyr); *III*, third ventricle of the cerebrum

of the electrode represents a stimulus, even if it is weak, caused in turn by the pulsations of the blood stream in the circle of Willis, which move the brain against the tip of the electrodes. The delayed appearance of adynamia would thus be understandable as a result of the summation of weak stimuli. In agreement with this interpretation is the observation of two cases in which adynamia appeared only after (weak elec-

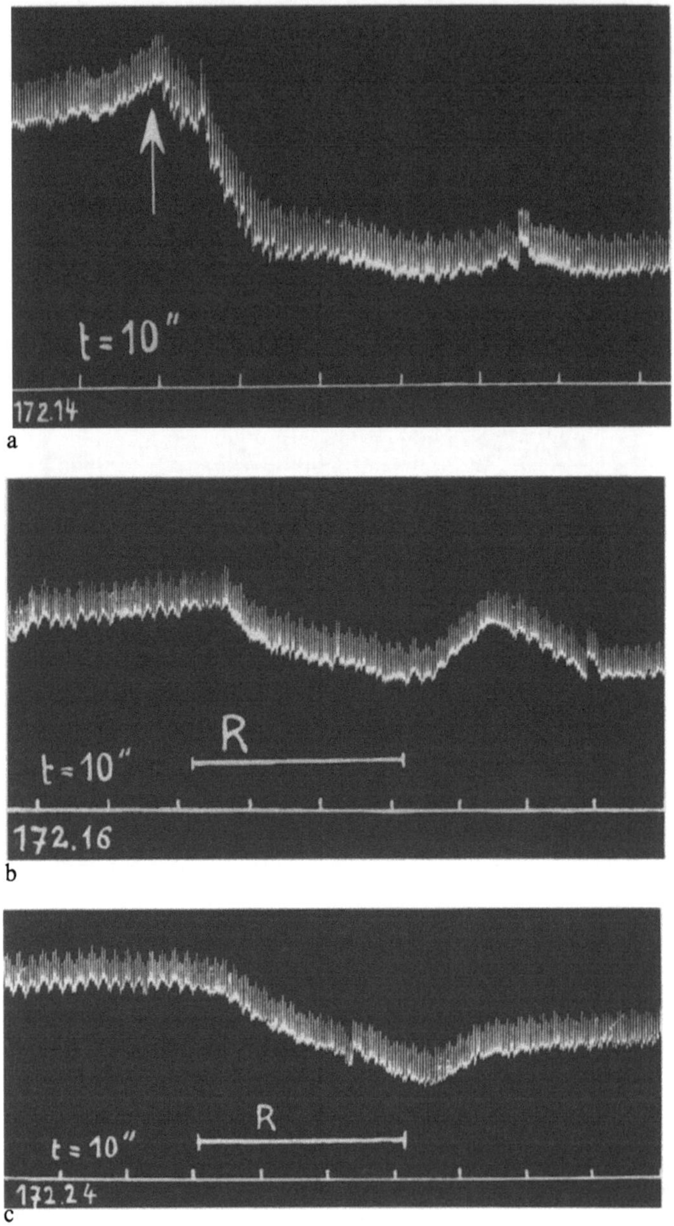

Fig. 10.5. a Blood pressure fall after inserting the electrodes into the hypothalamus (experiment 172). *b, c* Blood pressure decrease upon stimulation with 1.5 V and 3.0 V, respectively. In *b* a transient rebound follows the decrease (experiment 172). Note that the blood pressure remains at a lower level, exactly as in the experiments of Koch [9, 10]

trical) stimulation, or when the latter had been intensified. Thus, the adynamia described may be interpreted as resulting from inhibition. Indeed, the bilaterality of the effect is only explicable in terms of a stimulation action; it should *not* be regarded as a sign of a lesion.

In this connection the experiments performed by Koch [9, 10] are of great interest. In the course of an extensive investigation of the physiology of the carotid sinus, this author determined, among other things, that with sufficiently strong blood pressure increases in the carotid sinus preparation (as a result of a *pressor* stimulus), the spontaneous or in any way reflexly triggered movements of the *skeletal muscles are very much weakened, and most of the time even completely abolished.* He saw these effects, for example, in monkeys; one animal, which had previously always followed the movements of the investigator with its eyes, holding its arm extended and moving it continually, slowly allowed the arm to hang limp with increasing pressure also inside the carotid sinus. Spontaneous movements of the arm also ceased. The eyelids closed and the animal lay down as if sleeping. After release of the pressure, the condition was usually maintained, until after some stimulus the eyes were opened again and, with rapid increase of the blood pressure to the baseline level, the original behavior returned. Analogous symptoms appeared in the freely moving dog. With increasing pressure in the carotid sinus the animal was remarkably quiet, hung its head, closed its eyes, and lay down. With stimulation it reacted only slightly or not at all. The test procedure and special control studies excluded the possibility that this effect was due to a change in blood supply. From these observations Koch [9, 10] draws the conclusion that the pressoreceptor circulatory reflexes extensively influence the central nervous system and indeed, in an inhibitory sense, i.e., in the sense of diminishing excitability.

Thus, an unmistakable relationship is established with our observation described above and their explanation as an active inhibitory phenomenon. This may form the basis for instructive conclusions. It should be kept in mind that the *primary* (pressor) significance of the carotid sinus stimulation consists in the reflex induction of an *unloading (Entlastung) of the circulatory apparatus.* In the framework of circulatory regulation the inhibition plays the role of a protective mechanism which, at the same time, reinforces the economy of the circulatory operation. By irradiating as an inhibitory factor into the motor innervation the potentiated stimulus attests to the existence of a central relationship between circulation and motor system in conformity with the physiological demands of both systems. Thus, we see now that with intensified circulatory inhibition bodily performance is also slowed down, thereby effectively supporting the circulatory unloading reflex *(Entlastungsreflex)*. We stress that this correlation is not an interpretation but is based on factual findings. One can tentatively leave the question open as to whether it is the expression of a central *organization* or a *coin-*

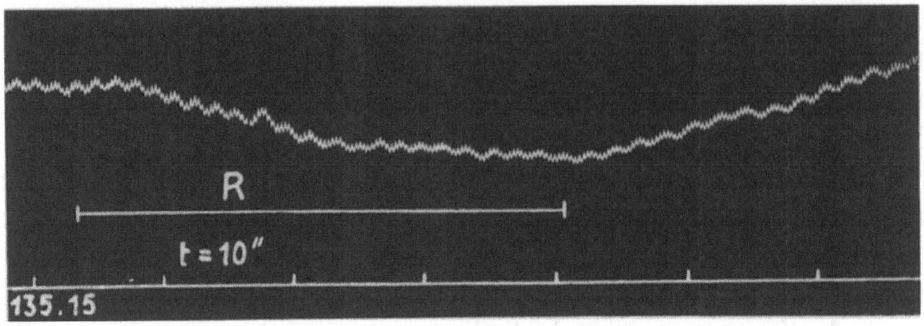

a

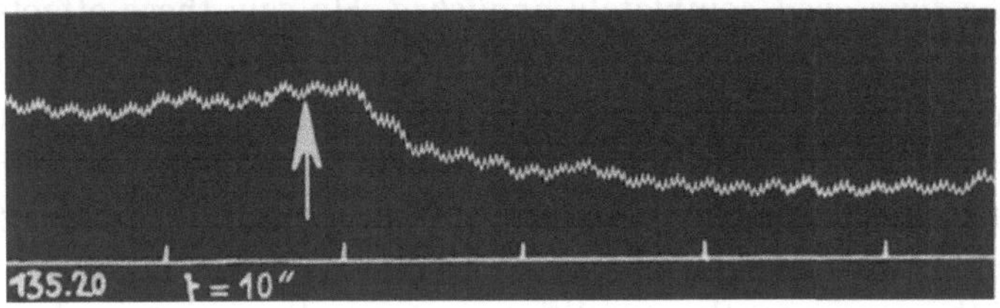

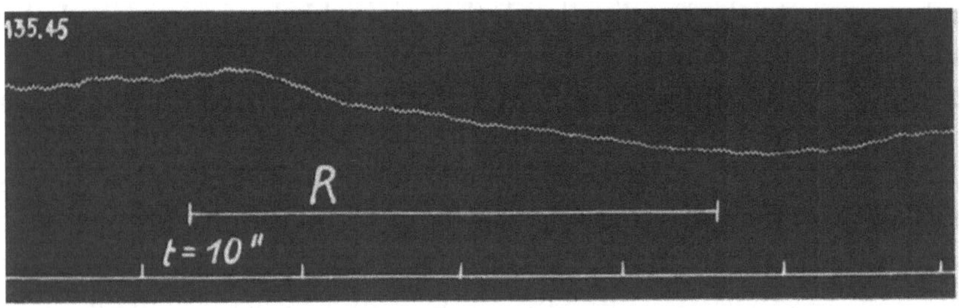

c

Fig. 10.6. a, c Blood pressure decrease after stimulation with 4.0 V (experiment 135); *b* blood pressure decrease after inserting the electrodes into the hypothalamus (experiment 135). *R*, electrical stimulation; *arrow* indicates the insertion of the electrode

cidental irradiation. In any case our attention is directed eagerly to the results of investigations reported earlier [7]. In a large series of studies the effect of electrical stimulation in the diencephalon was reexamined with respect to certain criteria of circulatory and respiratory behavior. It turned out that, among other findings, a hypothalamic area exists from which electrical stimulation consistently elicits a decrease in blood pressure, in part accompanied by slowing of the heart – in short, a typical depressor effect as it appears in the scope of the circulatory unloading reflex (Figs. 10.5, and 10.6). If we now compare the localization in our adynamic cases with the hypothalamic stimulation sites displaying depressor effects, we can conclude that in the former, as well as in the latter, the same area is involved

(Fig. 10.4). So the circle is closed; the results of investigations which had their point of departure from completely different viewpoints, may be recognized as parts of a *uniform mechanism*. As inhibition elicited by the intensified (peripheral) carotid sinus reflex spreads from circulation into skeletal musculature the same relationships are encountered when stimulation is applied to a circumscribed hypothalamic area. On the basis of the *central organization* an irradiation into the relevant substrate becomes manifested in terms of physiological *coordination*. The fact, moreover, that vegetative regulations and general behavior are *integrated by the hypothalamus* is confirmed by a new and instructive example; i.e., an unloading stimulus arising from circulation eventually invades the hypothalamus and spreads with synergic effect into the control mechanism of the skeletal musculature. In this context one should mention the observations of other authors [1–5, 8, 11–17] to gain insight into additional aspects of organizational principles. However, we shall limit ourselves here to reporting on the phenomenon of adynamia and the results of our own experimental work standing in immediate relationship to it.

The relationship of hypothalamic adynamia with circulatory regulation is consistent with the observation that adynamia is *not an irreversible* deficit sign, but rather a reversible expression of a disturbed balance of general excitability. It remains only to be explained how an *increased* central stimulation may overcome the inhibitory state. Here the investigations of circulatory behavior may help us again: when more intense central stimulation is applied, the depressor action gives way to a pressure effect and, indeed, the more *caudal the stimulation site* lies within the circumscribed area the more reliable the effect. If we move somewhat more posteriorly, the primary effect, i.e., a blood pressure increase accompanied at times by acceleration of the heart rate, appears even with weak stimulation. Here we refer to our comprehensive review of these findings [7]. From the circulatory unloading/muscle-relaxing area, stray currents exert their effects posteriorly and bring about an imbalance of the regulatory equilibrium; the ergotropic-regulating principle transmitted by the sympathetic chain opposes successfully, under given conditions, the less powerful endophylactic mechanism (transmitted by the parasympathetic elements).

In this context we should add some observations concerning the action of hypothalamic lesions, caused for example, by electrocoagulation or disease. When sufficiently large, they also lead to adynamia. However, in this case the symptom is the consequence of destruction, and therefore it represents a functional insufficiency, which may recede only in days or weeks, if at all (cf. [18]). Finally, the increased reflexes during adynamia in case 118, may be interpreted as secondary phenomena, namely as the consequence of the release from higher centers of motor innervation.

Summary

1. In a number of experiments the mere lowering of the electrodes or electrical stimulation (with low voltage) was followed by a pronounced adynamia, accompanied, to some extent, by a markedly diminished reactivity. This phenomenon is spontaneously reversible; it can also be caused to disappear, under some circumstances, by increased stimulation.
2. Adynamia was observed in cases where the electrode tips were located in the lateral hypothalamus up to anterior levels.
3. It is mentioned that this is in complete agreement with the inhibitory effect of an increased tension in the carotid sinus on the skeletal muscle system (Koch).
4. The functional relation between both inhibitory unloading phenomena is again reflected in centrally induced adynamia evoked from the segment of the hypothalamus, the stimulation of which triggers a typical circulatory unloading reflex *(Entlastungsreflex)*, i.e., depressor carotid sinus effect.
5. The dependence of higher levels of motor organization and circulatory regulation on a defined hypothalamic substrate demonstrates, on the one hand, the close relationship between centers of skeletal muscle function and circulatory regulation, and on the other hand, the role of the hypothalamus in coordinating vegetative and animal functions in the organization of a vitally important protective mechanism.

We wish to thank the Stiftung für wissenschaftliche Forschung of the University of Zürich and the Rockefeller Foundation, New York, for the support of this work.

References

1. Beattie, J.: Relation of tuber cinereum to gastric and cardiac functions (A preliminary note). Can. Med. Assoc. J. *26*, 278 (1932)
2. Beattie, J.: Hypothalamic mechanisms. Can. Med. Assoc. J. *26*, 400–405 (1932)
3. Beattie, J., Sheehan, D.: The effects of hypothalamic stimulation on gastric motility. J. Physiol. (London) *81*, 218–227 (1934)
4. Beattie, J., Brow, G.R., Long, C.N.H.: Physiological and anatomical evidence for existence of nerve tracts connecting the hypothalamus with spinal sympathetic centres. Proc. R. Soc. London Ser. B *106*, 253–275 (1930)
5. Billingsley, P.R., Ranson, S. W.: Branches of the ganglion cervicale superius, J. Comp. Neurol. *29*, 367–384 (1918)
6. Hess, W.R.: Beiträge zur Physiologie des Hirnstammes. I. Die Methodik der lokalisierten Reizung und Ausschaltung subkortikaler Hirnabschnitte. Leipzig: Thieme 1932
7. Hess, W.R.: Beiträge zur Physiologie des Hirnstammes. II. Das Zwischenhirn und die Regulation von Kreislauf und Atmung. Leipzig: Thieme 1938
8. Johnson, S.E.: On the question of commissural neurones in the sympathetic ganglia. J. Comp. Neurol. *29*, 385–404 (1918)

9. Koch, E.: Die Irrradiation der pressorezeptorischen Kreislaufreflexe. Klin. Wochenschr. *11*, 225–227 (1932)
10. Koch, E.: Weshalb sind die Aorten- und Karotissinusnerven als pressorezeptorische Kreislaufnerven bezeichnet worden? Z. Kreislaufforsch. *24*, 613–616 (1932)
11. Ranson, S.W.: An introduction to a series of studies on the sympathetic nervous system. J. Comp. Neurol. *29*, 305–312 (1918)
12. Ranson, S.W., Billingsley, P.R.: The superior cervical ganglion and the cervical portion of the sympathetic trunk. J. Comp. Neurol. *29*, 313–366 (1918)
13. Ranson, S.W., Billingsley, P.R.: The thoracic truncus sympathicus, rami communicantes and splanchnic nerves in the cat. J. Comp. Neurol. *29*, 405–439 (1918)
14. Ranson, S.W., Billingsley, P.R.: An experimental analysis of the sympathetic trunk and greater splanchnic nerve in the cat. J. Comp. Neurol. *29*, 441–456 (1918)
15. Ranson, S.W., Magoun, H.W.: Respiratory and pupillary reactions induced by electrical stimulation of the hypothalamus. Arch. Ges. Neurol. Psychiatr. *29*, 1179–1194 (1933)
16. Ranson, S.W., Kabat, H., Magoun, H.W.: Autonomic reactions induced by electrical stimulation of the hypothalamus. Am. J. Physiol. *109*, 85–86 (1934)
17. Ranson, S.W., Kabat, H., Magoun, H.W.: Autonomic responses to electrical stimulation of hypothalamus, preoptic region and septum. Arch. Ges. Neurol. Psychiatr. *33*, 467–477 (1935)
18. Stoll, W.A.: Beziehungen des Hypothalamus zur Temperaturregulierung. Helv. Physiol. Acta *1*, 329–357 (1943)

Chapter 11

The Subcortical Center of the Affective Defense Reaction*

W.R. HESS and M. BRÜGGER

Contents

In the experimental material collected by one of us since 1925 there are numerous cases in which carefully controlled stimulation of subcortical areas of the cat evoked symptoms of affective excitation [10–12, 15]. The centrally elicited reaction agrees extensively, as a rule completely, with the results we obtain under physiological conditions when the cat is confronted by a dog and, in this threatening situation, seeks to defend itself. Shortly after the beginning of electrical stimulation, one hears a warning growl which turns into hissing and spitting. At the same time there is piloerection of the tail, which soon gets thick and bushy; the hairs along the spine also stand up. The pupils widen and in some cases there is a rapid back-and-forth movement of the ears, recalling the flapping wings of a bat. Thus the entire effect at the climax of the affective defense reaction is particularly frightening and, as a rule, effective. The pronounced arching of the back,

* Hess, W.R., Brügger, M.: Das subkortikale Zentrum der affektiven Abwehrreaktion. Helv. Physiol. Acta *1*, 33–52 (1943)

11.1

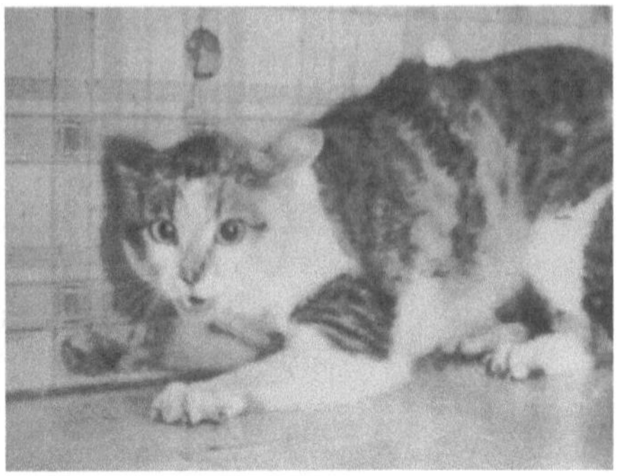

11.2

11.3

Fig. 11.1. Stimulation in the center for active defense with 1 V. Effects: arched back, piloerection, hissing directed at the camera man

Fig. 11.2. Stimulation as in *Fig. 11.1* with 1 V. Effects: piloerection, hissing, ear flapping, extension of claws, raising of forepaw to strike the camera man; marked dilatation of the pupils

Fig. 11.3. Stimulation as in *Fig. 11.1* with 2.5 V. Effects: bushy tail, hissing, ear flapping, directed at assistant standing nearby

which is rather striking in the physiological situation, is notably absent. Furthermore, it should be emphasized that the centrally excited animal passes from defense to attack, thus conforming to the normal defense tactics of the cat. This behavioral response is surprising in yet another respect. One would think that with central placing of the stimulus and the absence of an external trigger the symptoms would develop without a specific spatial direction. In fact, however, there is often a *goal-directed discharge*, as can be recognized from Figs. 11.1–11.3. Most often the cat in its defense/attack reaction turns against the nearest person participating in the experiment. Because the observer's behavior provides no reason for the cat's reaction, the attack can only represent the eccentric projection of a mood which touches off the defense situation; in other words, the central stimulus objectifies itself as an external event.

Fig. 11.4. Affective defense reaction due to natural stimulation (approach of a dog)

This behavior can be expressed so drastically that the cat springs viciously toward the observers within reach and claws the towel held for protection. Very impressive is the growing of this reaction to a maximum, which then suddenly stops when the electrical stimulation is discontinued. After less than 1 s one can see that all the symptoms of the excitation disappear. The return to normal clearly goes faster than the build up of the syndrome, which, depending upon the intensity of the stimulus, requires 5 10 or more s. In order to describe the effect of the subcortical stimulus, we present two short protocols:

Hissing 5 seconds after the beginning of stimulation; ears flapping like the wings of a bat: successively increasing effect. After 50 seconds of stimulation the cat remains with claws unsheathed ready to strike: in between, growling. Complete picture of defense (experiment 225).

Or:

Shortly after the beginning of the stimulation the tail becomes bushy; after 21 seconds hissing, panting, and movement of the ears develop only little by little, that is, with increase of the general excitation; urination after 23 seconds (experiment 243).

Not infrequently the affective syndrome is accompanied, as just mentioned, by urination and more rarely also by defecation. The typical posture taken under physiological conditions is thereby not assumed, as, e. g., following central stimulation in another area of the diencephalon. In contrast to a coordinated mechanism, the animal simply evacuates the urine or the feces. The movement occurs apparently involuntarily, that is, as a consequence of the excitation. As a result of its intensity it spreads beyond the boundaries which determine the manifestation of the *coordinated syndrome*.

An analogous sign of unspecific irradiation is the occasional occurrence of salivation and panting, both symptoms which have nothing to do with the active defense reaction – at least not primarily.

Topographical Localization of the Stimulation Area

For the localization it is important to consider the strength of the stimulus employed, because with increasing voltage the radius of the effect produced by each electrode increases, too, and the boundary of the active zone is less clear. Thus, it is imperative to give specific details about the stimulus voltage.

a) Stimulus Voltage. In correspondence with the regular procedure in our investigations, each stimulus site is excited with different voltages. According to our experience, the threshold effects are, under the best conditions, obtained when the stimulation voltage lies between 0.5 and 1.5 V. Based on our material, we consider this range as the low intensity level. The second level is taken – from a purely empirical standpoint – as 1.6–2.5 V. This permits potentiation of the symptoms and leads, under some circumstances, to new ones. The third level of intensity ranges from 2.6–4 V. What lies over this level *is not taken into consideration* in the analysis. With stronger stimuli one can often have very good and impressive effects, and one may have the chance to see something "interesting"; still, despite the interesting findings, as a rule the latter are not very useful for localization purposes, because stray currents and physiological irradiation contaminate the effect. One of *our* goals was to bring function and substrate into as close a relationship to each other as possible. Frequently, we even avoided exceeding 2.5 (−3) V. This was the case when very marked effects appeared even with the first and second intensity levels.

Besides the symptoms of the defense reaction, motor effects sometimes appeared, depending upon the site of stimulation, which prompted us to limit the intensity of stimulation. With regard to the speed and force of the motor responses whose character and diencephalic localization have already been extensively described [16]; see translation in this volume, p. 223, it was imperative for us not to exceed a certain voltage range. We also refrained from stimulating with the third voltage level when the effects were already clear cut with the lower intensities.

b) Technique of Localization. In all experiments the site of the blank electrode tips (1.5–2 mm long) was determined histologically in horizontal sections. The staining was according to Heidenhain and Nissl, which allowed identification of fiber tracts as well as cellular elements. Despite the fineness of the electrodes, made from hard steel, their anatomical position could be easily determined in serial celloidin sections, because there was always a trace of iron (marked by means of the Prussian blue reaction) in the area of the electrode tip. Our plan of investigation was not to localize specific effects but to explore the entire diencephalon and its bordering areas systematically, i.e., point-to-point, by means of electrical stimulation. In this way we wanted to gain insight into *the entire range of functions and principles of organization* of the diencephalon [10, 13]. These studies were therefore performed over a long period, i.e., between 1925 and 1942. This wide temporal "dispersion," however, proved to be no disadvantage, since the written protocols were enlarged upon extensively by means of movies allowing us to review and compare the stimulation effects at any time. The plotting of the data was carried out in the same way as in the work on the motor aspects of the diencephalon [16]. For each experiment the stimulation points, which were plotted in the photographs of the corresponding brain sections, were transferred to enlarged diagrams in which the typical landmarks were indicated at various levels. In the final analysis all the stimulation points from which the *affective* defense reaction could be elicited were selected and localized in cumulative charts. In order to establish the relationship between the 12 stimulation points belonging to one and the same experimental animal these points were connected with lines and, by adding the number of the experiment, the connection between histology, protocols, and films was maintained in each case. In order to symbolize the individual components of the defense syndrome we designed a "star" with six beams, each of which corresponded to a single effect as shown in Fig. 11.8. By plotting the appropriate "star" symbol at the point from which the corresponding effect was elicited, it was possible to determine at a glance not only which effects appeared at what site, but also what symptom was missing from the complete syndrome. Furthermore, the stimulus intensity, important for the definitive evaluation, was expressed in such a manner that symptoms which appeared with the lowest intensity were emphasized by a thickly drawn beam, while a thinly drawn beam designated the moderate and the dotted beam, the highest voltage level. The survey of the histological localization was finally facilitated by transposing all the stimulus points plotted originally at horizontal levels to a sagittal plane. This was possible without too great an effort, because of the preliminary work mentioned elsewhere [13].

c) Localization Data. With the aid of histological photographs (Figs. 11.5–11.7) and the corresponding maps (Figs. 11.9–11.11.), one obtains information about the position and distribution of the active points of stimulation with regard to the affective defense reaction.

The cumulative charts of the stimulation points with "positive" findings need to be complemented by plotting the "negative" findings (which also

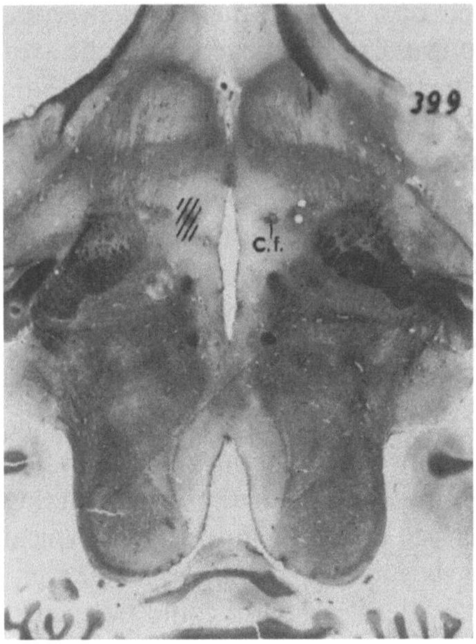

11.5

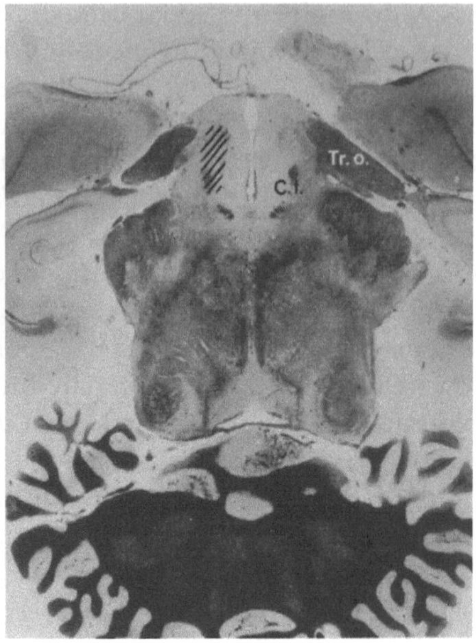

11.6

Fig. 11.5. Horizontal section 399 of the atlas [13]. *Shaded area* represents focal region for affective defense. Localization: intermediate hypothalamus in the region of the fornix. *C.f.*, fornix

Fig. 11.6. Horizontal section 449 of the atlas [13]. 1.5 mm ventral to section 399 *(Fig. 11.5)*. The focal area for affective defense is shaded. *C.f.*, fornix; *Tr. o.*, optic tract

have a "positive" aspect). It is important to identify the areas which fail to be involved in the symptoms under discussion, in particular, the points from which even with stronger stimulation a corresponding effect is lacking. Stimulation points which did not respond to the highest voltage step are designated by small solid circles; where the middle voltage step led to no particular symptoms, an empty circle is placed. The lowest voltage step producing negative findings has not been taken into consideration.

We shall consider first Figs. 11.9 and 11.10 in which the active stimulation points are mapped in the horizontal diagram. It is noteworthy that 80% of these points are crowded into a strikingly small area. We determined that the extent in a vertical and rostro-caudal direction amounts to just about 2 mm and even less in the transverse direction. This area is also notable in that the effects are particularly easily evoked, even with the weakest stimulation. Negative points – if any –involve animals under anesthesia. The excitability, in general, was considerably decreased in these cases. The localization of the highly sensitive region is depicted in Figs. 11.5–11.7. It involves the perifornical area of the anterior hypothalamus extending into the preoptic area and the ventro-caudal zone of the hypothalamus between the descending column of the fornix and the mammillothalamic tract (Vicq d'Azyr). Naturally it is advantageous to define the boundaries as exactly as possible. To this end the following is included.

d) Limits of Error in Localization. It is clear that the higher the voltage, the more one has to consider current spread. This source of error is somewhat larger with unipolar than with bipolar stimulation. In the latter case one must take into consideration that with rhythmic stimulation the effect often originates from the anode, which is 1.5 mm away. The error when one refers to the threshold effect of the cathode will be in this range. The physical and the physiological sources or error must be taken into consideration. These make the attempt to precisely define the boundaries of the responsive area difficult. Some of our experimental studies indicate that in certain substrates the excitation spreads by means of irradiation. In consequence, even areas specific for the affective defense reaction can be secondarily involved by *indirect* excitation.

Exactly for this reason it is of great importance to differentiate the stimulation effects according to the voltage steps and to place decisive emphasis on that site where effects appear at the lowest threshold level.

Still another physiological factor can, in our experience, impair exact localization. When a pronounced effect occurs in one animal, for example by stimulation in a focal area, the prior condition may often not return with the decline of the symptoms. Instead, a latent (physiological) excitation remains which expresses itself in a tendency toward reactivation of the stimulation symptoms. By stimulating with a neighboring electrode, one can make the "latent potential" apparent, since one can now producc the syndrome from an area which under normal threshold conditions is unresponsive. The boundaries of the active area thus appear to be widened. The opposite also occurs, namely, when the responsiveness is reversed after a marked excitation. Then it requires a stronger stimulus to evoke the earlier effects. Thus an electrode which lies, for example, on the edge of an active substrate may respond only to a high voltage as a result of a negative after effect. It appears as if one were farther removed from the focal area. Such threshold changes caused by the stimulation itself blur the boundaries between reactive and inactive areas. Nevertheless, the errors working in both directions balance one another in the total result, especially when higher voltages and therefore, stronger excitations are avoided and, as far as possible, longer pauses between stimuli are inserted. Finally, the individual temperament of the test animals should be considered, in that one is easily aroused and the other one less so. This fact must be included in the interpretation of the stimulation maps.

e) Further Differentiation of Localization. Additional precision in localizing the stimulation effects has been achieved by referring in the cumulative plots to the findings derived from each experiment. This was particularly revealing in those cases in which considerably different stimulation effects were obtained from neighboring electrodes or in which at least very different threshold voltages had been noted. By comparing such contrasting stimulation points, one obtains more detailed information about the boundaries of the active zone. The following is a differentiated analysis of this comparison.

The substrate, whose stimulation evokes the affective defense reaction, lies, as shown in Figs. 11.5 and 11.6, in the hypothalamus 1–2 mm from the midline. It is bilaterally symmetrical and corresponds to a *zone of central gray matter*, intermediate between the periventricular layers on one side and the lateral hypothalamus on the other. These facts are in agreement with the experiments with medio-laterally oriented stimulus positions (experiments 241, 243, 245, and 246) and more so with a rostro-caudal arrangement, on the left or right side (experiments 183, 231, 251, 262, 275, 276, and 285). Two of these experiments (243, 245) show particularly clearly how the excitation effect appeared completely within the intermediate zone, and indeed appeared with stimulation even at the lowest level. In contrast, the effect failed to occur more medially and laterally or occurred only with higher voltage and was then less pronounced.

In the ventral hypothalamus (plane 462) the substrate of the affective defense reaction extended more toward the midline. For example, in experiment 243, no excitation occurred at the medial electrode at the dorsalmost level (planes 412–449), while the complete effect developed even with the lowest stimulation voltage in the ventral level (planes 462–481). We could follow the responsive substrate down to plane 462 in two experiments (experiments 225 and 245), where it lies dorso-rostral to the ventromedial hypothalamic nucleus. That it still extends deeper in the tuber cinereum is not probable, but not to be excluded with certainty.

The dorsal border in the hypothalamus is clearly defined. It does not extend above plane 399 (experiments 227, 235, 237, 239). All stimulus positions in the area of the thalamus are negative in relation to our syndrome (see the sagittal section Fig. 11.11) A clear proof is delivered, for example, by experiment 241. Here the electrode tips reach on the *left* side down to plane 399, thus into the hypothalamus; on the *right* side, on the other hand, they reach only to plane 387, which still corresponds the thalamus. Thus while on the left the typical defense reaction is obtained, this is not the case on the right side. After advancing the electrode about 2 mm on the right side, the effect was promptly elicited.

Rostrally the responsive substrate extended into the medial preoptic area. Here the nearby supraoptic area proved to be inactive. This was proved by numerous cases (experiments 232, 234, and 239). Particularly decisive is experiment 234, in which the rostral electrode (A) in the supra-

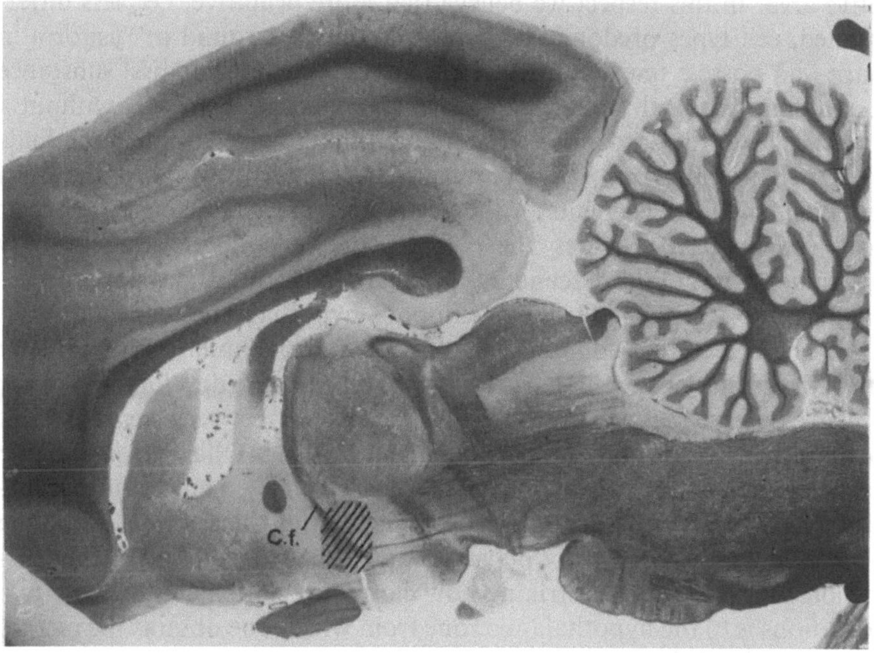

Fig. 11.7. Sagittal section 100 of the atlas [13], about 2 mm lateral to the median plane. Focal area for affective defense is shaded. *C.f.*, fornix

optic area shows no effect either on the left or on the right, not even by stimulation with the third voltage level. The caudally lying neighboring electrodes (B and C) yield, on the other hand, a distinct effect, and indeed even with the lowest voltage.

From the preoptic area the active substrate may be followed upward into the septal area, approximately along the descending column of the fornix; in three cases (235, 238, 253) we proceeded until the *dorsal level of the anterior commissure* was reached (horizontal section 334 and perhaps higher). In experiment 238 the electrodes did not reach below plane 324; thus for the positive effect elicited, no placing can be considered which lies more than 0.5 mm deeper.

Caudally the substrate of the affective defense reaction did not reach the mammillothalamic tract (Vicq d'Azyr). Experiments 207, 227, 231, 234, 235, 272, and 285 demonstrate that stimulation of the caudal electrode (C) in the vicinity or within the mammillothalamic tract gives no effect with lower and middle voltage. On the other hand the rostrally adjacent electrodes have a very impressive effect even with the lowest voltage level.

f) Histology. Although not possessing the experience of a professional brain histologist, we attempted to learn more about the fine structure of the active area. In this respect we believe that more primitive, i.e., less differentiated, cell types predominate here. The cells are round or fusiform in shape and possess few dendrites. They are pale with little Nissl substance accumulation at both poles; the nucleus is small and seemingly without a nucleolus. Still they are more differentiated than the cells which one finds in the neighboring zone of the periventricular gray surrounding the third ventricle. The fiber elements of the active area are characterized by their weak myelination. Besides these fibers one encounters the strongly myelinated tracts of the descending column of the fornix and the supraoptic commisure of Ganser and the myelin-poor hypothalamic-tegmental tract [22][1]. The descending column of the fornix and the anterior commissure can with certainty be excluded as beeing involved in the development of the affective defense reaction, because the excitation effect very clearly relates to the *pericolumnal* and the *pericommissural* areas. Olfactory and optic tracts are excluded for physiological reasons which need not be further discussed here. Whether fibers of the supraoptic *commissure of Ganser* and the *hypothalamic-tegmental tract* play a role may be subjects for further discussion. At any rate, it is striking how close the relationship of these two formations is to the hypothalamic zone from which one obtains the picture of the affective defense reaction. Close to the active zone, myelinated lateral tracts of the medial forebrain bundle pass by, as does the ansa lenticularis. Also at the lateral borders we find the H_1 field of Forel, the stria medullaris and the inferior thalamic peduncle. The external medullary lamina lies dorsally, the supraoptic commissure of Meynert ventrally and the mammillothalamic tract Vicq d'Azyr caudally. All these formations can be excluded from involvement in our syndrome because numerous cases document the fact that their stimulation remains without a relevant effect. The same holds for the periventricular layers of the gray substance of the third ventricle.

Functional Arrangement of the Center

After determining that there is a strikingly circumscribed area from which the affective defense reaction symptoms can be consistently elicited, the question arises whether this area can be further divided into segments to each of which the specific elements of the entire syndrome can in turn be assigned. The method according to which our evidence was obtained permits this question to be answered without difficulty. Referring to Figs. 11.9–11.11, we can indeed conclude that not all the stimulation positions appear to be equivalent. Some produce all or almost all of the individual

1 The little-known fibers of the hypothalamic-tegmental tract originate in the dorsolateral part of the ventromedial hypothalamic nucleus and course mainly caudally to the medial border of the H_1 field of Forel; a small part courses to the zona incerta [22]

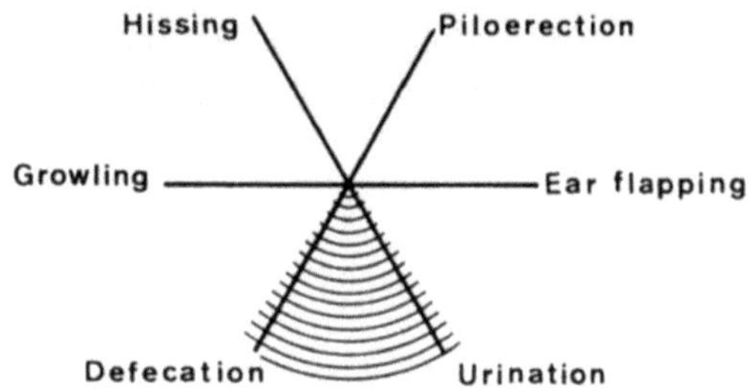

Fig. 11.8. Symbol for the syndrome of the affective defense reaction. The main symptoms are arranged as single beams of the six-pointed star and are utilized in the localization diagrams in *Figs. 11.9–11.11.* Urination and defecation are the result of an (uncoordinated) irradiation during very strong excitation

symptoms; in others, with voltage near the threshold, one or the other symptom is missing. An arrangement in the sense of *functionally specific sectors* is, however, not obvious. On the contrary, we can clearly exclude such a surmise. A certain regularity occurs only insofar as piloerection can be elicited from a somewhat wider radius; it thus appears alone, especially when near the ventral border, suggesting that this symptom has generally a more extensive area of representation. The opposite is the case with flapping of the ears. The entire syndrome is given a special emphasis in that it only appears in the presence of the full-blown defense reaction. To this corresponds the observation that this symptom appears mainly with stimulation of the core of the active area, from which the climax of the reaction is obtained at lowest threshold. In the border zones, on the other hand, a stronger stimulus is needed, or only individual symptoms appear, in the following rank order (growling, piloerection, hissing, flapping of the ears).

The stimulus sites for urination and defecation are inconsistent. One can understand this because they involve a generalized spreading of excitation to the vegetative function areas, which have nothing to do with the *coordinated* defense reaction. In fact, urination and defecation are a simple consequence of a strong emotional shock with unspecific irradiation. Naturally the condition of the peripheral target organs plays a decisive role. For example, a full bladder facilitates the evacuation of urine; similarly defecation is dependent on the condition in the colon. In summarizing what is known, it can be shown that from a well-defined region a number of single functions are released which are integrated into a goal-directed response, namely protection by defense. Because this relationship between substratum and performance is not duplicated in another area of the subcortex with nearly the same consistency and completeness, there is no doubt that we have here a *subcortically coordinated and graduated apparatus for the defense reaction.* Thus, according to localization and relative arrangements it may be considered a *center*. In evaluating the observations with the lesion experiments as known from the literature, we may be able to support our conclusions still further.

Centrifugal Transfer of Excitation

We shall now turn to the question of how excitation arising in the area in which by far the most active stimulation points lie, is transmitted to the effector organs so that these are set into action. The piloerection effect shows that a very close relationship to the sympathetic nervous system must exist; this is confirmed by the widening of the pupil [14], which is always very pronounced and up to now has not been mentioned because *it is not specific for the defense syndrome*. The same is valid for the activation of blood pressure and respiration which has already been described in detail elsewhere [13]. Hissing represents a peculiar form of facial expression combined with phonation. It belongs to the categories of automatic motor activities induced by subcortical innervation. Growling and ear flapping are analogous mechanisms with different peripheral effector organs.

Because these symptoms require the *coordinated action* of numerous muscles, it is not reasonable to assume that excitation flows directly to the corresponding motoneurons; rather, it must be conducted through a supranuclear "organizer," i.e., a neural apparatus which is subordinate to the center of the affective defense reaction. Without further taking up the question of dividing the entire reaction into individual functions, it is still interesting to us because certain experimental observations are directly related to it. Consider, for instance, stimulus sites which were situated outside, and even quite a distance from the center and from which certain symptoms could be elicited, in part isolated and in part combined with other effects (Figs. 11.9–11.11). To this point we can make the following statements.

Beyond the area established as the center, multiple stimulation points with symptoms of the affective defense reaction have been determined which are spread out caudally across the boundary between the diencephalon and mesencephalon (Fig. 11.11). From these one can obtain growling, piloerection, and hissing; not seen when stimulating these points was ear flapping in the form typical for this syndrome. The effective stimuli belong almost without exception to the higher voltage levels. Thus the threshold is clearly higher than in the center. The stimulation effects were not as consistently seen in the region between the center and mesencephalon down to the rostral border of the pons (isthmus) as in the focal area. Often there were silent placements near the active stimulus points which showed no effect with stimuli up to 4 V. The positive stimulus points were found to be situated without exception near the midsagittal plane. In such areas we

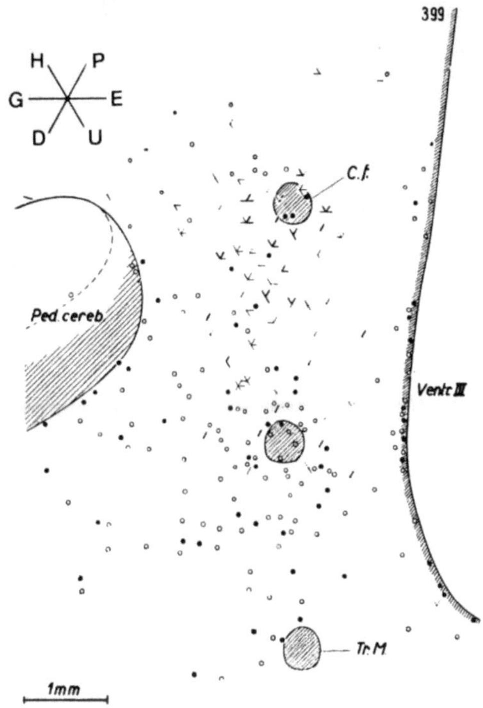

Fig. 11.9. Diagram corresponding to horizontal section 399 (*Fig. 11.5*). Localization of individual symptoms of the affective defense reaction. *C.f.*, fornix; *Tr. M.*, tract of Meynert; *ped. cereb.*, pes pedunculi cerebri; *Ventr. III*, third ventricle of the cerebrum. Symbols see *Fig. 11.8*

found them at the lower border of the massa intermedia (experiments 71, 221, and 236), in the central gray matter at the rostral entrance of the aqueduct of Sylvius (experiments 230, 236), and still further caudally (experiment 125). From the periaqueductal region we obtained in one case hissing, piloerection, and urination from points which lay relatively far laterally, namely at the lateral border of the central gray matter (experiment 130). But in this case *faradic* currents had been applied which have a strong tendency to spread far out and which may have excited layers lying near the midline.

From these data one must conclude that even the excitation induced in the center itself finds its way to the effector organs, i.e., to the supranuclear controls along a pathway which lies near the ventricles. The higher threshold voltage required for the infracentral stimulation points suggests that this pathway does not involve a compact fiber tract. Rather, the mediating elements are distributed over a relatively large "road," and within this the weakest stimulation at any one point always affects only a relatively small fraction of them. Therefore the excitation is not sufficient to activate the

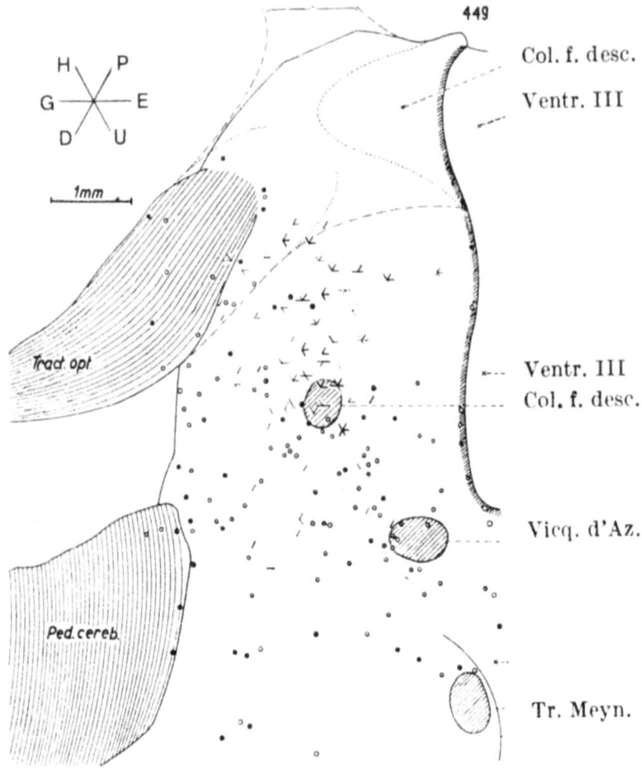

Fig. 11.10. Diagram corresponding to horizontal section 449 *(Fig. 11.6)*. Localization of individual symptoms of the affective defense reaction. *Col.f.desc.*, fornix; *Ventr.III*, third ventricle of the cerebrum; *Tract.opt.*, optic tract; *Ped.cereb.*, pes pedunculi; *Tr.Meyn.*, tract of Meynert; *Vicq.d'Az.*, mammillothalamic tract (Vicq d'Azyr). Symbols see *Fig. 11.8*

entire affective defense reaction. For each excitatory state the affect-free condition induced is dependent upon an equlibrium whose displacement requires a certain amount of excitatory forces. So it is finally understandable why "ear flapping" is not observed outside the area of the center. This effect is the *last* element in the rank order of the symptoms belonging to the affective defense reaction. It appears only when the excitation has reached its high point and also only in certain cats; with others the release of this impressive fright-inducing mechanism is not seen because (in the defense situation) the cat proceeds quickly to attack. With reference to individual differences, the striking experience was noted that outside the center the affective defense reaction was obtained mainly in young animals.

Observations of Other Authors

Since the investigations of Karplus [18] and Karplus and Kreidl [19, 20], it has been known and confirmed many times that effects produced by hypothalamic stimulation resemble those obtained by applying electrodes in the peripheral sympathetic pathways. Thus for the first time certain functional relationships were established which interest us here. Besides this we also find reports about effects of hypothalamic stimulation which correspond to electrical stimulation of parasympathetic pathways. Such findings have already been described [13]. In this connection we should like to mention the investigations of Beattie et al. [4, 5], Fulton [8], and Kabat et al. [17], which provide detailed information on the localization concerning blood pressure, respiration, pupil, and nictitating membrane, i.e., factors indicating changes in the *general* excitatory state. Data related to the affective defense reaction we find, apart from our own communications [10–12, 15], in the works of Kabat et al. [17]. These authors speak of "rage-like reactions." The animals search around with wide pupils, the respiration is markedly accelerated, the hairs are erect, the claws unsheathed; urination occasionally occurs; and the activity of the gastrointestinal tract is inhibited. In one case rhythmic hissing was seen (with stimulation in the region of the fornical part of the medial forebrain bundle). As accompanying signs, vigorous movements characteristic of marked emotional reactions were observed. The authors localized the region of the "rage-like reactions" in the lateral *hypothalamus*.

The conclusions drawn from the stimulation results can be further elaborated by adding the data from lesion experiments, especially those studies in which the symptoms of the affective defense reaction were presented after *ablation of brain areas* at different levels. These results are important not only for the localization, but also for the understanding of the organization of behavior.

First we shall mention Goltz [9], Rothmann [23], and Dusser de Barenne and Magnus [7], who were able to elicit by peripheral stimuli hissing, growling, piloerection, and defense reactions in dogs, and to some extent, also in the cat after removal of the telencephalon. From this, we must conclude that these symptoms are essentially based on a subcortical mechanism. Woodworth and Sherrington [24] proved that in the decerebrated cat, that is, after removal of the cortex and diencephalon, with stimulation of an afferent nerve, the so-called pseudo-affective reflex occurred. Yet, even with a very strong peripheral stimulus a coordinated defense reaction in the sense of attack of flight never occurred. Cannon and Britton [6] studied the behavior of the sympathetic-adrenal system in the decorticated cat. In the acute experiment after decortication and recovery from anesthesia a typical excitation syndrome with piloerection, dilated pupils, sweating of the paws, rise of blood pressure, and increase of pulse rate and blood sugar levels was obtained. This condition was designated as "sham rage." In studies by Bard [1], in which the cortex and with it the corpus striatum and the cranial part of the diencephalon was removed, mild peripheral stimulation elicited "sham rage" which was described as an intensive affective excitation with

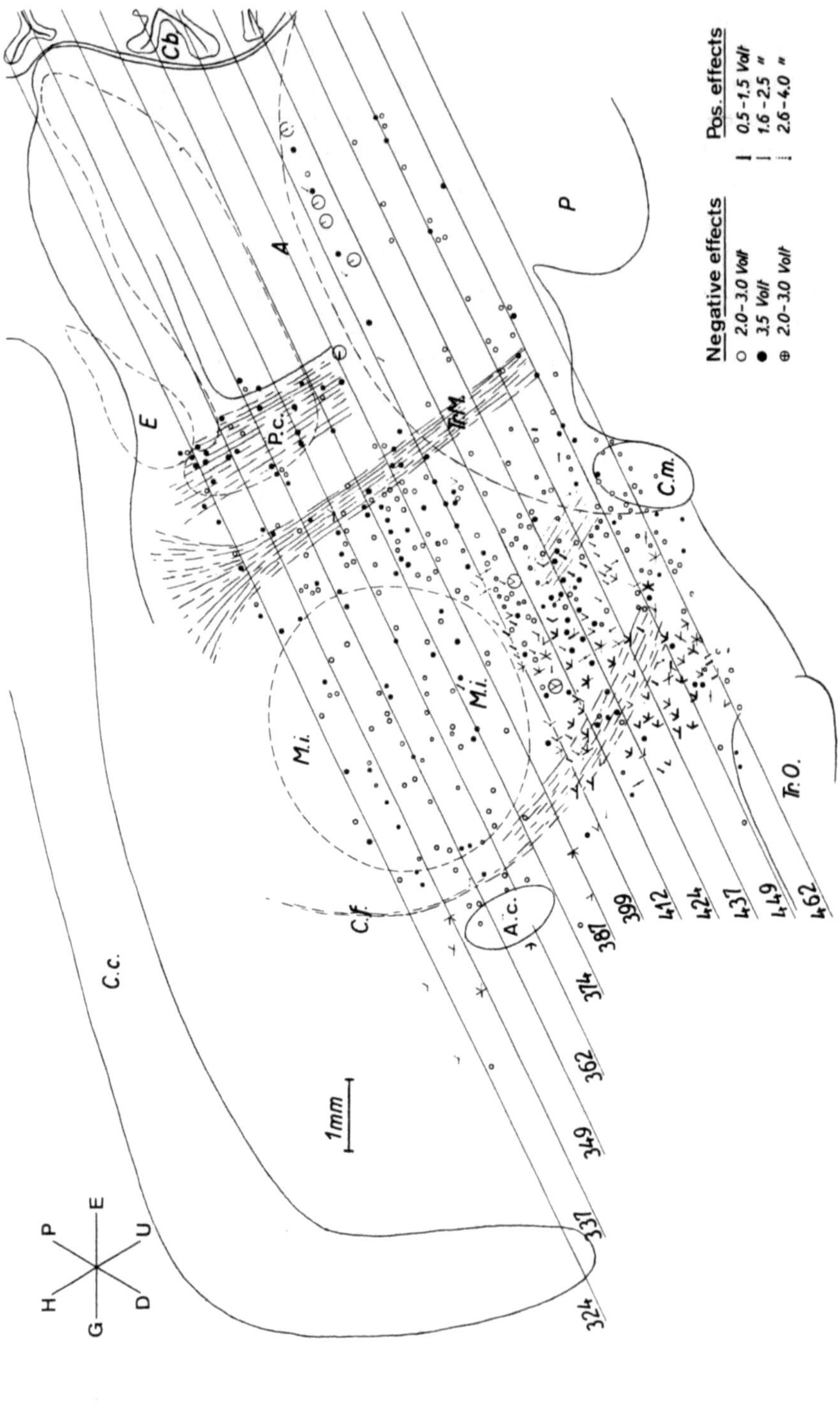

Fig. 11.11. Diagram corresponding to sagittal section 100 in *Fig. 11.7.* Symbols see *Fig. 11.8.* *A*, aqueduct of Sylvius; *A.c.*, anterior commissure; *C.c.*, corpus callosum; *Cb*, cerebellum; *C.f.*, fornix; *C.m.*, mammillary body; *E*, epiphysis; *M.i.*, massa intermedia; *P*, pons; *P.c.*, posterior commissure; *Tr.M.*, tract of Meynert (habenulo-interpeduncular); *Tr.O.*, optic tract. Approximately × 10

piloerection, sweating of the paws, contraction of the nictitating membrane, exophthalmus, blood pressure increase and increase of the pulse rate. Powerful (although nondirected) attack movements were also seen. Otherwise the response was similar to that of a normal animal when made angry, quite in contrast to the pseudoaffective reflex of the midbrain animal. For closer localization of the diencephalic "rage mechanism" Bard [1] sectioned the diencephalon at different levels along the rostrocaudal axis. He concluded that sham rage can still be easily elicited after transection at a frontal plane through the optic chiasm, but also after section between the optic chiasm and the pituitary stalk and still midway through the tuber cinereum. However, if the section was made between the anterior two-thirds and the posterior one-third of the mammillary bodies or between the rostral part of the midbrain, the effect designated as sham rage was no longer obtained. Bard [1] therefore localized the substrate responsible for sham rage in the caudal part of the hypothalamus and in the most ventrocaudally situated regions of the thalamus. Further studies [2, 3] led him to a limitation of the localization in the caudal hypothalamus, thus excluding the thalamus. In the final interpretation Bard [2, 3] stated that the hypothalamus is a "center of excitation and integration of sympathetic reactions", as one sees with anxiety and rage. Keller [21] argued in another manner from his ablation studies performed on decerebrated cats. The section was made through the middle of the mesencephalon. The periphery was stimulated (patting, rubbing of fur, change of the external temperature) and the responses involved growling, hissing, exophthalmus, dilated pupils, piloerection, and striking with the paws in a nondirected fashion. If the section was made sharply at the rostral end of the midbrain, these symptoms were much more difficult to evoke than when the section was placed 1–2 mm more caudally. According to the author [21], the absence of a coordinated defensive reaction demonstrated that "no central connection exists between the localized afferent and somatic efferent impulses in the midbrain animal." Thus is seemed conceivable that such a mechanism could lie further rostrally, possibly in the hypothalamus.

Discussion

Our stimulation findings seem to be in surprisingly good agreement with ablation experiments. Among the points we can confirm is the conclusion reached by Keller [21] on the basis of his ablation experiments. On the one hand, the symptoms following the ablation suggest that a supramesencephalic mechanism must be involved which forms the connection of afferent with somatic efferent impulses; on the other hand, our stimulation studies demonstrate that this coordinating apparatus exists within subcortical regions. Furthermore, we now know where it is localized. Of particular interest in terms of its activity are the contributions by Bard [1–3] and Keller [21]. These authors emphasize that after section of the *cranial diencephalon*, and even after complete transection of the *mesencephalon*, not only can the physiologically induced syndrome, which is transmitted to the peripheral target organs by the sympathetic pathways, still be elicited, but also the midbrain preparation is even capable of inducing aimless pawing and clawing movements. On the other hand, the connection between the defense reaction and a *concrete object* in the environment was absent. In our experiments the *affect, once discharged, directs itself clearly toward a target*. The (artificially) induced endogenous drive is superimposed and matched with the visual perception. Under the given circumstances only the visual system, including the intracerebral processing through which the optically perceived picture is identified, can be the directive in this target-oriented *organization of the defense reaction*. Sensory organs and cerebral cortex thus appear *auxiliary to a subcortically organized performance*. The functional relationship between coordination and perception is, however, *not in one direction* only. Under physiological conditions it is predominantly the eye from which the determination of the dangerous situation arises and which gives the stimulus to the defense reaction! We could not definitely decide whether the occasional positive stimulation sites in the dorsal region of the septum are related to the reciprocal transfer of excitation to cortex. Rather, we limited ourselves to a reference to observations of Karplus and Kreidl, who mentioned that excitatory states were elicited from the frontal cortex [18–20]. Future research must elucidate this issue. With respect to subcortical organization, which is our special field of interest, further clarifications are desirable and also *possible*. It seems important that from the infracentral sites, symptoms of the affective defense reaction can be activated by peripheral stimulation even after ablation of the brain segment in which the center lies. This can only mean that the descending pathways can exercise not only directing but also, in part, coordinating functions. While the latter play only a subordinate and fragmentary role in the intact animal, their competence increases after the abolition of the essential parts of the system. In any case, long-standing experimental experience shows that, in an attempt to compensate the functional deficit, subordinate, even rudimentary, regulatory functions come into more pronounced action after the disappearance of the primary control. In this respect, our findings are fully consistent with this role.

Summary

1. The preliminary observations on affective excitation elicited by hypothalamic stimulation were reported some time ago [10]. These data have

now been systematically analyzed together with the experimental material accumulated since. Using threshold stimulation we were able to define certain areas reaching from the septum into the hypothalamus from which a reaction can be elicited that one sees under physiological conditions when the cat is in a defense situation.

2. Almost 80% of these stimulation points are grouped bilaterally in a circumscribed area of about 2 mm in height and length, and less in transverse diameter. With higher sensitivity of the substrate, the discharge of the reactions is target directed, despite subcortical triggering. The animal passes from defense to attack on nearby persons. Thus, endogenously induced mechanisms are excentrically projected and fused with exogenous (visual) perception. In this way unity of the total behavioral pattern is established.

3. The focal area of the stimulation points concerned with a complete and goal-directed reaction, from which the different threatening defense mechanisms are mobilized, possesses all the qualities of a coordination apparatus oriented toward a defined performance; that is, it possesses the qualities of a center.

4. The extracentral stimulation points yielding individual, in part also combined symptoms of defense, indicate descending pathways of the excitation running in the central gray matter. Complete agreement exists between the results of stimulation studies and those of the ablation experiments of other authors. The findings as a whole provide, in addition to the localization of a center (controlling and coordinating vegetative as well as somatic components, including psychological behavior), the physiological proof of a well-defined interaction between cortex and subcortex, whereby the latter assumes *control in the executive phase*.

We wish to thank the *Stiftung für wissenschaftliche Forschung* of the University of Zürich and the Rockefeller Foundation, New York, for the support of this work.

References

1. Bard, Ph.: A diencephalic mechanism for the expression of rage with special reference to the sympathetic nervous system. Am. J. Physiol. *84*, 490–515 (1928)
2. Bard, Ph.: On emotional expression after decortication with some remarks on certain theoretical views. Part I. Psychol. Rev. *41*, 309–329 (1934)
3. Bard, Ph.: On emotional expression after decortication with some remarks on certain theoretical views. Part II. Psychol. Rev. *41*, 424–449 (1934)
4. Beattie, J., Brow, G.R., Long, C.N.H.: The hypothalamus and the sympathetic nervous system. In: The vegetative nervous system. Time, W. et al. (eds.), pp. 249–316. Baltimore: Williams & Wilkins 1930. (Association for research in nervous and mental disease IX)

5. Beattie, J., Brow, G.R., Long, C.N.H.: Physiological and anatomical evidence for the existence of nerve tracts connecting the hypothalamus with spinal sympathetic centres. Proc. R. Soc. London Ser. B. *106*, 253–275 (1930)
6. Cannon, W.B., Britton, S.W.: Studies on the conditions of activity in endocrine glands. Am. J. Physiol. *72*, 283–294 (1925)
7. Dusser de Barenne, J.G., Magnus, R.: Beiträge zum Problem der Körperstellung. III. Die Stellreflexe bei der großhirnlosen Katze und dem großhirnlosen Hunde. Pflügers Arch. Ges. Physiol. Menschen Tiere *180*, 75–89 (1920)
8. Fulton, J.F.: New horizons in physiology and medicine: The hypothalamus and visceral mechanisms. N. Engl. J. Med. *207*, 60–68 (1932)
9. Goltz, F.: Der Hund ohne Großhirn. 7. Abhandlung über die Verrichtungen des Großhirns. Arch. Ges. Physiol. Menschen Tiere *51*, 570–614 (1892)
10. Hess, W.R.: Bewegungsanalytische Filme. Ber. Ges. Physiol. Exp. Pharmakol. *42*, 547–575 (1928)
11. Hess, W.R.: Filmdemonstration zur Physiologie des Zwischenhirns. Verh. Ver. Schweiz. Physiol. (June), pp. 18–19. Basel: Schwabe 1935
12. Hess, W.R.: Hypothalamus und die Zentren des autonomen Nervensystems: Physiologie. Arch. Psychiatr. Nervenkr. *104*, 548–557 (1936)
13. Hess, W.R.: Beiträge zur Physiologie des Hirnstammes. I. Die Methodik der lokalisierten Reizung und Ausschaltung subkortikaler Hirnabschnitte. Leipzig: Thieme 1932. II. Das Zwischenhirn und die Regulation von Kreislauf und Atmung. Leipzig: Thieme 1938
14. Hess, W.R.: Pupille und Zwischenhirn. Klin. Monatsbl. Augenheilkd. *103*, 407–413 (1939)
15. Hess, W.R.: Beziehungen des Hypothalamus zu Trieb- und Stimmungsäußerungen. Verh. Ver. Schweiz. Physiol. (July), pp. 5–9. Basel: Schwabe 1939
16. Hess, W.R.: Charakter der im Zwischenhirn ausgelösten Bewegungseffekte. Ein Beitrag zur extrapyramidalgesteuerten Motorik. Pflügers Arch. Ges. Physiol. Menschen Tiere *244*, 767–786 (1941)
17. Kabat, H., Anson, B.J., Magoun, H.W., Ranson, S.W.: Stimulation of the hypothalamus with special reference to its effect on gastro-intestinal motility. Am. J. Physiol. *112*, 214–226 (1935)
18. Karplus, J.P.: Die Physiologie der vegetativen Zentren. Dtsch. Z. Nervenheilkd. *106*, 213–238 (1928)
19. Karplus, J.P., Kreidl, A.: Gehirn und Sympathikus. I. Zwischenhirnbasis und Halssympathicus. Arch. Ges. Physiol. Menschen Tiere *129*, 138–144 (1909)
20. Karplus, J.P., Kreidl, A.: Gehirn und Sympathicus. II. Ein Sympathicus Zentrum im Zwischenhirn. Pflügers Arch. Ges. Physiol. Menschen Tiere *135*, 401–416 (1910)
21. Keller, A.D.: Autonomic discharges elicited by physiological stimuli in mid-brain preparations. Am. J. Physiol. *100*, 576–586 (1932)
22. Rioch, D. McK.: Studies on the diencephalon of Carnivora. III. Certain myelinated-fiber connections of the diencephalon of the dog (Canis familiaris), cat (Felis domestica), and aevisa (Crossarchus obscurus). J. Comp. Neurol. *53*, 319–388 (1931)
23. Rothmann, H.: Zusammenfassender Bericht über den Rothmannischen großhirnlosen Hund nach klinischer und anatomischer Untersuchung. Z. Ges. Neurol. Psychiatr. *87*, 247–313 (1923)
24. Woodworth, R.S., Sherrington, C.S.: A pseudoaffective reflex and its spinal path. J. Physiol. (London) *31*, 234–243 (1904)

Section IV

Organization of Motor Systems

Chapter 12

Critical Considerations on the Concept of a Nervous Center*

W.R. Hess

Although the concept of a nervous center frequently appears in connection with physiological and clinical questions, opinions about the nature of such centers are quite divergent. A good start has been made toward clarification, especially as regards localization, on the one hand, and functional considerations, on the other. The work of von Monakow can be mentioned as an example. Centers are often spoken of although it is quite obvious that each author lends his own meaning to the concept and content of the term. This leads to misunderstandings which make themselves felt in discussions and create confusion in research. Thus, critical considerations on the concept of a nervous center are justified. Such considerations should, however, define what is meant by the term and how it should be used.

For this purpose I will have recourse to a term in current use to critically illustrate a concrete example: *the vagus center*. As a nerve, the vagus is a morphological structure with a nucleus which is also defined by *morphological* criteria. However, a center explicitly denotes the regulation of a physiological *function*. It indicates a central organization which regulates a peripheral event having its own aim and purpose. If this contradiction makes it clear that the term "vagus center" is scientifically incorrect and is a hybrid concept composed of morphological and functional elements, we must consider the possibility that the term was adopted earlier and has been subsequently linked, in spite of being incorrect, with a precise idea. This possibility seems likely, for instance, when we speak of a *"cardiovagus center."* The inhibitory effect of the vagus on the heart is substituted by the term "vagus." What we mean by cardiovagus center is in reality the center of cardioinhibitory function. But even then, such an interpretation of the term is not acceptable, because it diverts us from a correct approach to center functions. We are reminded that inhibition of cardiac activity occurs in

* Hess, W.R.: Problèmes de biologie et de médecine, pp. 43–49. In: *Problems Relating to Biology and Medicine*, volume dedicated to Prof. Lina Stern on the occasion of the 30th anniversary of professional activity in the field of scientific and social education. Leningrad, Moscow: Governmental Publications in Science and Medicine, 1935

response to artificial stimulation of certain areas of the vagal nucleus. This effect, however, should not be identified with the function of a center, but, at the very most, with its final result. The function of the center (expressed as cardiac inhibition) is to mobilize the cardioinhibitory vagus elements at the right time and in the proper degree of intensity. The mechanism is not likely to be in the vagus nucleus itself; rather, the vagal nucleus represents the output stage of a superimposed complex of nervous action, a centrally located auxiliary device. Thus we reject the term "vagus center" in this usage, not merely because the expression itself is incorrect, but also because it is based on erroneous concepts.

In a way, the situation is similar when circumscribed cell complexes are designated as centers in the morphological description of certain areas of the brain, e.g., in the thalamus. Even though we are quite obviously not dealing with nerve nuclei here, but rather with structures belonging functionally to a higher level, the relationship with a definite function is absent.

Here is yet another example of a common use of the term "center:" we speak of a *seizure center* in the medulla oblongata. This term is related to a characteristic phenomenon. Nevertheless, the term "seizure çenter" is a conceptual absurdity. Physiological function is based on *order*, and order is based on the above-mentioned central organization which calls the individual muscles of the musculoskeletal system into action in the right combination, in the right order, and with the appropriate tension. The regulatory center is thus the very essence of order. This order is the goal *(Leistungsziel)* for which the muscles are set in action. A seizure, however, does not express a physiological order, nor does it attain this goal. It is therefore simply incorrect to speak of a center in this connection. It makes no difference whatever if, in a specific case, a seizure can be precipitated by stimulation of a locally defined substrate. The concept of a center must not be identified with that of localization. It may well be that clearly defined areas in the central nervous system are related to equally clearly defined activities. However, it need not be so, and often is not.

Here is yet a third usage which we shall examine critically. A number of autonomic centers are assumed to exist in the hypothalamus, e.g., a *center for water balance*. In this instance, we correctly think of the regulation of a specific physiological function. Nevertheless, we must challenge the supposition of a center for water balance. Here is how the matter stands. We have already pointed out that the order projected by a center into the periphery is distinguished by a clearly defined purpose or goal. One of the activities connected with water balance is the *regulation of body temperature*. To prevent congestion of heat, the sweat glands are activated (evaporation of water from the body's surface reduces the body's temperature). Another characteristic activity is the elimination of water-soluble waste products. In this *process* of *excretion*, the elimination of water via the glomeruli is a partial act, in which the eliminated water assumes the role of a vehicle. Elimination of water plays a similar role in the secretion of saliva, in the production of gastric juices, etc. Here too water is a vehicle, in this case for the physiological agents serving *digestion*. Regulation of temperature, elimination, and digestion are well-defined physiological activities. As we have seen, they all have an effect on the water balance because water acts as the *instrument* of these activities. There is, however, no reason to suppose that the elimination of water through the sweat glands, the glomeruli, and the gland cells of the digestive tract implies a common source. A center for these three functions is hypothetical, to say the least. In one respect, there might be some reason to assume a center for water balance. As the role of water in temperature regulation, urine elimination, etc., makes greater or lesser demands on the water reserves of the body, it becomes physiologically necessary to *replenish these reserves*. We may imagine this regulation as operating in the following way: depending on the water level, stimulation of a specific nervous substrate is generated which impresses itself on consciousness as thirst; this acts as a stimulus to water ingestion. If this is indeed the case, we are dealing with a *center for the regulation of water intake*, i.e., a uniform function as regards aim and purpose. The concept of a center for water balance would, however, only be meaningful if this same central organization could activate mechanisms which, in some way or another, eliminate water from the body – purely for the sake of regulating the water level. As the evidence is not irrefutable, it seems out of place to speak of such a center as a foregone conclusion. We have similar objections to assuming the existence of a center for protein metabolism, for carbohydrate metabolism, etc. *Metabolism* as such is not an activity, but rather a phenomenon in the service of a wide variety of activities organized according to purpose. One example is locomotion which, by making demands on the muscles, affects the source of energy in the muscle tissue. Nevertheless, we cannot categorically reject the assumption that a center for some "metabolism" or other exists as the present state of our knowledge does not allow us to give a definite answer.

We now come to a fourth example in which the center concept is under discussion. I quote a sentence from Bethe [1]: "We must reject the idea of coordination centers in the functional and the anatomical sense."

Bethe categorically denies the existence of functionally specific organizations in the central nervous systems. Further on he writes: "The ancient building of the theory of centers, often bolstered by auxiliary hypotheses, lies in ruins." By rejecting the functional specificity of the neural substrate, the fundament on which the theory of centers was based, Bethe is seeking a new approach to nervous activity. He opposes his concept of "gliding innervation" *(gleitende Innervation")* and the "plasticity" of the central nervous system to the supposition of fixed relations between function and regulating central substrate.

One argument which at a first glance really speaks against the existence of functionally specific centers is that a particular stimulus, depending on the circumstances, is not always answered with the same responses. A well-known example is that of the cat which reacts to the tip of its tail being pinched by moving it to the right if the tail is bent to the left. If the tail is bent to the right, the very same stimulus causes the cat to move its tail to the left. The reflex also varies depending on whether the leg is stretched or retracted; if it is bent, the leg is then stretched. The relativity of the reaction seems clearly to suggest that there is no connecting structure with functionally specific properties between the afferent and the efferent sides of the reflex arc; otherwise one would have to assume that identical stimuli always have an identical effect. This simplistic conclusion fails, however, to do justice to reality. There do indeed exist cases of a necessary relation between stimulus and response, for instance, where the reacting organ has only a limited or a single possibility of reaction. One example is the lid reflex when the cornea is stimulated. It is a different matter with the skeletal muscular system, which must carry out a large variety of movements; depending on the position of the body and the limbs, the action of the individual muscles may be completely different. No movement can be made without displacing the mutual position of origin and insertion of the individual muscles. Our previous criticism of the center concept now gains a positive note, because we will show that the principle of a specific relation between substrate and function can apply to the skeletal muscular system, even though the forms of activity are highly varied and the action of each individual muscle is dependent on the momentary posture. We shall become acquainted with a mechanism which explains the "plasticity" of nervous activity from the very functional specificity of the nervous substrate. Our conclusion, as will be seen, not only removes all doubts concerning the existence of centers, but also provides us with guidelines for a precise definition of the concept.

We shall start by considering peripheral activity and its functional organization. We realize that the prerequisite for every movement, for every functional expression of an organism, is preparedness. The stimulus inducing the reflex, an external influence, only produces a reaction if the body in general and, more particularly, the nervous system is prepared for a reaction. For instance, the sleeper does not react to a sound to which the waking person would immediately react by turning the head in the direction from which the sound came. In deep sleep, the muscle does not even answer the patellar reflex of its tendon, which it would otherwise unfailingly respond to. In the waking state there are degrees of preparedness, and various functional systems operating at the same time may not necessarily react in the

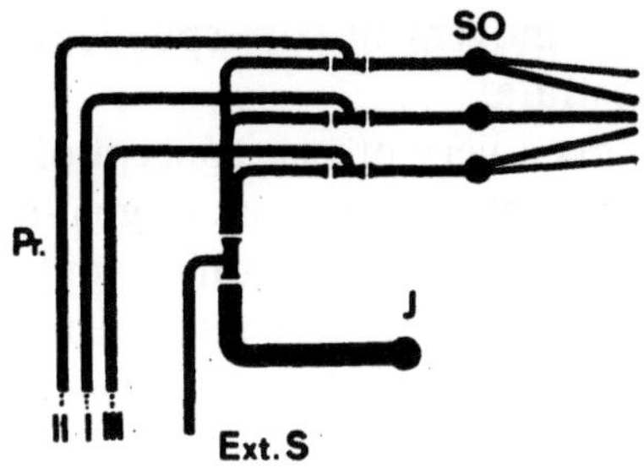

Fig. 12.1. Representation of the organization of a purposive action. *J*, initiating impulse (expressed as motor preparedness); *Ext. S*, external stimulus guiding the impulse toward its effectuation; *Pr.*, proprioceptive afferents; *I*, *II*, *III*, paths of control signals (selecting from the multitude of effector combinations the one corresponding to the momentary biomechanical constellation); *SO*, selection organ (carries out switchboard function)

same fashion. An observer following a process closely with his eyes is liable to "fail to hear" an otherwise distinct acoustic stimulus. On the other hand, someone paying close attention to an anticipated event will very quickly react to its occurrence. These phenomena can be explained by different degrees of tension (due to particular circumstances) seeking expression. They remain latent until an external factor is added, e.g., an auditory, visual, or tactile stimulus. With this stimulus, the nervous activity enters into a new phase. The stimulus which has occurred makes the latent tension manifest. Depending on the nature and intensity of the stimulus, the tension develops in a particular direction. A goal is aimed at, either positively (turning the eye, ear, hand, etc., toward) or negatively (turning away, flight, avoidance, etc.). We have already seen that achieving this end is complicated by the fact that, apart from the simplest forms of response the reaction of the individual muscles depends on the position of the body and limbs. The stimulus-induced movement thus encounters not a few but a multitude of possibilities when the intended movement is to be effected. Each of these possibilities corresponds to another, always well-defined constellation in the effector system. The intended action can be successfully carried out only if, from the multitude of possibilities, a single one (corresponding to the constellation) is selected. Thus, the third phase in carrying out the intended movement is that of selection, a nervous process which provides the manifested state of tension with certain effector combinations. This selection function represents a specific task of proprioceptive control.

I have included a diagram (Fig. 12.1) to illustrate this segmentation of functions which applies in principle to all purposeful reactions and actions. The segmentation can be highly differentiated or reduced to a uniform mode of reaction. The view is upheld that the functional organization of the center governing the movement corresponds to a structural arrangement of the neural substrate.

Figure 12.1 represents a very precise concept of a center, which is distinguished by its responsibility as an organ of goal-oriented order. Attaining the objective depends on the motivating impulse (J) originating from the autonomic sphere of the central nervous system. This impulse is guided by a specific external trigger signal (Ext. S). Achieving the goal, however, is also dependent on the constellation existing at the moment of stimulation. A control stimulus originating from the selection organ directs the appropriate message to the effector combination and lends the general intentional innervation its particular form. This approach excludes any statement on the localization of the centers in the topographical organization of the central nervous system and on the local relations between the structures corresponding to the individual phases of center functions.

In closing, we would like to add that not providing an answer to this question is part and parcel of the center concept, but we are far from being at the end of the path we have begun to travel. Analysis of peripheral activity can tell us a great deal more and serve as a guideline to understanding the structure of the entire central nervous system. In the final analysis, the peripheral activity which occurs as a result of the activation of the effectors reflects the functional order of a central organization.

References

1. Bethe, A.: Plastizität und Zentrenlehre. In: Handbuch der normalen und pathologischen Physiologie, Vol. 15: Correlationen 1/2. Bethe, A. (ed.), pp. 1175–1220. Berlin: Springer 1931

Chapter 13

Critique of the Hering-Breuer Theory of Self-regulation of Respiration*

W.R. Hess

Since the classical studies by Hering [13] and Breuer [1], the influence of the vagus nerve on respiration is thought to operate as follows: the expansion of the lungs, caused by inspiration, delivers the stimulus which inhibits inspiratory innervation and "switches" over to expiratory innervation. At the end of expiration, deflation of the lungs is assumed to cause a stimulus which inhibits expiration and permits inspiration. Occasionally, compared with the control mechanism of a steam engine, this mode of operation of the respiratory center, *active at every change of the respiratory phase* and operating from the periphery, is the core of the Hering-Breuer theory. It has been established on the basis of the well-known experiments in which air was blown into and withdrawn from the lungs. This demonstrated that every filling or emptying of the lungs triggered the opposite movement in the diaphragm.

A prolonged discussion was begun on whether there perhaps existed only an inspiration-inhibiting "switch" reflex *(Schaltreflex)*. Inhibition of expiration would then be brought about solely by means of the disappearance of the inspiration-inhibiting reflex connected with expiratory emptying of the lungs [10, 18–20, 23, 24]. We will not contribute to the discussion at this time. What interests us in this connection is the fact that here, too, the decisive role is played by a "switch" reflex occurring with every breath.

In addition to the above-mentioned impressive stimulus effects elicited by inflation of the lungs, observations have also been made which do not readily fit into the theory of self-regulation of respiration, as it has been briefly outlined. Hering and Breuer had already realized that the vagus nerve also exercises certain tonic functions. They speak of these functions when they mention that the stimuli dependent on expansion are not the only ones communicated by the vagus nerves, but that there is also a maintained excitatory influence. Head [11, 12], who worked in Hering's laboratory,

* Hess, W.R.: Kritik der Hering-Breuer'schen Lehre von der Selbststeuerung der Atmung. Pflügers Arch. Ges. Physiol. Menschen Tiere *226*, 198–211 (1930)

also refers to a tonus function of the vagus nerve for which he offers an explanation. He imagines that the inhibition of inspiration, which is repeated with every breath, leads, as it were, to the retention of an inspiratory innervation which is not discharged. The remaining excitation which is prevented from being discharged is manifested under suitable conditions as tonus (inspiratory tetanus is also mentioned). Later investigators, e.g., Loewy [21], C. Heymans [15], and J. F. and C. Heymans [16], came back to the question of tonus and sought a connection with peripherally acting chemical stimulus components and with influences from the cardio-aortal system.

Registration of action currents, either in the centrifugal respiratory nerves [22, 26, 27] or in the respiratory muscles [2, 3, 25], has led to renewed interest in the tonus phenomenon involved in respiratory innervation.

In particular Wachholder and McKinley [25] showed in their investigations that at the end of expiration, i.e., in the relaxation period, clear cut action currents can still be recorded from the diaphragm. The same applies to the intercostal muscles.

The findings, made by means of action current recordings, reveal important insights but do not clarify the role played by the tonus phenomenon in respiratory innervation. The results are also not clear cut in every respect. The investigations of Dittler [2, 3] have shown, for instance, that innervation is maintained during the whole expiration only under certain conditions. He thus concludes that there are two types of breathing. In the first, the action current derived from the diaphragm displays an alternation between an active and a passive stage. In the second, the passive stage is absent. As to the conditions necessary for the appearance of the second type, Dittler can only say that it regularly occurs in unilateral pneumothorax. Wacholder and McKinley's [25] comments also reflect the lack of clarity in the interpretation of prolonged innervation. In their view the fact that action currents can be recorded from external intercostal muscles throughout the entire expiration phase, makes it probable that the function of these muscles involves more than inspiration movements.

The prolonged innervation of the inspiratory muscles cannot be related so easily to the tonus function ascribed to the afferent vagus nerve. Wachholder and McKinley found that the two phenomena are interrelated after all. The remark by Wachholder suggesting another function of the *inspiratory* thorax muscles is thus well founded.

A thorough study of respiratory regulation [14] gave me the definite impression that, apart from the points raised here, the mechanism of self-regulation is not as clear cut as is generally supposed. This has prompted some supplementary experimental observations. The result was a surprise insofar as it showed that the tonus effect of the vagus nerve and of a certain part of deep sensibility from the thorax, which has been overshadowed and excluded by the current concept of self-regulation of respiration, must be viewed as primordial factors in the physical regulation of respiration. The "switch" reflexes, on the other hand, are of secondary importance. This result, announced in advance, means that we abandon the generally accepted view presented in textbooks. It has always seemed very obvious, but is essentially incorrect, as the following experimental results show.

Experiments

We opened the thorax of an anesthetized rabbit by removing the sternum and about half the ribs. The lowest costal arch with the sternal xiphoid process is, of course, left for the sake of diaphragm attachment. Care must also be taken to have a clear view of the diaphragm.

Respiration is artificially raised somewhat to bring about the well-known apneic effect (Fig. 13.1). During apnea, the lung is only slightly filled. Air is then blown in (approximately 5–10 cm^3) through a tube fixed to a lateral branch of the tracheal cannula. If the artificial respiration required to bring about apnea is not too strong or maintained too long, the diaphragm (upper curve) carries out the movements interpreted by Hering [13] and Breuer [1] as expiration. It becomes somewhat arched. Now comes an important point, which has apparently not been noticed or at least not correctly assessed. If more air enters to fill the lung, the diaphragm rises even more. When the air is withdrawn, the opposite effect is seen. The diaphragm gradually returns to its original position (strikingly flattened) as a result of the reflex movements. We would like to stress one aspect which has hitherto been incorrectly interpreted and which is the subject of much discussion. It is the fact that these movements of the diaphragm also occur when there is no respiration (i.e., during apnea). This would apparently contradict some of the data given in the literature, according to which the reflexes described by Hering and Breuer are absent in apnea. Hoffmann and Keller [17] were unable to demonstrate any reflexes on electrographic records. This seemed very important to the authors as they drew certain conclusions on this basis about the nature of the reflexes and their relationship to the stimulus mediated by the blood. The curves reproduced in Fig. 13.1 illustrate how reflex tension appears in the diaphragm during apnea. This is certainly not possible without action currents. The differences between the two sets of results can probably be attributed to the manner in which apnea is produced. It should be mentioned that chemical apnea (i.e., that induced by lowering the stimulus mediated by the blood below the stimulus threshold) is involved in extremely few of the cases described in these experiments. Usually apnea is conveyed rather by the vagus nerve, since inhibition of the respiratory center occurs when the stimulus mediated by the blood is still above the stimulus threshold. If artificial ventilation is

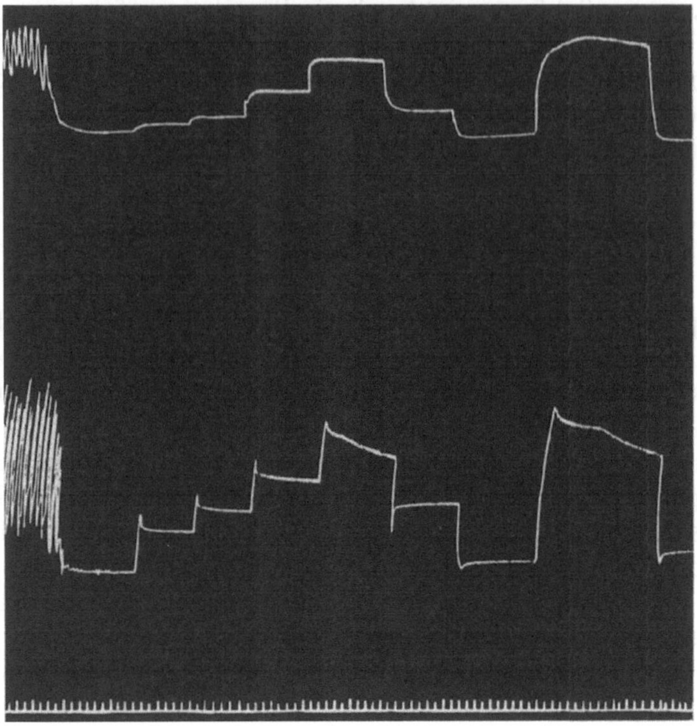

Fig. 13.1. Change in position of the diaphragm *(above)* following a change of lung inflation *(below)*, measured by inflation pressure. The curves begin with the recording of artificial respiration. When this is discontinued, apnea ensues. As the lung is inflated stepwise, the diaphragm gradually arches in the thoracic cavity. When air is withdrawn, the diaphragm retracts. It rises once again when the lung is refilled. Diaphragm movements have been reduced to two-fifths. The small waves are due to heart movements. Time scale: 1 s

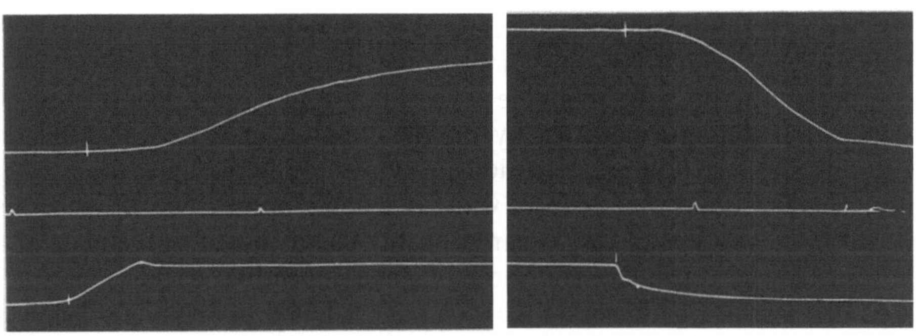

a b

Fig. 13.2a, b. Apneic diaphragm, tonus recorded at rapid drum speed. *a* When air is blown in *(lower curve)*, the diaphragm becomes slack with a latency of 0.22 s and rises; *b* when the air is withdrawn, the latency for retraction is 0.14 s. Both curves belong to the same record; 3.1 s of recording have been left out between the two samples given excessively, especially if the lungs are fully inflated, the reflexes in the subsequent apnea are absent. The reason, in our view, is that a strong overstimulation by the reflexes of the vagal apparatus had already taken place. After a brief pause, the reflexes return, usually before spontaneous respiration begins. The possibility of separating the diaphragm reflexes from respiratory activity is of utmost importance.

Another fact must be considered in assessing this reflex effect: the alternation of tension in the diaphragm has a fairly rapid onset, but requires a good deal of time until a new equilibrium is achieved (cf. Fig. 13.2). Furthermore, it seems quite striking that it is possible, by varying the amount of air in the lungs, to impose any desired position on the diaphragm from maximum relaxation to almost complete contraction. After impartial consideration of these facts, we must doubt whether the Hering-Breuer reflexes are really movement effects (i.e., expiration and inspiration). They may, in fact, represent changes in the tonus of the diaphragm. We have recorded the movements of the diaphragm to better demonstrate any changes easily perceived with the naked eye. A light lever was placed on the top of the diaphragm, with the mediastinum open and without damaging the foremost branches of the phrenic nerve. The lever permitted all movements and changes in position to be transmitted pneumatically to a Marey capsule. The internal pressure of the trachea, whose rise and fall causes the various degrees of inflation of the lungs, was recorded at the same time. Figure 13.1 and 13.2 depict changes in position of the diaphragm triggered by a reflex as a result of a change in lung inflation during apnea. What is involved is not merely diaphragm movements opposite to the change in the lung's inflation, but gradual tonic changes of diaphragm *position*. The time sequence is more easily recognizable in Fig. 13.2a and 13.2b because of the greater recording speed of the drum. The interval between stimulus and effect is 0.22 s when the lung is being filled and 0.14 s when it is being emptied. Other measurements have yielded times between 0.07 and 0.11 s and 0.07 and 0.08 s, respectively. The duration of the response of the diaphragm until a new equilibrium is established is 4 full seconds, when tension is decreasing. When the tension of the diaphragm increases, this time is considerably shorter (as can be seen from the curve).

The relatively long duration of the mechanical response of the diaphragm contradicts to some degree the electrographic findings of Hoffmann and Keller [17] on the high speed of impulse transmission. The onset of the reflex can be less than 20 ms. It is obvious that the recording of mechanical effects must yield considerably longer times. It is also evident that the time sequence of the mechanical effect – not only its onset but its entire course – is decisive for fitting the reflex responses into the time course of the cycle. Nevertheless, these questions of timing call for further study. There may be individual differences dependent on the state of the respiratory apparatus. Our recent findings suggest this to be the case.

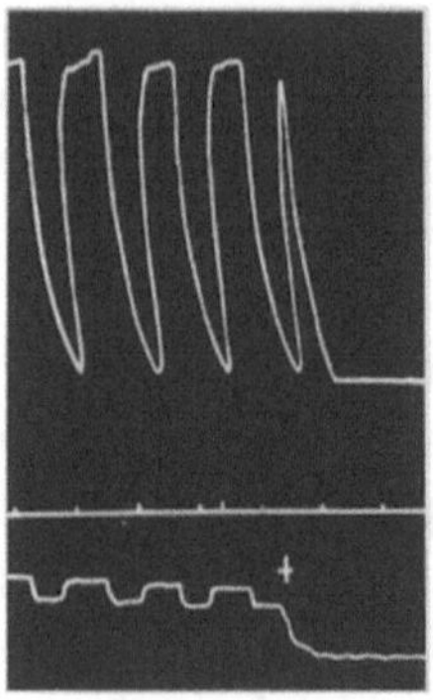

Fig. 13.3. Withdrawal of air from the lung during spontaneous breathing. The intratracheal pressure *(below)* shows the time relation between artificial changes in inflation and spontaneous breathing. The change (+) occurs at the end of inspiration. Spontaneous breathing continues and completes almost the entire next cycle before the diaphragm reacts. Time scale: 1 s

We also generated intercurrent inflation and deflation while spontaneous breathing was in progress to clarify the time lapse between a change in lung inflation and the corresponding response of the diaphragm. One of the results is shown in Fig. 13.3. The tracheal pressure curve (below) indicates the tension change of the diaphragm as related to the respiratory phase. At +, inflation begins at the end of expiration. The mechanical responses of the diaphragm (above) show that three-quarters of the next respiratory phase are completed before the respiratory cycle is disturbed by the arrival of the reflex excitation. It is difficult to conceive, in the presence of such a considerable phase shift, that these reflex responses could represent the mechanisms by which the changes in respiration cycle are generated, although this corresponds to the classic concept. However, as already mentioned, we would like to refrain for the time being from any further evaluation of the timing problem and investigate the fact that controlled changes in lung inflation produce gradual displacements of diaphragm positions. What is the significance of these reflex-induced position changes? We have a clue when spontaneous breathing returns after apnea with successively increasing amplitudes (Fig. 13.4). Every change in tension is related to changes in the degree of expiratory relaxation of the diaphragm. The tonus of the diaphragm controlled by lung inflation determines the final point of the expiratory movement and thus the initial point for the following inspiration. We have now discovered the explanation why action currents can be recorded from the diaphragm during expiration: the diaphragm is only relaxed to a certain resting tension which can be viewed as the basic tonus on which breathing movements are based. We now understand why this basic tonus is not always present. As a result, Dittler

[2, 3] distinguished two types of respiration, depending on the presence or absence of expiratory action currents. The decisive factor is the inflation of the lung. If the latter is adequate the resting tension of the diaphragm, which persists even when the lung is only moderately or slightly filled, gives way to a complete relaxation of the diaphragm in expiration.

These observations not only inform us about the occurrence of tonus phenomenon in the mechanics of respiration; they also reveal the significance of the phenomenon. The more the lung is inflated, the higher the dome of the diaphragm is arched into the thoracic cavity in order to relieve lung tension. On the other hand, the more the lung collapses, the more the diaphragm is tonically innervated and withdrawn from the thoracic cavity to give the lung more room for expanding. These displacements are not inspiratory and expiratory movements in the least. They represent the postural regulation of the diaphragm on which the respiratory movements are superimposed. What we see is a change in the basic tension (tonus) on which are superimposed the tension fluctuations causing the breathing movements. This is one finding of a general nature. The second finding has to do with the quantitative relationship between postural regulation and movement regulation. Tonus is the factor determining the range of respiratory movements of the diaphragm. The higher the basic tonus, the shorter the path taken by the diaphragm in completing its respiratory movement. The tonus can become so strong that it completely prevents respiratory movement.

It has hitherto been assumed that the reflex transmitted by the vagus nerve inhibits inspiration in particular. As was stated in the introduction, some doubt was attached to the "switch" function of the vagus nerve in expiration. However, observation shows that once respiration is properly functioning after apnea, the diaphragm, like the heart muscle, always undergoes almost complete contraction. The amplitude of the movement is limited not by blocking inspiration when the lung has been filled to a certain point (as in the Hering-Breuer theory), but by limiting expiratory deflation.

Our description, based on observations of the diaphragm, seems to contradict the conclusions drawn from the well-known pneumogram after vagotomy, which shows that the amplitude of respiration expands somewhat, and sometimes even exclusively toward inspiration. Thus, it seemed justified to assume that the inspiratory movement had previously been inhibited by the "switch" reflex transmitted by the vagus nerve. However, this conclusion left out one point. The pneumogram represents the activity of the diaphragm and thorax in both static and dynamic terms. When the diaphragm loses its tone as a result of vagotomy and is arched high in the thoracic cavity, the thorax can prevent the collapse of the lung by an appropriate change in position. Gad [10] has already pointed out that the thorax

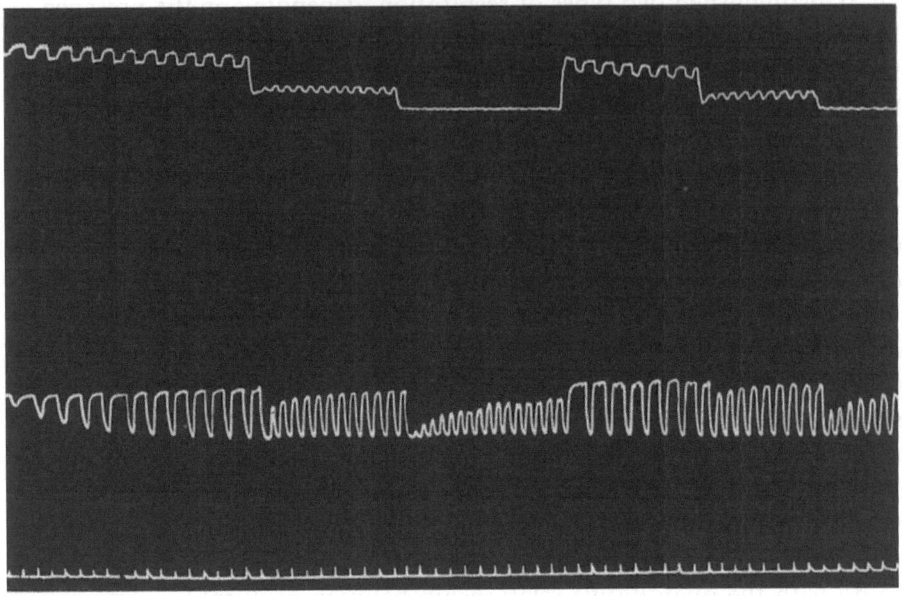

Fig. 13.4. Changes in position of the diaphragm *(below)* as a consequence of changes in lung inflation *(above)* during spontaneous breathing. The inspiratory movements (downward) take their point of departure from the various tonic levels of the diaphragm which, as the basic position of respiratory movement, is dependent on the various degrees of inflation of the lung. The inspiratory movements all reach the same amplitude at maximum contraction of the diaphragm. Diaphragm movements have been reduced to one-fifth. Time scale: 1 s

of a vagotomized animal is unmistakably positioned for inspiration. However, this has nothing to do with respiratory movement but rather is the postural regulation (substituting for the slackening diaphragm tone) of the thorax. At any rate, the immediate consequence of this change in position of the thorax (the mechanism governing the change requires closer study) is that the lung remains filled almost as much as initially (i.e., prior to the vagotomy). Respiratory movements are then superimposed on the inflated lung. They are far more extensive than before the vagotomy, because the exspiratory relaxation of the diaphragm is greater and also because the thorax expansion brings about a broadening of the surface covered by the diaphragm. Moreover, the space taken up by diaphragm movements is expanded in three dimensions. The apparent contradiction of the vagotomy experiment is thus solved, but we have also had an opportunity to throw some light on the differentiation between postural regulation and movement regulation in the mechanics of respiration.

Returning to the curves in Fig. 13.4, there is one further point which has not yet been discussed, but which fits in perfectly with what has already been said. A specific respiratory rate corresponds to every tonic position of the diaphragm and thus to every size of amplitude. This phenomenon is very impressive when the experiment is carried out in such a way that artificial ventilation is not interrupted for too long, thereby causing secondary stimulus factors, with a central site of action, to appear and distort the picture. To ensure an adequate response of the experimental animal, too strong an inflation stimulus must be avoided. It is also advisable in preparing the experimental animal to ensure that the stomach is not too full, for otherwise the movements of the diaphragm would be impeded. If these precautions are reasonably well observed, the correspondence between strong diaphragm tone and high respiratory rate, on the one hand, and weak diaphragm tone and low respiratory rate, on the other, will most certainly be seen. Slight freedom of movement is correlated with a short contraction time, and great freedom of movement with long contraction time. Respiratory rate, like diaphragm tone, is independent of lung movement. It is most pronounced even when the lung is completely still (i.e., no "switch" reflexes). It is not true, as has been supposed, that small variations in pressure, visible at the curve of intratracheal pressure (they are the result of a simple mechanical transmission of diaphragm movements to the lung lying in the open thorax), somehow play a role in regulating amplitude. This hypothesis is disproved by an experiment in the curarized animal, which was conducted for other reasons and will be described later.

One question which must still be answered is how the relations between tonus function of the diaphragm and respiratory rhythm are mediated. It is hardly necessary to explain the significance of this question. At first we thought that respiratory rate is automatically regulated by the fact that tonus limits the freedom of movement of the diaphragm by innervatory as well as by mechanical means. It would mark the instant at which the "switch-over" of the cycle occurs. This is the moment when the expiratory movement of the diaphragm has reached the tension corresponding to the basic tone. The mechanism involved here would be analogous to that supposed operating in the Hering-Breuer self-regulation of respiration. The only difference is that the vagi have an indirect (tonus-regulating) function and the actual control stimulus comes from the diaphragm (with vagal tone) and back again. This seemed to be contradicted by the findings of Hoffmann and Keller [17], who were unable to detect characteristic stretch reflexes *(Eigenreflexe)* of the diaphragm. Hoffmann concluded that this muscle occupied a special place. On the other hand, we were familiar with the investigations of Fleisch [4–9], who described the existence of such reflexes for the whole of the respiratory muscle complex. These reflexes would perhaps be suitable for transmitting the suggested control of the respiration cycle. We sought a clear answer which we found. Diaphragm movements, or still better – with curare – all movements of the respiratory apparatus, must be blocked and the nerve impulses coming from the respiratory center observed at the proximal phrenic nerve stump. This experiment was already successfully applied to prove the self-regulatory nature

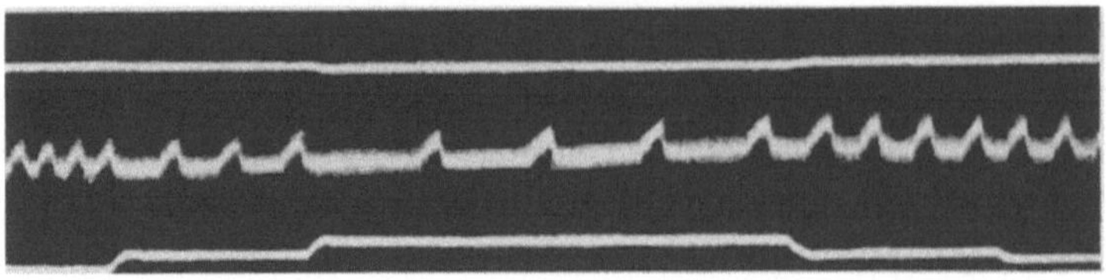

Fig. 13.5. Action currents of the phrenic nerve *(middle curve)*. Change in lung inflation *(lower line)*. Upward movement indicates inflation. Movements of diaphragm: *upper line*; their registration is represented by the silhouette of the lever used for the kymograph recordings. The rabbit has been given curare. The respiratory movements of the diaphragm have stopped. The slight displacements of the *upper line* are caused by the pressure of the more or less inflated lung on the diaphragm lever. The action currents of the phrenic nerve, although there is absolutely no respiratory movement, display the same changes in rhythm as the diaphragm of the normal animal: gradual decrease in respiratory rate together with gradual increase in lung inflation, and vice versa

of the respiratory center [26, 27]. Our new experiments, however, include a new element which arose from the results of our above-mentioned investigations.

The lungs must be filled to different degrees during the recording of the action currents from the phrenic nerve; Fig. 13.5 shows the result. We see that the degree of lung inflation (lowest curve) governs the time allotted to each individual breath even in the absence of any movement in the respiratory system. Even if we were entirely unacquainted with previously discussed findings, this last fact suffices to prove that physical regulation of respiration works differently from the way in which it is portrayed by the theory of self-regulation. The decisive factor in controlling respiratory rate and depth are not "switch" reflexes, but a *reflex tone* which determines amplitude and rate of respiration.

Insofar as switch reflexes are involved in respiratory movement, their role corresponds to that played by proprioceptive afferents for goal-directed movements in general. Their task is to modify a movement whose intensity and duration has already been determined centrally. It does so by "fine regulation" of time and intensity. We do not need to go into further details since we deal with this question in another place [14].

Our observations identify the lung as the place of origin of the tonus-generating stimulus and the diaphragm as the reacting organ. It is obvious that any further development of this new finding must also take the thorax into account. We have referred briefly to its behavior when discussing the vagotomy effect. Experiments in which half-amputated ribs are held at the free end and slowly but vigorously forced apart, give some information on their involvement as a reflexogenic zone. The vagotomized animal not treated with curare displays an inhibitory effect analogous to that in lung inflation (Fig. 13.6). The rib joints are most likely responsible for this reflex which is associated with the reflex coming from the lung. We must think

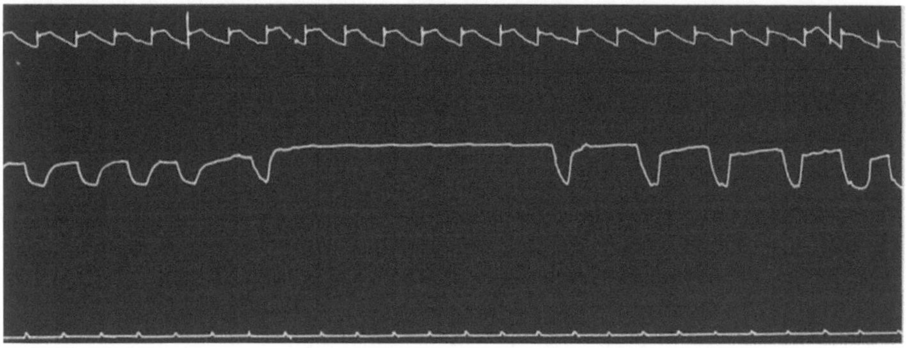

Fig. 13.6. Changes in rate of spontaneous breathing (*lower curve*, diaphragm) when rib stumps are forced apart (between two vertical strokes on the uppermost curve): vagotomized rabbit. Artificial respiration (tracheal pressure, *above*) during the experiment. There is pronounced inhibition of spontaneous breathing as with lung inflation. Time scale: 1 s

of this when interpreting the findings of Wachholder and MacKinley [25], namely that the expiratory tone of the diaphragm cannot be made to disappear by vagotomy. More investigation is needed on the coordination between the postural regulations of the diaphragm and the thorax wall.

Summary

Tension changes in the diaphragm, which are observed in conjunction with changes of lung inflation (Hering-Breuer reflexes), show that the responses of the diaphragm caused by reflexes are not to be considered as inspiratory and expiratory movements but as graded changes of the basic tension (tonus) of the diaphragm. The postural tone correlates with the amplitude of the respiratory movements by limiting the expiratory relaxation of the diaphragm. Thus, the regulation of respiratory amplitude by the vagus nerve is generated by tonic innervation and not by a self-regulatory switch reflex. Regulation of respiratory rate is connected with the strength of the reflex tone determining the amplitude. This is demonstrated when the action currents of the phrenic nerve are recorded in the curarized animal, in which respiratory movements are eliminated. Every change in lung inflation is matched by a change in frequency of the respiratory rhythm even when no switch reflexes can be generated because of the paralysis of the motor apparatus. The regulatory reflex tone is regarded as partly of thoracic origin.

I wish to thank Miss M. Liechti for her aid in the experiments.

References

1. Breuer, J.: Die Selbststeuerung der Atmung durch den Nervus vagus. Sitzungsber. Akad. Wiss. Wien *58/II*, 909–937 (1868)

2. Dittler, R.: Über die Innervation des Zwerchfelles als Beispiel einer tonischen Innervation. Arch. Ges. Physiol. Menschen Tiere *130*, 400–443 (1909)
3. Dittler, R.: Über die Aktionsströme des Nervus phrenicus bei natürlicher Innervation. Arch. Ges. Physiol. Menschen Tiere *131*, 581–588 (1910)
4. Fleisch, A.: Propriozeptive Atmungsreflexe. Pflügers Arch. Ges. Physiol. Menschen Tiere *219*, 706–725 (1928)
5. Fleisch, A.: Bahnung und Hemmung der propriozeptiven Atmungsreflexe durch die Atmungsphase. Am. J. Physiol. *90*, 346–347 (1929)
6. Fleisch, A.: Über die Eigenschaften der propriozeptiven Atmungsreflexe. Pflügers Arch. Ges. Physiol. Menschen Tiere *222*, 12–25 (1929)
7. Fleisch, A.: Über Eigenreflexe. Schweiz. Med. Wochenschr. *59*, 737–739 (1929)
8. Fleisch, A.: Beeinflussung der propriozeptiven Atmungsreflexe durch Adrenalin und Atropin. Pflügers Arch. Ges. Physiol. Menschen Tiere *224*, 390–402 (1930)
9. Fleisch, A.: Bahnung und Hemmung der propriozeptiven Atmungsreflexe. Pflügers Arch. Ges. Physiol. Menschen Tiere *223*, 509–533 (1930)
10. Gad, J.: Die Regulierung der normalen Atmung. Arch. Anat. Physiol. (Physiol. Abt.) 1–32 (1880)
11. Head, H.: On the regulation of respiration I. J. Physiol. (London) *10*, 1–70 (1889)
12. Head, H.: On the regulation of respiration. II. Theoretical. J. Physiol. (London) *10*, 279–290 (1889)
13. Hering, E.: Die Selbststeuerung der Atmung durch den Nervus vagus. Sitzungsber. Akad. Wiss. Wien *57/II*, 672–677 (1868)
14. Hess, W.R.: Die Regulierung der Atmung. Leipzig: Thieme 1931
15. Heymans, C.: Über reflektorische Einflüsse auf das Atmungszentrum. Verh. Dtsch. Ges. Kreislaufforsch. 92–105 (1928)
16. Heymans, J.F., Heymans, C.: Sur le mécanisme du tonus respiratoire vagal. C.R. Soc. Biol. (Paris) *99*, 633–635 (1928)
17. Hoffmann, P., Keller, Ch.J.: Untersuchungen über Atemreflexe mit Hilfe der Aktionsströme. Ber. Ges. Physiol. Exp. Pharmakol. *50*, 296 (1929)
18. Keller, Ch.J., Loeser, A.: Der zentripetale Lungenvagus. Z. Biol. *89*, 373–395 (1930)
19. Lewandowsky, M.: Die Regulierung der Atmung. Arch. Anat. Physiol. (Physiol. Abt.) 195–248 (1896)
20. Lewandowsky, M.: Über Schwankungen des Vagusstromes bei Volumänderungen der Lunge. Arch. Ges. Physiol. Menschen Tiere *73*, 288–296 (1898)
21. Loewy, A.: Über den Tonus des Lungenvagus. Arch. Ges. Physiol. Menschen Tiere *42*, 273–281 (1888)
22. MacDonald, J.S., Reid, E.W.: Electromotive changes in the phrenic nerve. A method of investigating the action of the respiratory centre. J. Physiol. (London) *23*, 100–111 (1898/99)
23. Schenck, F.: Über den Einfluß der Apnoë auf die vom Vagus ausgelösten Atemreflexe. Arch. Ges. Physiol. Menschen Tiere *83*, 99–119 (1901)
24. Schenck, F.: Über die Bedeutung der Lungenvagusfasern für die Atmung. Arch. Ges. Physiol. Menschen Tiere *100*, 337–347 (1903)
25. Wachholder, K., McKinley, C.: Über die Innervation und Tätigkeit der Atemmuskeln. Pflügers Arch. Ges. Physiol. Menschen Tiere *222*, 575–588 (1929)
26. Winterstein, H.: Die automatische Tätigkeit der Atemzentren. Pflügers Arch. Ges. Physiol. Menschen Tiere *138*, 159–166 (1911)
27. Winterstein, H.: Die Regulierung der Atmung durch das Blut. Pflügers Arch. Ges. Physiol. Menschen Tiere *138*, 167–184 (1911)

Chapter 14

The Nature of the Movements Elicited by Electrical Stimulation of the Diencephalon*

A Contribution to the Problem of Extrapyramidal Motor Control

W.R. Hess

Contents

In our previously published analysis of motor effects the entire material was divided into three groups, depending on the type of musculature involved (face, extremities, or axial musculature) [2–4]. We shall retain this division, which was employed to achieve clarity of presentation, and shall now examine more closely the "motor innervation patterns," starting with the axial muscles, i.e., body, head, and eyeballs, with regard to quality and character. Because extensive film material was available, it was possible to achieve a degree of accuracy in the analysis of the movements equivalent to that of the histological localization of stimulation sites. The results of 12 years of experimental work, recorded pictorially, are now available for systematic analysis. In spite of the considerable flexibility offered by cinematographic records, the extensive and exceptionally varied material remained quite difficult to process. No memory can retain all the essential details. Hence, it was necessary to arrange the results of the analysis thematically with text and graphic illustrations and to supplement these data with a vast number of photographs from the film. This enabled us to study and com-

* Hess, W.R.: Charakter der im Zwischenhirn ausgelösten Bewegungseffekte. Pflügers Arch. Ges. Physiol. Menschen Tiere *244*, 767–786 (1941)

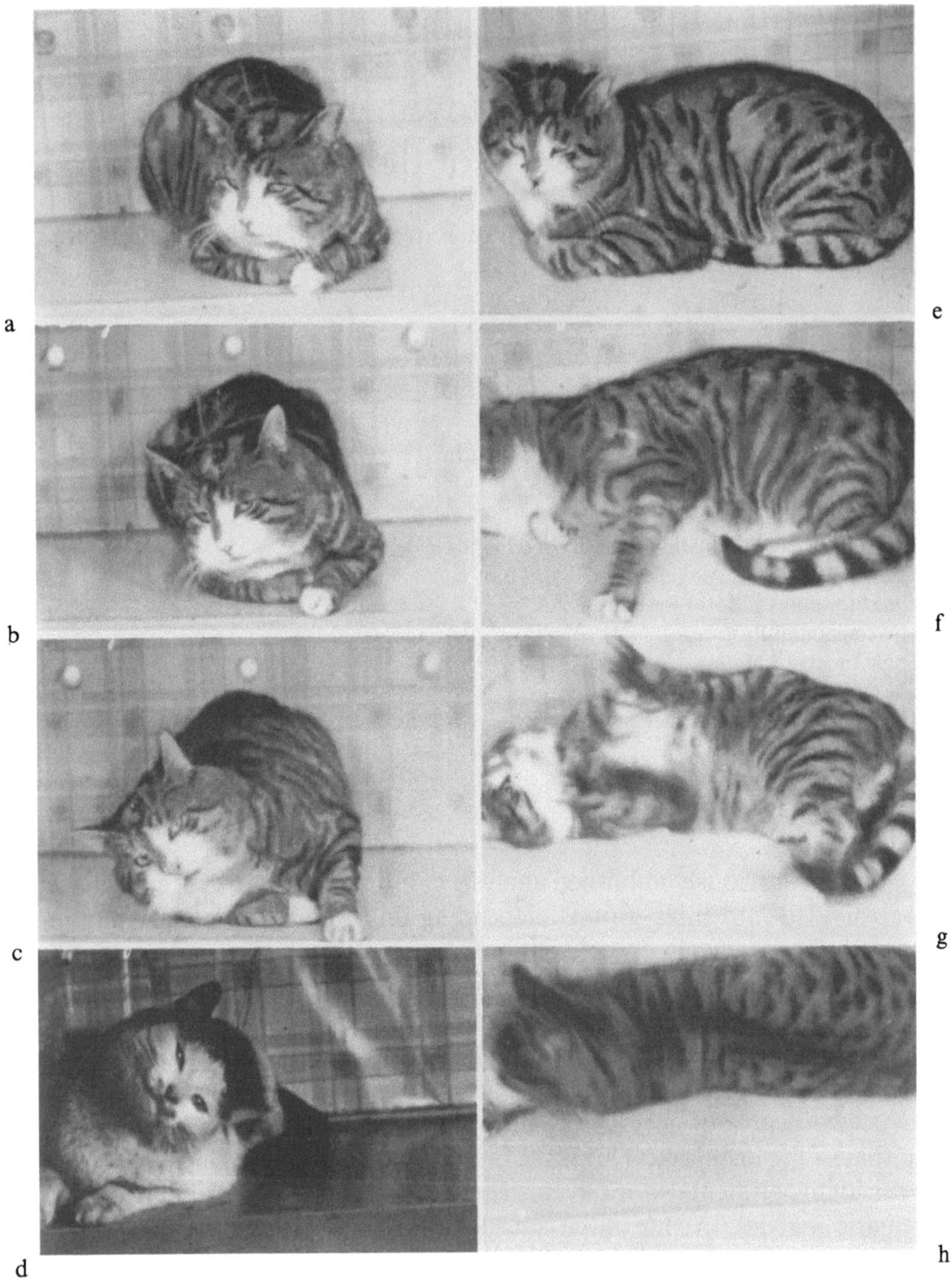

Fig. 14.1. a–c. Still pictures from experiment 218: *rotation* about the sagittal axis, stimulation of the right diencephalon. *a* before stimulation. *d* Still picture from experiment 184: rotation about the sagittal axis, stimulation of the left diencephalon. *e–h* Rotation of head and body in various successive stages (experiment 218). *e* Before stimulation

pare the individual cases from every angle and to detect contrasts between them. By selecting individual sections of the film for further study, a documentation was produced which revealed the principles underlying the manifold stimulus phenomena.

Axial (Rotational) Movements

The various motor effects are identified by reference to a spatial coordinate system; in the case of rotational movements (as this is the most frequent type of stimulus effect, it will be dealt with first), the coordinate system is represented by the usual three vertical rotational axes; the longitudinal (fronto-occipital), the transverse (bitemporal), and the vertical (dorsoventral). The corresponding movements were all observed very frequently, often in pure form (i.e., defined by a single axis), but usually in complex form (i.e., with components of various rotational planes).

The motor effect may be restricted to the head and, in spite of continued stimulation, only reaches a certain point at which it then stops. The centrally induced *movement* is thus transformed into a *posture* which deviates from the primary (head straight) position. The head reverts to its initial position when the stimulation is discontinued. It is also possible to *grade the deflection* by slightly changing the voltage. In this manner, every possible position between "vertex straight-up" and "vertex almost-horizontal" (to the stimulated side) can be achieved in rotation. With increased voltage and correspondingly strong rotation, the effect spreads from the head to the body, the latter being "carried along," as it were. The head rotation turns into a rolling of the whole body. Raising the head leads to raising the anterior part of the body until the animal falls over backwards. Turning the head sideways becomes a circling movement. In isolated cases, the body axis is bent while the animal remains standing. These phenomena are illustrated in Figs. 14.1 and 14.2.

A slight rotation of the head about the longitudinal axis is, of course, almost always present in vertical movements, i.e., about the transverse axis, when stimulation is one sided. The side contralateral to the stimulation is somewhat raised. The behavior of the eyeballs deserves special attention. When head and body are unimpeded, a central stimulation produces nothing particularly worthy of observation, except when the tips of the electrodes are in the area of the intracerebral root or the intracranial path of the oculomotor nerve. In such cases, however, we are dealing with a peripheral mechanism which is of no further interest. When the head, as a result of central (i.e., diencephalic) stimulation, is about to carry out a rotation but is prevented from doing so by being held in place, the movement is switched to the eyeballs, which deviate from their normal position (in relation to the

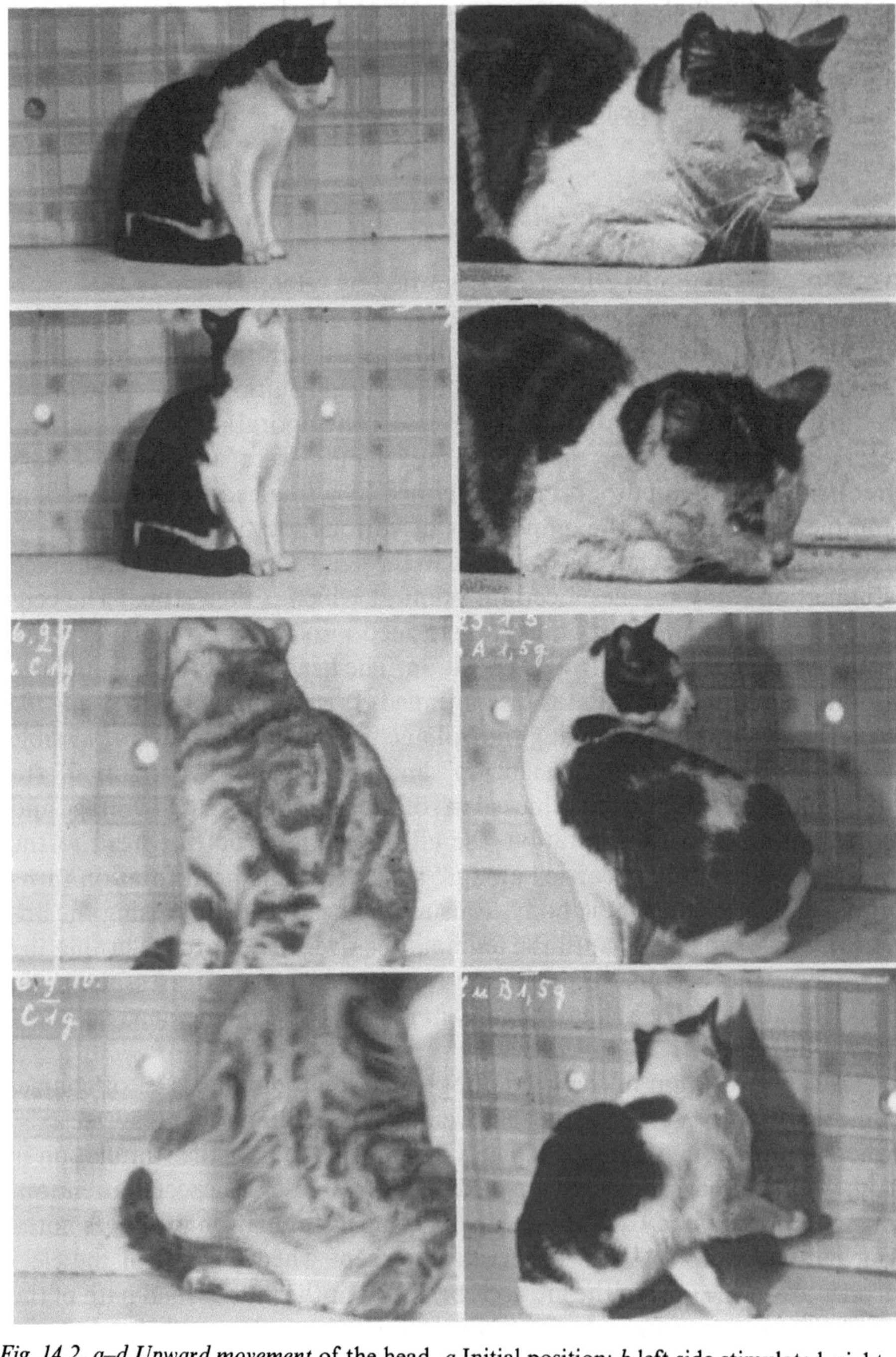

Fig. 14.2. a–d Upward movement of the head. *a* Initial position; *b* left side stimulated, right side rises first (experiment 210); *c* right side stimulated, upward movement is more pronounced on the left; *d* later situation – tipping over backwards (experiment 216). *e–f Downward movement* of head. *e* Before stimulation; *f* right side stimulated (experiment 93); *g turning* to the *contralateral* side; *h turning* to the *ipsilateral* side (experiment 223)

head). The relation between head and eyeballs is particularly striking in these impeded rotational movements: the slit-shaped pupils of the cat are inclined. If the head is then released or allowed to make a passive rotation, the motor effect caused by innervation of the eye muscles disappears entirely (Fig. 14.3). A similar relation between the head and the eyeballs under central stimulation is also observed when the head is raised. The situation is analogous to a perfectly balanced posture of the head when compulsory deflection of the head, e.g., from the primary posture, is answered by a compensatory eye movement. If the position is a result of diencephalic stimulation, it is, in accordance with what has already been said, "corrected" by a corresponding movement of the eyeballs.

The *direction* of rotation about the longitudinal axis is, without exception, an upward movement toward the side of the stimulus. The direction of movement in the sagittal plane is also nearly uniform. The anterior pole of the body rises, and the head is bent backwards. A *lowering of the head* is obtained from only very few characteristic stimulus sites; the head is lowered until the nose touches the table. With turning in the horizontal plane the situation is different. There are two groups here: one group turns to the stimulus side, the other in the reverse direction. This difference is not accidental, but is systematically dependent on the localization of the stimulus sites. More will be said about this later.

Hitherto the discussion has centered on cases in which deviation is about a single axis. However, it has already been mentioned that combined movement about various axes is more frequently evoked, particularly the combination of upward movement and rotation. However, rising with simultaneous turning in the horizontal plane and rotation, turning and lifting the head and possibly the anterior part of the body are also quite often found in association. Figure 14.4a–d illustrates such combined movements about more than one axis.

The relation between rotation in the horizontal plane and upward movement merits special attention. Rotation is carried out in such a way that only one-half of the head or body is raised. This mechanism is often accentuated by the appearance on the corresponding half of the face of certain motor phenomena such as raising of the upper lip and the cheek. Depending on the stimulus site, purely one-sided (i.e., contralateral) raising may be associated with an ipsilateral component; the rotating head is simultaneously pulled upward. Certain stimulus sites cause such a strong ipsilateral rising movement that the forces on either side are almost equal, with the result that the anterior part of the body shoots up nearly simultaneously with the contralateral side. Thus the boundary between rotation and rising is not sharp. This is the functional expression of an innervation apparatus which governs rotation both in the transverse and in the sagittal planes. Since stimulation on the contralateral side inevitably produces mirror-

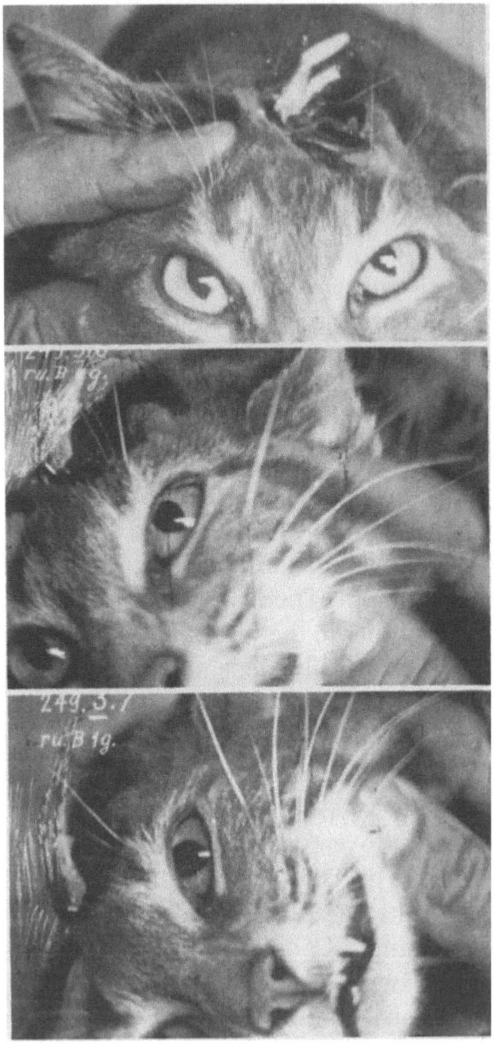

a

b

c

Fig. 14.3a–c. Relation of head to eyeballs. *a* Head is held fast and prevented from making rotation induced by diencephalic stimulation; the eyeballs rotate. *b* and *c* The head is turned passively. The eyeballs readjust – normal relation to head (experiment 249)

image movements, the reactive elements situated in both halves of the brain must be regarded as connected "half apparatuses" which work together, particularly in stabilizing a certain head or body posture. This view will be presented in greater detail in connection with the relation between movement and posture. For further clarification refer to Figs. 14.4e–g and 14.5.

The Influence of Stimulation Intensity and Duration

If a near-threshold stimulus causes a distinct rotation of the head and a slight augmentation of the stimulus intensifies this rotation, the effect will spread to the anterior part of the body if the voltage is raised. Further intensification of the stimulus causes the posterior part of the body to move. As a result, the whole animal rotates about the longitudinal axis (rolling movement). The head is thrown back, if the effect is restricted to the head. If those regions whose excitation spreads to the body even at low voltages are stimulated, the anterior part of the animal rises, with the head more or less in advance of the body. The forepaws are stretched out in typical fashion while the toes are spread wide. Intensity and duration of the stimulus are the factors determining whether the rising of the anterior part of the body will develop into a complete rotation about the transverse axis, with the animal falling over backwards. In turning about the vertical axis, the head usually leads, but simultaneous movement of the head and anterior part of the body has been observed as the first response of the motor phenomenon. The result is that the orocaudal axis formed by the head and the body is not bent. Spreading of the effect can often be obtained by prolonging the stimulation while maintaining the same intensity. The movement, which at first is observed only in the head, is then gradually (i.e., after 20 or more seconds) seen in the anterior part of the body. This phenomenon is apparently the consequence of the cumulative effect of stimuli which are at first prethreshold for the body. The same principle is in operation when a stimulus effect begins with a rotation about one axis and, upon stimulation being continued, joins the movement about a second axis. This "drag" is most easily observed in turning in the horizontal plane. This corresponds to our experience that this constitutes differences in the time separating stimulus and response, between turning on the one hand and rotation and upward movement, on the other.

Latency Period and Aftereffect

Seen with the naked eye, rotation and upward movements begin immediately on stimulation. To obtain a more exact understanding of these temporal relations, a number of cases were photographed in slow motion (128 frames per second). Study of the film shows that, under favorable conditions, the time which elapses between the movement of the magnetic stimulus signal and that of the head is that between two frames. In other cases, especially where the stimulus is weaker, the interval is two to three frames. This delay is quite rightly attributed to the fact that the inertia is less rapidly overcome by the weaker stimuli and, hence, movement begins later. In other words, the shortest times measured are crucial. Of course, for calculating the true latency, the inertia of the magnetic signal must be known, as it adds up to the time lag between stimulus and response. To rule out this source of error, we tested the inertia of the magnetic signal by comparing it with current of a pulsating light bulb discharge released by the stimulation. The light signal and the magnetic signal were both projected onto the moving film. The test showed that the movement of the magnetic signal was visible within $^{1}/_{1000}$ of a second. This time lag is of no importance in our experiment, and it can be stated that the interval between stimulus and response (that is, latency time) is less than $^{1}/_{50}$ of a second. The same does not apply to turning in the horizontal plane. Whether this is so in all cases, however, is not yet certain. As a rule, a few seconds pass before movement commences. Later in such cases, there are also occasional pauses of several seconds during which the animal stands still and only continues its movement afterwards. A pronounced positive aftereffect is often associated with delayed onset. The head remains in the deviated position for a few seconds longer, or else the circling movement is continued for a quarter or half a circle. In certain circumstances, the consequence of the longer latency and aftereffect is a combined movement, beginning with rotation or raising of the head and finishing with a turn.

Time Structure of Motor Responses

A relatively low stimulus frequency was employed in order to analyze the time factor as thoroughly as possible. For further details, the reader is referred to a description of the method [1]. We were guided by the consideration that only sufficiently long pauses between the individual impulses would enable us to observe whether the central excitations remain more or less discontinuous on the way to the reactive substrate or whether they pass through substrates with summation power or even substrates with their own rhythm. Experience has demonstrated that a frequency of six to ten impulses per second permits this distinction to be made. In rotation and upward movements, the responses are always clearly manifested in the form of jerks; a return phase may even intervene between the individual jerks. The return phase is, however, always shorter, and the result is a one-sided saccade movement. The situation is similar to that involving peripheral muscle stimulation with frequencies producing an incomplete tetanus. The comparison is also valid in that, in the course of a stimulus series, the initially stronger jerks are less accentuated and the movement is rapidly evened out as the stimulus frequency increases. The fragmentation of the movement into individual perceptible units gives it a rather nonphysiological appearance. However, as soon as the frequency reaches beyond 20–30 per second the response emerges in a perfectly smooth manner. The jerky

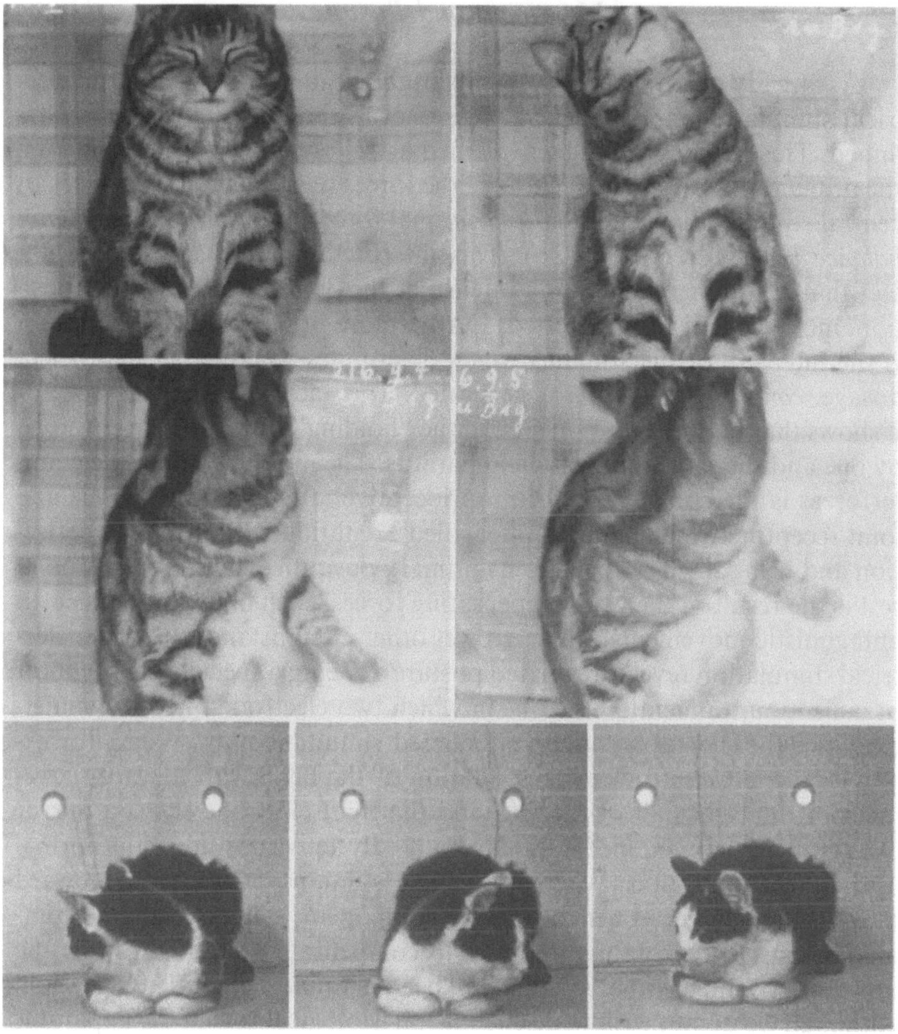

a b c d e f, g

Fig. 14.4a–g. Movement with combined deviations in more than one plane. *a* Before stimulation; *b* upward movement of the head with slight rotation; *c* and *d* turning in the horizontal plane is finally added to the syndrome (experiment 216). *e–g* Simultaneous stimulation of bilaterally symmetrical points. *e* Left side stimulated, head turns to the right; *f* right side stimulated, head turns to the left; *g* simultaneous right and left stimuli: responses cancel each other out (slight predominance of right turn) (experiment 119)

movements stand in opposition to smooth motions which arise from the start at eight impulses per second. This is the rule for turning in the horizontal plane. Finally, it should be mentioned in passing that there are also types of movement whose rhythm is completely unrelated to that of the stimulation. We shall return to this subject when discussing automatisms evoked from the diencephalon.

Movement and Posture

As already mentioned, movements of the head induced by central electrical stimulation can suddenly "freeze" even though stimulation is continued. This behavior may be explained as follows: when deflection from the primary position is great enough, the forces released in the periphery by central stimulation encounter oppositional forces, thus creating a *dynamic balance* (Figs. 14.1c, d, 14.2b). This interpretation is supported by the fact that additional stimulation intensifies the deflection; in the new position, movement is again arrested. Furthermore, it has been proved that it is the intensity of the stimulus which determines whether, at a sufficient dosage, *continued movement* is generated. This fact is very significant, for it shows that both the movement and the "holding" functions are governed by one and the same innervation apparatus. The mobilization of the counterforces is the consequence of proprioceptive afferents from muscle and joint receptors which undergo intensified excitation with increased deflection and bring the head back to its primary position. Thus, the position at rest is the result of the two compensating forces, each one creating its own antagonistic movements. There are yet other circumstances in which electrical stimulation reveals regulated posture as being a special case of motor organization, to mention the one in which two electrodes placed symmetrically in the diencephalon are stimulated simultaneously. When, for example, the left electrode evokes rotation of the head with upward movement of the right half of the face and the right electrode elicits a mirror image effect, the combined effect is a markedly active posture in the normal upright head position. Whereas a single stimulus produces an upward movement of the head with a rotation component, symmetrical dual stimulation causes the rotating impulses to compensate each other, while the upward movement is potentiated: the head is raised without any rotation. Although we do not wish to discuss the physiological role of the apparatus set in motion by the artificial stimulus before the effects on the extremities and the face have been described, it should be mentioned that in our acute (stimulation) experiments, we are dealing with innervation mechanisms whose failure, as revealed, e.g., in extrapyramidal syndromes, is of interest to the clinicians.

Tonic Innervation

When the stimulus electrode is used for electrocoagulation, as has been already described [1], thus creating a small focus in a motor substrate, permanent deflection results which is particularly pronounced in the head. The direction of the deflection from the normal position is diametrically op-

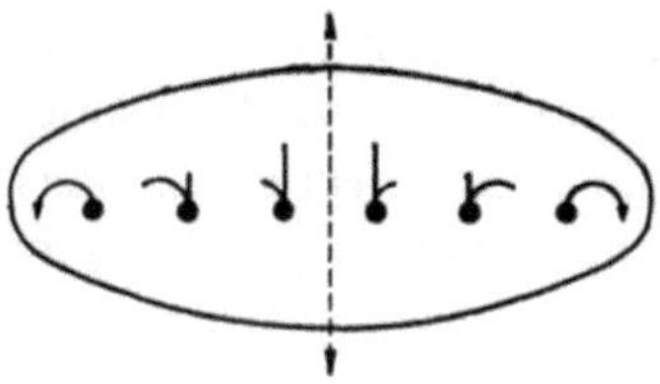

Fig. 14.5. Representation of relations between rotation and upward movement. Rotation and upward movement combined in different proportions, depending on the localization of the stimulus site. The *right half* of the field supplies a mirror image of that of the *left half*. The interplay of both sides fixes head and body posture in a definite position (see *Fig. 14.6*, numbers 1, 2, 6). The *dotted line* corresponds to the midsagittal plane (plane of symmetry)

posed to that caused by feeble or moderate stimulation. This behavior proves that the elements which responded to stimulus in our experiments are normally in a tonic excitatory state and thus have a decisive influence on the posture of the head. Two facts should be stressed in this connection. The postural defect does not primarily involve a specific muscle or group of muscles; it is oriented according to a spatial vector. Thus, we are dealing with "direction-specific" atonia, or hypotonia. The second fact concerns movement, for we note that active movement is reduced along the same vectorial components as are revealed by the impairment of tonic innervation. This is dealt with elsewhere [5] because of its special neurological interest (due in part to the pathological focal symptoms). Here it should only be mentioned that these observations confirm the conclusion drawn here that the influence mediated by the diencephalon on postural mechanisms is identical in nature with that mediated by the extrapyramidal system on direction-specific movements. Posture is a special case of regulated "movement."

Localization Findings

Our present task is to establish the relationship between the stimulation sites established by histological examination of the experimental brains and the stimulus effects described above. For reasons of clarity, symbols have to be given to the various phenomena. These symbols are entered on schematic representations of horizontal brain sections, especially sections 362, 399, 424, and 462 of the horizontal atlas [1], and are explained in the legend of Fig. 14.6. It is essential that they be easily understood and suitable for representing complex motor phenomena. Nevertheless, one point requires special explanation. When identifying a right or left turn, the usual practice is to orient oneself according to the object observed. However, it is also common that data on the direction of a rotation are given from the observer's point of view: vertex to the right (or clockwise) means right from the observer's point of view. The direction is, of course, reversed as seen by the animal. One has to imagine one's own head in place of the animal's or the head of the animal with its viewpoint (observed from behind). The common inconsistency regarding point of view usually does not affect pure description. However, where the aim is to understand the organization of the control apparatus and outwardly visible processes are to be correlated with central control, only the point of view of the subject carrying out the movements can be valid. Thus, clockwise rotation indicates the direction in which the head represents the hand of the clock, and the rotation is seen as if one were standing in the animal's place (or behind it). This method of representation is mandatory. One need only think of the confused situation which would arise if, in the combination of rotation and turning to the left or to the right, the former were seen from the observer's viewpoint and the latter from the animal's.

In the original version, colored signs were used to enter the symbols on the drawings of the brain sections. This allowed the stimulus intensity of a particular effect to stand out. By correlating stimulus site, stimulus voltage and stimulus effect, an attempt has been made to assemble the material necessary for answering the localization question to which the following remarks can be added (see Figs. 14.7–14.10).

Apart from certain special cases, motor response sites in the diencephalon are first encountered – moving from top to bottom – only within the horizontal plane which passes approximately through the middle of massa intermedia. The position of this horizontal plane (section 362) is further defined by the fact that it intersects the anterior and posterior commissure. From here we encounter motor activity as far down as the horizontal plane (section 462) which touches the dorsal pole of the mammillary body. The substrate from which weak stimuli elicit movement is just 4 mm in depth. Medially, the response field is sharply defined by an area which fails to give movement effects. At lower levels, the medial area with unresponsive sites decreases so that it is only 1.5–2 mm in diameter at the level of the supramammillary commissure. The lateral extension of the response area can be less well defined. In this direction the responses involved the extremities or the face (Fig. 14.8) rather than the trunk. For purposes of lucidity a cumulative plot in a single plane has been prepared to summarize the results contained in the individual charts, including those with effects on extremities and face, a subject which has not yet been broached and therefore not closely defined. Even sites which are unaffected by a voltage of more than 2.5 V have been included. Representing stimulus intensity by using different colors had to be abandoned for the printed text, very much to the detriment of clarity. Instead, this goal was achieved by using direc-

1 2 3 4 5 6 7 8 9 10 11 12 13 14 15 16 17 18

Fig. 14.6. Explanation of symbols used in the localization charts: *1*, rotation of the head, direction of movements relative to the animal. *2*, upward movement; *3*, downward movement; *4*, turning in the horizontal plane to the stimulated side; *5*, turning to the contralateral side; *6*, rotation and upward movement; *7*, rotation, upward movement of the head with circling to the contralateral side; *8*, head movement, not definable; *9*, head and eye movements; *10*, head with anterior part of body; *11*, head with entire body; *12*, turning with longer latency; *13*, turning in saccades; *14*, turning in circles with pauses; *15*, motor effect in the contralateral half of the face; *16*, effect on extremities; *17*, *small dot:* stimulus sites with no effect at second voltage level; *18*, *large dot:* stimulus sites with no effect at third voltage level. *Line* indicates movement in corresponding direction until new balance is reached. *Arrow* indicates movement if continued, e.g., rolling or falling over backward or circling. The effects on face and extremities are not differentiated here. *Heavy line* means that the response is elicited by first voltage level (0.5–1.5 V). *Thin line* refers to phenomena of second voltage level (1.5–3.0 V). A *short line* indicates a weak effect, a *long line* a strong effect. All signs are for left-sided stimulus

tional symbols of varied thickness. Finally, the entries distributed over 17 individual charts were summarized in four planes (each 1–2 mm distant from each other) and restricted to effects on the axial musculature, and to stimulus intensities at first and second voltage levels (Figs. 14.9, 14.10).

No absolute rostral and caudal limits can be drawn for the active region because the effect of stimulation at various sites very much depends on relatively small differences in stimulus intensity. There is a limited field from which responses of the axial muscles can be evoked without exception and with strikingly low voltage. This zone in which the best effect is achieved is intersected by the plane (section 399) which lies below the massa intermedia and parallel to the commissural plane (section 362). Horizontally, the zone lies between the mammillo-thalamic tract and the habenulo-interpeduncular tract (Fig. 14.7).

Having dealt with the localization of motor effects in general, we must now ask whether any relationship can be established between the quality of the phenomenon and the stimulus site. In fact, such a relationship does exist. On stimulation with the first voltage level (not higher than 1.5 V), the whole of the body or even its anterior part, can, with very rare exceptions, only be set in motion if the stimulus site is behind the mammillo-thalamic tract (i.e., about the level of section 399). Further forward, usually only head movements can be achieved by intensifying the stimulus up to the second voltage level (up to 2.5 V). The whole body responds better to more ventral stimulation (section 424) of the active region. Effects on the eyeballs are clearly observed only when the head is firmly held.

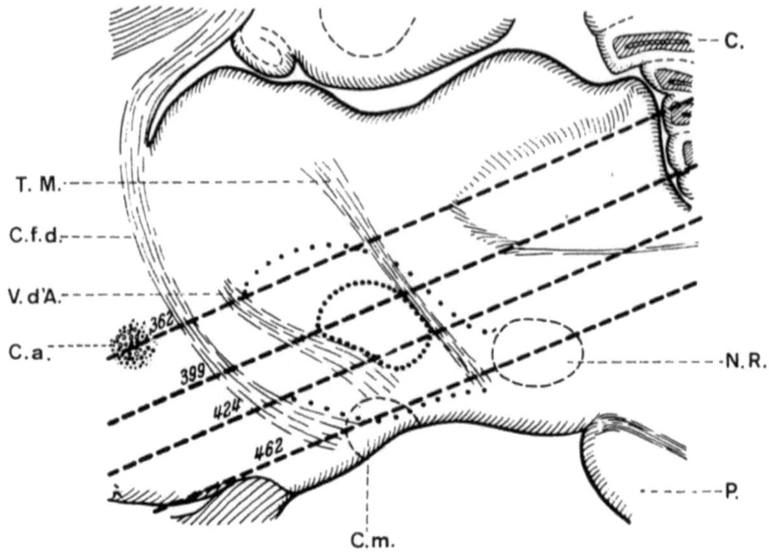

Fig. 14.7. Sagittal section through the diencephalon, partly schematic, 2 mm from the midline. The *broken lines* represent the position of the horizontal projection planes (Hess [1]). The zone defined by *closely spaced dots* between the mammillothalamic tract *(V.d'A.)* and the habenulo-interpeduncular tract *(T.M.)* corresponds to the projection of the area in which the active stimulus sites are most dense and the stimulus thresholds lowest. The zone defined by *loosely spaced dots* encompasses the area comprising a thin spread of active stimulus sites at high stimulus intensities. The field is open rostrally and caudally. *T.M.*, tract of Meynert; *C.f.d.*, fornix; *V.d'A.*, mammillothalamic tract (Vicq d'Azyr); *C.a.*, anterior commissure; *C.*, cerebellum; *N.R.*, red nucleus; *P.*, cerebral peduncle; *C.m.*, mammilary body

With regard to the axes of the different types of movement, the dorsalmost part of the active region is exclusively associated with rotation around the sagittal axis involving the eyes, head, and body. Increasing the stimulation strength from the first to the second level usually intensifies the rotation. In very rare cases, a second movement component appears. The substrate for pure rotation can be followed somewhat deeper laterally than medially (Fig. 14.9 a, b). Otherwise the rotation is joined by an upward movement which will be discussed later. Further ventrally, the rotation effect gradually disappears; here it is limited to the lateral and, less so, rostral regions. Finally, two cases of pure rotation of the head were evoked from the red nucleus or adjacent site. Here too the rule obtains that rotation is in a certain direction so as to cause the contralateral side of the head to rise (the vertex thus inclines toward the stimulus side).

The movements about the transverse (i.e., bitemporal) axis usually take the form of an *upward movement*. The degree of symmetry of the response depends on the distance of the stimulus site from the midline. As the distance increases, the contralateral side is increasingly in advance when ris-

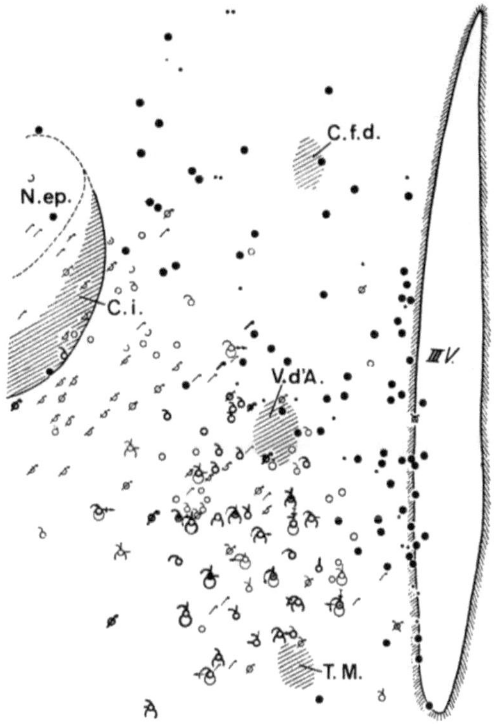

Fig. 14.8. Horizontal projection plane at the level of section 399 with plotted stimulation points (the corresponding histological section can be found in the atlas p. 100 [1]). *C.a.*, anterior commissure; *C.f.d.*, fornix; *C.i.*, internal capsule; *N.ep.*, nucleus entopeduncularis; *T.M.*, tract of Meynert; *V.d'A.*, mammillothalamic tract (Vicq d'Azyr); *III V*, third ventricle

ing. This is equivalent to an accentuation of the rotating component. Another fact is the relative increase of the vertical component when the stimulus moves from the level of section 362 to deeper planes of the active region. This type of effect appears only when the electrodes on their way down have reached at least the plane passing through the lower limit of massa intermedia. At this level the effect is only moderate, whereas 1–2 mm deeper it dominates the picture. The maximum upward response in its purest form is obtained at the level of section 399 (Fig. 14.9b) where it is restricted to a relatively small area, the diameter of which corresponds to the distance between the mammillo-thalamic tract and the habenulo-interpeduncular tract. One mm deeper, the area is somewhat more extensive in all directions (section 424, Fig. 14.9c). Here is where we encounter the most medial of the stimulus sites affecting the motor system. These sites respond only at the second voltage stage so that electrical irradiation from just

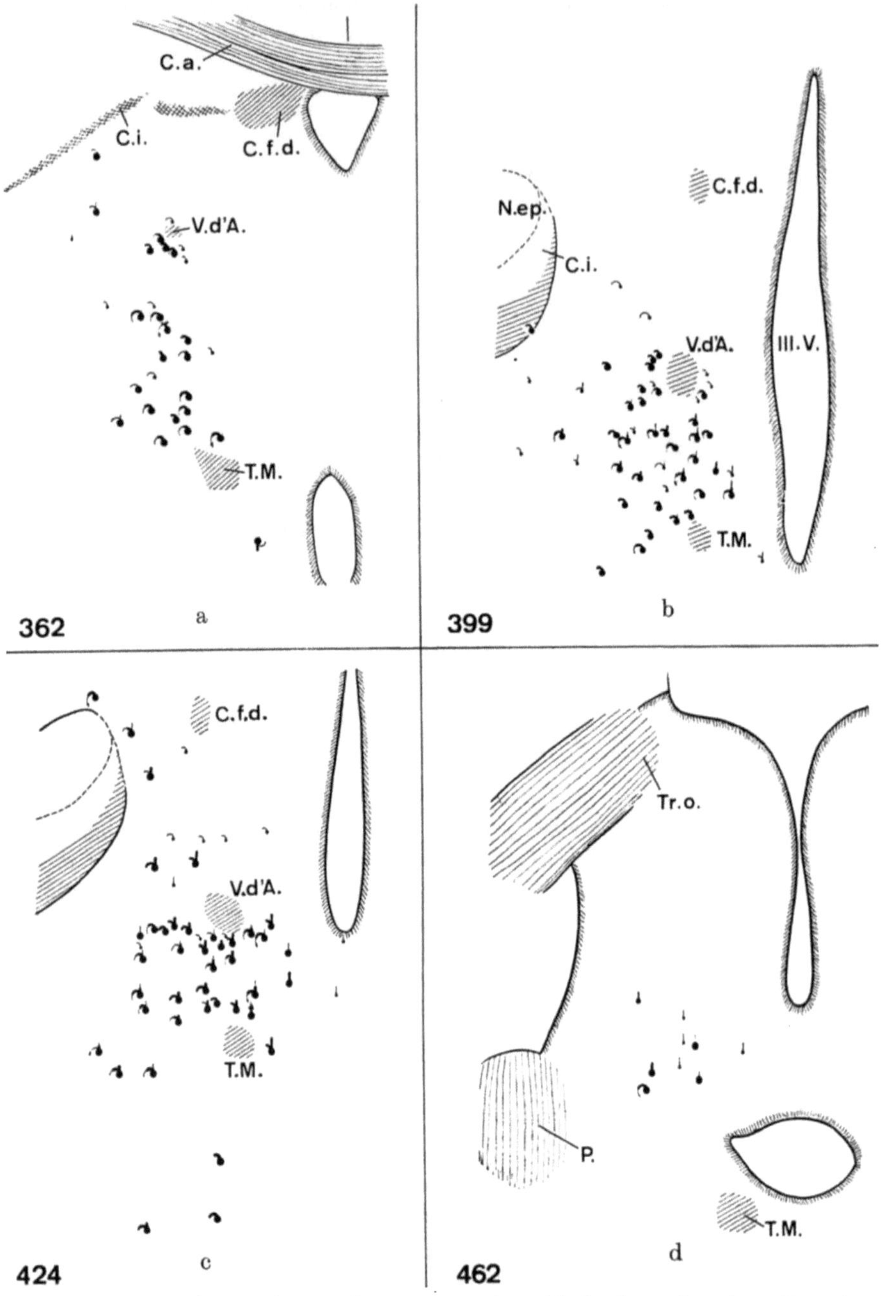

Fig. 14.9a–d. Rotations and upward movements plotted in horizontal sections. Dorsalmost plane contains (with rare exceptions) only stimulus sites with rotation. The upward component appears further down, more pronounced medially than laterally. Isolated stimulus sites are found quite far rostrally. The densest accumulation is in the area between the mammillothalamic tract *(V.d'A.)* and the habenulo-interpeduncular tract *(T.M.)*. *a* Level of section 362; *b* level of section 399; *c* level of section 424; *d* level of section 462: horizontal section through hypothalamus (tuber cinereum). *C.a.*, anterior commissure; *C.i.*, internal capsule; *C.f.d.*, fornix; *V.d'A.*, mammillothalamic tract (Vicq d'Azyr); *T.M.*, tract of Meynert; *N.ep.*, nucleus entopeduncularis; *III.V.*, third ventricle; *Tr.o.*, optic tract; *P.*, cerebral peduncle

1.5 mm cannot be ruled out; 1.5 mm further down, (section 462) the substrate responsible for the upward movement is again more restricted (Fig. 14.9 d).

Earlier, some rare exceptions were mentioned. These are stimulation effects in which the *head is lowered* (Fig. 14.2 f). The four stimulus sites from two different animals are largely consistent. They are found at the transition from the diencephalon to the mesencephalon at the level of sections 362 and 399 (Fig. 14.9 a). Three of these sites lie somewhat further back than shown in our diagram. The striking point about the localization of the stimulus sites causing *turning* in the horizontal plane is their relatively great dispersion. The substrate responsible for turning reaches more dorsally than any other and also rather ventrally. The turning effect is also associated with stimulus sites in the rostral- and caudalmost parts of the active region and reaches far to the side. Only medially is sharp delimitation maintained, a fact which also holds for the other forms of movement. In the light of our knowledge about rotation, it is striking that turning to the left and to the right can be generated on either side. Direction, however, is not a matter of chance but is dependent upon localization. Stimulus sites causing the animal to turn to the stimulated side (the ipsilateral side thus precedes in the circling movement) are situated rostrally. Those for movement in the opposite direction are at the caudal end. The transition line (at which one direction gives way to the other) depends on the level. The deeper the stimulus site, the more the line is displaced. Hence, contralateral turning is elicited only by very rostral stimulus sites on the dorsalmost level (section 362). At the next level (section 399), the border runs more or less through the middle of the active region. At the level of section 424, contralateral turning dominates the entire region except for one point far to the caudal end. At the most ventral level (462), only contralateral turning occurs. Furthermore, the borderline does not run at right angles to the midline, but deviates somewhat obliquely. This situation is represented in Fig. 14.10 a–d, where the border is entered as a wavy line. In the transition area, one direction or the other may predominate, depending on the stimulus intensity.

The Identity of Active Elements

It has been pointed out that the active motor area is congruent with a sector of the diencephalon, through which fronto-occipital fibers run, converging toward the caudal end [2, p. 642; 3, p. 681]. By classifying motor effects according to the direction of movement, we have been able to determine the relationship between the site of the stimulus and the type of effect. Thus, we can project the borderline separating ipsilateral from contralateral turning onto the entire motor response area. Our present results confirm the observation that the sites from which identical effects are obtained ad-

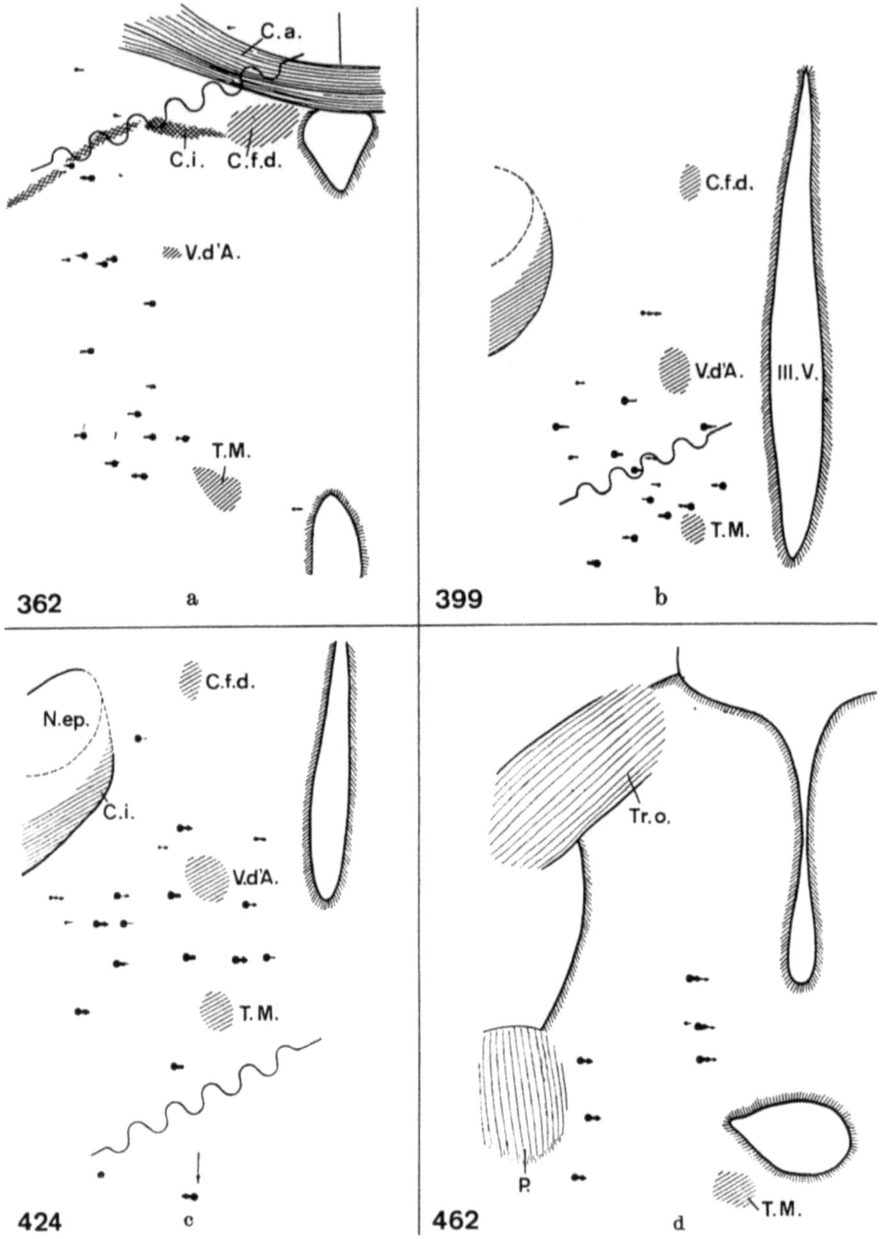

Fig. 14.10. a–d Stimulus sites representing turning in the horizontal plane. Same horizontal sections as in *Fig. 14.9*. *Wavy line* defines the boundary between turning to the contralateral or to the stimulated side. This boundary is far to the front in plane 362, passes through the middle of the active area in section 399, and is at the rear in section 424. Only section 462 has turning to the contralateral side. *C.a.*, anterior commissure; *C.i.*, internal capsule; *C.f.d.*, fornix; *V.d'A.*, mammillothalamic tract (Vicq d'Azyr); *T.M.*, tract of Meynert; *III.V.*, third ventricle; *N.ep.*, nucleus entopeduncularis; *Tr.o.*, optic tract; P., cerebral peduncle

here to the course of the above-mentioned fibers. One point should be made in this connection.

The relative number of active stimulus sites decreases toward the rostral end of the exploration field. This reduction is most pronounced with respect to the responses of the axial muscles. By intensifying the stimulus, more effects, particularly from the rostral areas, can be produced, without causing a further spread of current. The most obvious explanation of these findings is that the reflex apparatus is by and large found in the diencephalon itself, specifically, where the positive sites are most dense and the stimulus threshold lowest. A smaller percentage of stimuli operate via fibers which establish connections with more rostral substrates. This is depicted in Figs. 14.8–14.10. The fact that very jerky movements predominate is due, in our view, to the involvement of efferent fibers. This applies particularly to rotation as well as to up- and downward movements of the head. The situation for turning in the horizontal plane is probably more complex; in spite of the low stimulus intensity, the movements are often completely smooth. The question arises as to the origin and termination of the transthalamic fibers [4]. The corpus striatum is the most likely choice; however, this cannot be definitely established at the present time. On the contrary! A good deal of material, mostly relating to the extremities and the face, remains for further detailed analysis. Furthermore, we have recently combined stimulation with electrocoagulation and appropriate histological investigations of the reactive sites in the hope of establishing contact with the advanced knowledge concerning the morphology of the brain assembled in recent years. We must also wait with the answer to the question about the physiological role played by motor phenomena activated by artificial stimuli and about the functional relations to motor activities which are organized entirely at mesencephalic levels (e.g., red nucleus). This will provide us with the opportunity to examine the most important experimental findings of other authors.

Summary

The analysis of motor responses from the diencephalon, in particular axial movements, was undertaken with generous use of the documentary film, and with special reference to rotational axis and direction. The results of the study, seen in relation to the localization of the corresponding stimulation sites, lead to the following conclusions:

1. Within the motor response area, there is a definite grouping according to rotational axis. The relation between sites and effects of stimulation are diagrammatically represented.

2. The movement effects in the frontal and sagittal planes represent cooperative regulatory mechanisms. The responsible substrates are topo-

graphically continuous and their functions bilaterally integrated. Intensifying the stimulus not only increases the response but causes it to spread to additional parts of the body which are all moved about the same axis ("syntelic" movement). Increasing the duration of the stimulus also extends the effect. The latency for rotational and upward movements is less than $^1/_{50}$ of a second; that for turning in the horizontal plane not infrequently exceeds one to several seconds. The stimulation frequency, in rotation and upward movements, is reflected in a saccade response (at eight per second). Jerky movement is occasionally observed with horizontal turning, too, but most frequently the movement is smooth.

3. Movement and posture of the head and body are organized by the same central apparatus. Posture corresponds to a dynamic balance between simultaneously and symmetrically released central excitations which are under proprioceptive control. Focal destruction involving certain areas of the diencephalon leads to deficits which have a mirror-image relation to the motor stimulation responses obtained from the same site. The functional defect affects both posture and movement functions, atonia or hypotonia referring to a spatial vector which is dependent on the localization of the focus.

References

1. Hess, W.R.: Beiträge zur Physiologie des Hirnstammes. I. Die Methodik der lokalisierten Reizung und Ausschaltung subkortikaler Hirnabschnitte. Leipzig: Thieme 1932
2. Hess, W.R.: Zwischenhirn und Motorik. Pflügers Arch. Ges. Physiol. Menschen Tiere *243*, 634–650 (1940)
3. Hess, W.R.: Motorische Symptome im Gesicht bei Reizung im Zwischenhirn. Pflügers Arch. Ges. Physiol. Menschen Tiere *243*, 678–685 (1940)
4. Hess, W.R.: Diencephale Reizsymptome und Körperstamm; Beziehungen derselben zu Bewegungseffekten im Gesicht und an den Extremitäten. Pflügers Arch. Ges. Physiol. Menschen Tiere *243*, 741–747 (1940)
5. Hess, W.R.: Experimenteller Beitrag zur Frage der extrapyramidalen Motorik. Z. Ges. Neurol. Psychiatr. *172*, 639–642 (1941)

Chapter 15

The Biomotor System as an Organization Problem*

W.R. HESS

Contents

Whoever has had the pleasurable experience of watching a startled deer on its flight through the open wood can scarcely withhold a feeling of admiration for its harmonious movements. The smooth flow in the dynamics, the rhythm, and an unexcelled surefootedness are the causes underlying the aesthetic impression. To the physiologist the viewing of movement provides a mental stimulation beyond the pleasure of observation. It suggests that behind it is the creative power of a neural machinery whose exploration provides a tempting and almost presumptuous problem, for it unquestionably involves a very complex event which, as such, remains invisible to the eye and is only recognizable in its outward projection, that is, in the form of the motor performance. Three arguments may help, however, to overcome our hesitations: firstly, just as with the development of morphology, function has its phylogenesis from simpler to more highly organized levels. With this a pathway to the study of the organization of movement is marked out, so that hope of eventually arriving at the goal is justified. Secondly, we can draw from the data of previous research which elucidate at least partial aspects of this overall problem. Finally, the systematic application of modern recording techniques, especially cinematography, seems promising for the analysis of *the organization of movement*.

* Hess, W.R.: Biomotorik als Organisationsproblem I und II. Naturwissenschaften *30*, 441–448, 537–541 (1942)

The Problem

First we shall turn our attention to conditions which are valid for the lower forms in the animal kingdom. As a concrete example let us look at the movement of a pike in quiet water as it rushes out from its lurking position toward its prey. What takes place here is an oriented movement toward a visual target. The mechanism which is set in motion receives, as far as can be recognized externally, its impulse from a retinal image whose specific projection determines the direction of the movement. This results from a specific activity pattern of muscles which are responsible for propulsion and navigation. In whichever way the relationship between sensory stimuli and movement may be produced, *the correspondence between spatial order in the optical and the motor phase of the total performance is clear*, because otherwise the prey could not be hit and captured. Identity of spatial order is only possible on the basis of inherent relationships between activity of the (visual) receptors and the (motor) effectors. Our problem, brought to its simplest formulation, is to learn these relationships. It is already expressed in the apparent quiet of the fish lying in wait. In fact, with the motionless fish the locomotor system is steadily functioning to compensate for forces of hydrostatic origin and thus to "hold" the fish in a suitable position for launching the attack. The character of dynamic equilibrium is still more impressively demonstrated when the fish remains motionless in *running water* while lying in wait. Translatory and rotatory forces coming from outside continuously force the fish out of position. Thus, remaining motionless requires active control of significant forces. When under corresponding environmental conditions the trout jumps relatively accurately at a mosquito moving above the surface of the water, this means functional control of a relatively complex system of forces. The same holds for the bird aiming in stormy weather at its booty, e.g., a seagull swooping down on a small piece of bread floating on the water. This performance is even superior to that of the trout from the point of view of coordination. With a fish the effect of gravity is almost negligible; in the flight of a bird, gravity involves the full weight and tends to divert it from the goal. The successful act can only be achieved by a mechanism which is adequately incorporated into the whole organization of movement. Gravity also plays an important role for terrestrial animals and man. Gravity tends likewise to impair the visually induced movement pattern. When, despite this, the goal is reached, it proves that the visually organized combination of forces is mixed with automatic impulses which counteract gravity and permit the target-directed forces to operate as if in a gravity-free field.

We have drawn attention to the ability (of the organism) to counteract external forces quantitatively and in the proper direction and have pointed out that the responsible mechanisms must form an integrated part of the target movement. In order to grasp the dynamic relationships in their entire multiformity, an account of the actual findings must be added regarding muscular mechanics. In other words, the relationship between the vectorial system of forces on the one hand, and the muscular system as a biological apparatus, on the other, should be discussed. This is not accomplished by merely recognizing that muscle tension delivers the necessary forces. For a qualitative and quantitative grasp of the entire problem, it is important to realize that the forces are oriented according to a system of spatial coordinates, while the direction of pull of individual muscles in the course of movement, depending on their attachment sites on the bones, changes continuously. What this means for the organization of a target-directed motor performance may best be explained on the basis of a given object. The typical example is the directional movement of the eye, because the eyeball is moved in its socket in all planes and, within rather wide limits, possesses complete freedom of movement. Thus, the physical requirements of eye movements can be easily analyzed. The eye is also ideal because the passive mobility is realized with the minimum of muscle activity, namely, through three muscle pairs acting perpendicularly to one another.

Easily understandable is the relationship between muscle action and direction of movement when the visual stimulus, which induces the movement, strikes the eye in its primary position, that is, when gaze is directed forward. As soon as an object appears, for example, in the temporal field, the tension of the lateral (external rectus) muscle is activated and the turning of the eye is completed. In an analogous way stimulus objects arising in any sector of the visual field are sighted with the same accuracy. As a rule, however, a single muscle does not lead to the target as occurs in the simplest case; rather it requires a muscle combination whereby the different forces are mobilized, and their action leading to the target-directed result is constituted according to the parallelogram of forces. These relationships are shown in Fig. 15.1, where the direction of gaze must be deviated upward and outward to visualize the stimulus object. In order to attain this result a balanced coordination of the lateral muscle and the elevator of the eye (external and superior rectus muscles) is required. That this combined action takes place automatically and immediately, that is, without prior to-and-fro searching, means a remarkable performance! In reality the capability of the central motor apparatus surpasses considerably the combination act that is seen in this movement. Consider the following situation: our eye can not only sight objects from the *primary position* in a radius of more or less the peripheral field of vision; the gaze movement is also precise when the eye *in an already deviated position* picks up the image of a stimulus object. In order to appreciate this faculty correctly, the mechanical relationships of the eye muscle apparatus will have to be analyzed somewhat closer, because the *direction of pull of each individual eye muscle* is *different*

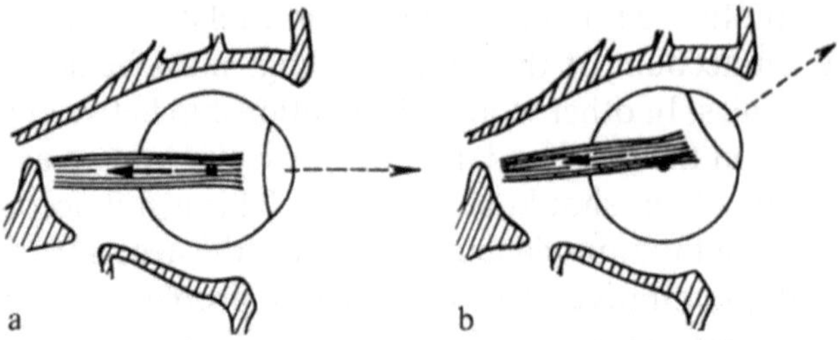

Fig. 15.1. a Pull of the external rectus muscle from the primary position of the eye (gaze straight forward). It develops a purely abducting effect; *b* pull of the external rectus muscle with the eye looking upward. Abduction is now combined with an elevating and slight rotating component. In analogous fashion, the effective movement of each eye muscle changes in a geometrically defined way with the eye position

according to the position of the eye. Neither in direction nor in quantity does it represent a "constant" factor which could be included as such in the central processing. On the contrary! As a result of the displacement of the eye relative to the starting point, the action of any muscle projected into space changes considerably, e.g., when a lateral deviation exhibits an elevating and a rotating component (Fig. 15.1 a, b). This *limitation of the relationships* between a certain muscle and its projected direction of pull in a spatial coordination system complicates the target-directed form of a movement; nevertheless the organization functions are accomplished with great precision. This does not mean that for a target-directed movement certain muscles are combined but rather that their vectorial forces are combined. From phase to phase, muscles are utilized whose direction of pull continuously changes. In each phase they are grouped anew so that the resultant, defined according to direction and magnitude, projects into the space defined by fixed coordinates.

For the final formulation of the problem it remains to be added that, as a rule, the organizing process is not limited to the oculomotor apparatus, but in most instances includes that of the head and also, under some circumstances, that of the trunk. Basically, this extension brings no new aspects to the problem, since it involves the same principle within a system of muscle groups functioning in parallel. Thereby the motor forces responsible for the eye, head, and trunk movements are integrated with respect to space and time toward a goal-directed movement.

Experimental Investigations

Up to this point we have limited the discussion to the problem of the central organization of motor performances taking place under physiolog-

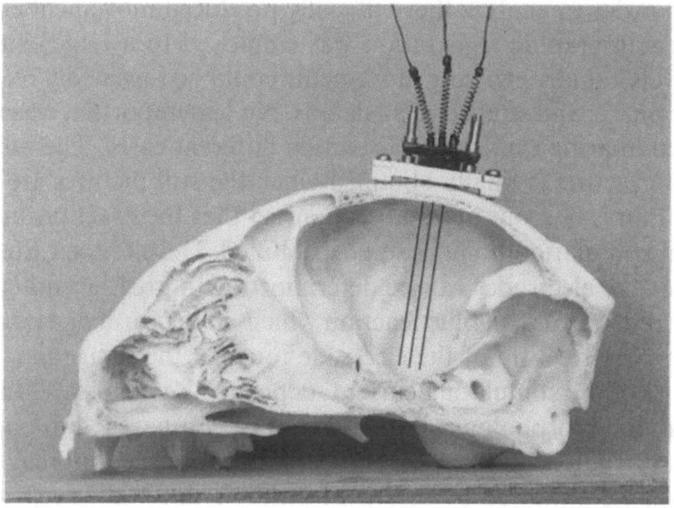

Fig. 15.2. Two double electrodes in situ. The sockets for the electrodes are firmly attached to the electrode carrier which is screwed to the skull

ical conditions and we have cited pertinent observations. Now we should like to tackle the problem from the experimentalist's point of view. By referring to previous original contributions from this laboratory we should like to give a summarizing report on a series of experimental investigations [1–6] (see translation of [5] in this volume, p. 223).

a) Electrical Stimulation Experiments. The project developed out of the necessity to stimulate electrically an insufficiently investigated part of the brain, the diencephalon, and to examine more closely the motor effects and to bring them into relationship with motor symptoms known from other, already better investigated parts of the brain. The *experimental technique* was designed to meet the following requirements: The application of electrical impulses was carried out in such a way that the free movement of the experimental animal was not impaired. Thus the forces activated by the artificial stimulus, if they became visible, appeared as deviations from the normal motor behavior. This implantation of electrodes was carried out under nitrous oxide anesthesia. An extremely light electrode carrier was designed which could be screwed firmly to a short segment of the exposed skull to which a block with three very fine steel electrodes was fastened by means of small clips. The small diameter of the individual electrodes (0.25 mm) and their high elasticity made it possible to insert them with a minimum of injury (Fig. 15.2). In fact, after lowering them into the diencephalon, no or only a very brief change in behavior of the animal was observed. In order to analyze the artificially produced movement effects reliably, cinematographic registration was employed to a large extent. Thus the relatively rapidly evolving movements could be repeatedly reexamined, phase for phase, and studied in all details. No less important was the possibility of comparing similar effects seen in different cases. The full value of the motion picture is finally obtained when the still pictures are available and the brain is laid out in serial sections so that the exact position of the individual stimulation points can be examined. Despite their fineness, the position of the electrodes can be determined with great accuracy, namely by means of the Prussian blue reaction which, in the form of a point, shows where the bare steel tip of the electrode touched the tissue and, owing to the passage of the stimulating current, deposited an iron trace. In this way functional and morphological findings can be compared with one another directly and controlled from case to case. Because the different effects of stimulation were plotted by means of symbols in an atlas representing a standard series of brain sections (corresponding to equidistant planes of section), certain organization features became readily evident. In completing this description of the technique it should be mentioned that the stimulus intensity was held close to the effective threshold, so that the artificially induced movement remained in the physiological range and, if possible, without secondary effects.

To check and enlarge upon the observations, control stimulations were repeated each time with the same and with different voltages. The stimulation was accomplished by direct current impulses with a frequency of 6–8 per second. The low frequency was chosen because information about the *temporal pattern* of the movement induced by a single stimulus impulse provided important conclusions. With individual animals the number of electrically stimulated points was limited to 12. The cat, which possesses a well-developed target-oriented motor system, served as experimental animal. The small range of variation in head size made it possible, moreover, to standardize the technical instruments. Of primary interest were the stimulation effects which concerned the movements of the eyes, the head, and the body. The head responds most readily. Stimuli near the threshold cause a small deviation from the normal position. A slight increase enlarges the angle of deviation and, for the duration of the stimulation, a new equilibrium obtains in the deviated position. If the stimulation is still further increased, the body follows the movement of the head in the same plane and direction. Even now the movement can stop during the stimulation, while the head and the anterior part of the body retain the position of deviation until the stimulation is interrupted (Figs. 15.3–15.6). All this makes a completely natural impression; the posture is not forced and spastic. Finally if one goes still one step higher with the stimulus intensity, the *deviated position* goes over into a *continual movement* which stops only with cessation

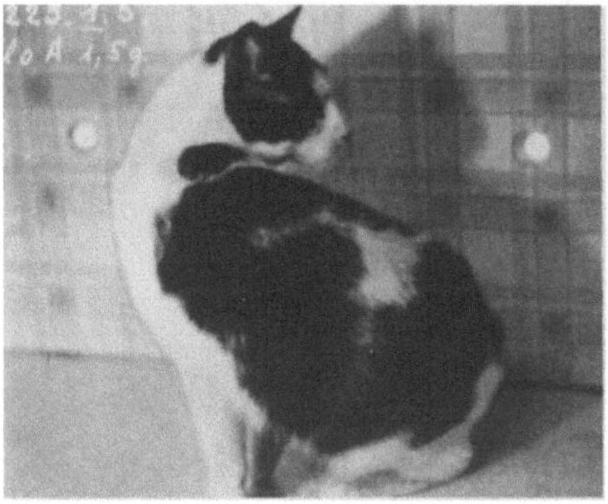

15.3

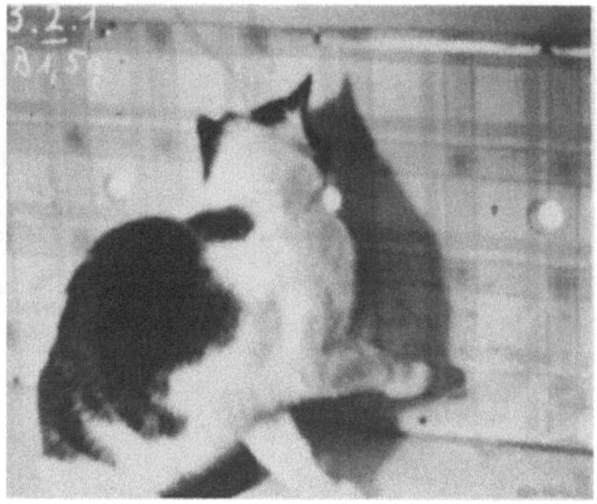

15.4

Fig. 15.3. Turning toward the opposite side
Fig. 15.4. Turning toward the stimulated side

of the stimulation. In pure cases, the movement remains within the same plane and direction. It is noteworthy that the *same stimulation* which causes a change in the tonic activity of those muscles controlling the posture of head and trunk may elicit phasic stepping movements of the extremities which cause turning of the body in a given plane.

A further aspect concerns the behavior of the eyes, which also can be involved in deviation from the primary position. Such effects are more striking when head and body are immobilized during stimulation. In the case of head rotation, for example (that is in the transverse plane), the eyeball preserved its initial position relative to the head. If one hinders, on the other hand, the movement by holding the head tightly, or by turning it back into the initial position, the eyes roll and remain "rolled" as the stimulation is continued. If we relate the effects of stimulation to the localization of the corresponding brain sites, it turns out that *topographically defined areas are responsible for the direction of movement and the related deviation.* This finding is of particular interest when we compare it with the organization of the motor zone of the cerebral cortex. Here we find the muscles, i.e., the muscle groups, represented according to their peripheral anatomical arrangement. The substrates responsible for the motor effects of the diencephalon are, on the other hand, geometrically oriented in their arrangement, i.e., relative to a coordination system through which the space in which the animal moves is defined. The substrate predetermines the *plane and direction* in which the eye-head-body axis is deviated to a new stationary position or under some circumstances is led to continued movement. More precisely, stimulation slightly lateral to the midline in the posterior third of the diencephalon, at a middle level, causes lifting of the head; with an increase of the stimulus the anterior body may rise until the animal falls backwards. Somewhat lateral from the position mentioned, an upward movement makes itself appar-

15.5

15.6

Fig. 15.5. Still picture from a film: rotation, stimulus left

Fig. 15.6. Raising of head, stimulus left: contralateral side precedes the other side somewhat in the upward movement

ent more unilaterally so that the head and the half of the body opposite the stimulus side are raised more strongly than on the side of stimulation. If we move more laterally, the contralateral head and body half alone are raised. At the same time the ipsilateral side is lowered so that a pure *rotation* occurs (Fig. 15.5). With increasing stimulus the latter goes over into a continued rolling movement. Through the choice of the site of stimulation, the movement may be placed in any desired plane between vertical-longitudinal and vertical-transverse. In the described effect, both halves of the brain complement each other as a unit and thus exert a perfect control over the two planes of space named. The vertical plane movement is only represented by *a lifting of the forepart of the body in the area described* (Fig. 15.6), while downward movement is elicited in a specific segment lying further posteriorly.

Rolling movements and upward movement in the sagittal plane have other common features. The latencies are extraordinarily short – less than $^1/_{50}$ of a second – and thus the movement thrusts triggered by single stimulation impulses (six to eight per second) are rather sharply separated from one another so that continued stimulation leads to sharply accentuated saccades. With prolonged stimulation, however, the movement thrusts gradually merge into one another so that a more or less smooth movement with deviation from the primary position takes place. The time-locked saccades and the very short latency indicate that the responding substrate stands in relatively *close relation* to the (peripheral) effector organs. Hence the more remarkable the demeanor of the experimental animal, for the artificially induced interference does not appear to be perceived as an imposition from outside sources, but is taken as a natural event. One recognizes this because the animal during the turning pays attention as usual to the events in the environment. After a rolling movement it is neither surprised nor dismayed; it acts as if nothing special were happening, that is, as if a normal regulatory mechanism were in play. So everything indicates that the phenomena described are to be interpreted as an expression of a displacement in the neurodynamic balance. In the contest of antagonistic innervations which maintain the normal posture, the electrical stimulation seems to affect the system of muscular forces by introducing additional excitations.

The same interpretation is also valid for movements in the horizontal plane which can be elicited from certain areas of the diencephalon. They differ from the movements in the vertical plane in that, despite the low frequency of stimulation, the turning of the head and body as a rule are smoother due to summation of the individual responses. This suggests that the relationship of the electrically stimulated substrate to the peripheral effector apparatus is rather indirect, i.e., that elements with pronounced summation property and a more tonic form of discharge are inserted between stimulation site and muscle. Consistent with this notion is the observation that movements in the horizontal plane have a relatively long latency, often several seconds, and that they exceed the period of stimulation. In contrast, the onset of movements in the vertical plane occurs almost immediately with the onset and breaks off suddenly with cessation of stimulation. A further peculiarity of the movements in the horizontal plane involves the topographical position of the stimulation sites, which are distributed evenly in a rostrocaudal direction through the diencephalon while the stimulus points controlling the vertical planes are grouped in foci. The former are not completely absent in the rostral area but there they are much rarer than the responses evoked in the posterior diencephalon. Also, higher voltages on an average are required in the rostral regions. From all these findings one must conclude that the "organizer" of the horizontal turning movement lies further rostral while for rolling and upward movements in the sagittal plane, *one* apparatus is responsible which at least, to a considerable extent, lies nearer to the midbrain and thus to the effector system. The relevant relationship between stimulus position and character of the motor effects is schematically shown in Fig. 15.7. Whether the stimulation affects cells of the "central," "precentral," or "postcentral" tracts is a question of its own. With respect to the diencephalon the answer is of little consequence because the different possibilities give practically the same results. With respect to movements in the horizontal plane the arrangement of the corresponding stimulus points along the already-mentioned fiber tracts is so manifest that doubt about their relationship to the motor effects cannot exist. Recent data obtained by combining brain stimulation with the Marchi technique are completely in agreement with this interpretation [8]. Two fi-

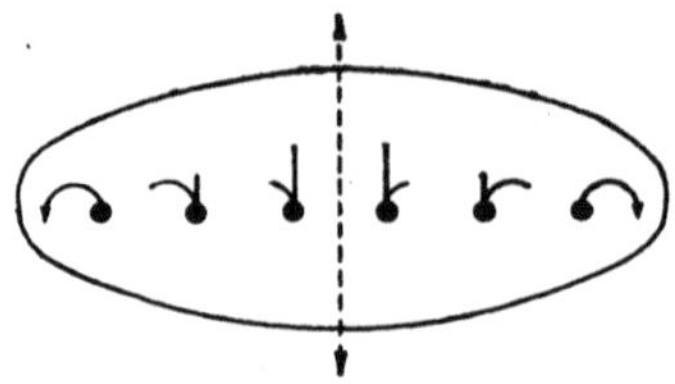

Fig. 15.7. Relationships between rotation and elevation. According to the localization of the stimulation site, rotation and elevation combine with each other in different proportions. The right half of the substrate causes a picture symmetrical to that of the left half. Through the interaction of both halves, head and body postures are fixed in a definite position. Explanation of symbols: ↷, rotation; ↓, head and anterior body upwards; ⤵, combination of rotation and upward movement. The *dashed line* corresponds to the midsagittal plane (of symmetry)

ber layers can be clearly distinguished, a deep one whose excitation brings about turning toward the contralateral side and a less deep layer whose effect induces ipsilateral turning.

With these observations we have characterized the main experimental observations which needed to be communicated here. To be added is only the fact that, depending on the stimulation site, complex movements may occur in which components from different levels work together. Thus, different types of combinations arise as resultants, such as turning with rotation or rotation with lifting of the head. Also noteworthy is the observation that each increase of the stimulus intensity leads not only to a potentiation of the motor effect but also to a widening of the stimulation area. New symptoms occur. They remain – as long as not too strong irradiating stimuli are employed – consistently oriented to movement in the same plane, i.e., in the same direction, a behavior which we wish to designate as "syntelic coordintion" (*telos*, target).

In order to obtain still closer insight into the mechanism of diencephalically induced movements, we also tested the effects when two separate points are stimulated. Here we chose the most logical paradigm, namely, simultaneous stimulation of symmetrical points, one of which, for example, elicits head-turning toward the right, the other head-turning toward the left. As a result of simultaneous stimulation, the opposing turning effects annul each other. It is not, however, that the two systems of muscles are simultaneously contracted so that the *peripheral effectors* are in competition, from which a more or less rigid middle position would result. On the contrary, the head remains in a flexible equilibrium despite the double stimulation. The stimuli are balanced at the level of the *innervating organization*, that is, they summate there with oppositely directed patterns and neutralize each other. The quantitative processing of the stimulation effects is reflected in the responses to a slightly asymmetrical stimulation which are clearly asymmetrical in the direction of the stronger stimulus. Additional insight is provided by the following observation: a stimulus which triggers lifting of the right forelimb is combined with a stimulus which – employed alone – causes lifting of the left forelimb. Simultaneous stimulation of both points results in mutual inhibition, i.e., the animal moves neither the left nor the right extremity. This behavior is the more striking because the responses of one extremity do not exclude responses of the others, as is the case, for example, with turning of the head toward the left and toward the right. Indirectly, however, the limb movements compete insofar as the body would lose support if both forelimbs were at the same time lifted from the floor. Thus, the postural mechanism is included; it even gains the upper hand, but only as long as one works with stimuli with low intensity. If the voltage is increased, the behavior changes. Now *both* forelimbs are lifted with the anterior part of the body, so that the postural support breaks down. From this it is clear that the artificial stimulus is integrated into the physiological play of excitations and that in the experimental situation one can demonstrate quantitatively that the "antagonism" between directional forces represents the guiding principle of motor organization.

b) Lesion Experiments. The experimental technique goes one step further when an elimination of the responding substrate is added to the stimulation experiment, since deficit signs represent a new parameter. The experiment is carried out in the following way: After the main stimulation effects have been observed and cinematographically recorded, one can use the two most active stimulation electrodes for the application of a high frequency coagulation current. Even with very weak currents the substrate lying between the electrodes is inactivated by the electrothermic effect. If it involves a substrate from which significant movements had been elicited, one can see well-defined impairments in the motor performance. Important deviations from the normal posture of the resting animal appear. This deviation is a mirror image of the response produced by stimulation. The *new* position is evidently stabilized. For, after having been passively moved into normal position, the head will immediately return to the deviated posture. This means that a component of the directional force which normally contributes to the excitatory balance for the maintenance of normal posture has been eliminated. As a consequence, a corresponding shift in the equilibrium of regulating forces has been established. This conclusion is strengthened by yet a further observation: the animal with deviation of the head in the horizontal plane can hardly turn it toward the right, while it can easily turn it in any direction to the left (compare Fig. 15.8a, b). Such behavior means that the diencephalic *regulation of posture* and the diencephalic *organization of movement* are implemented by the *same apparatus*. Actively maintained posture can also be understood in the physiological sense as an expression of a dynamic

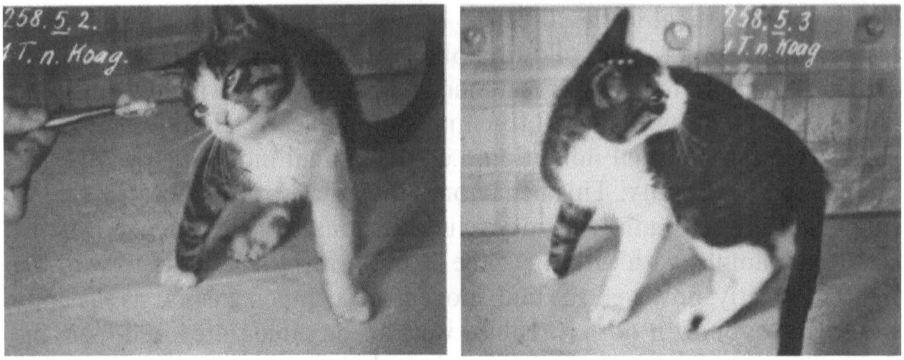

a b

Fig. 15.8a, b. Limited freedom of movement toward right *a*, but not toward left *b* (see text for explanation)

balance within the organizer, which represents potential forces mobilized by the body in the form of its muscles. At the level of the diencephalon it is vectorially represented. The pattern of excitation within this representative "field of force" projects into the effectors through which corresponding forces are released. The result becomes apparent to us as a controlled movement, i.e., as a regulated posture. The previously described behavior of the eyes is in complete agreement with such a concept.

Before evaluating the results of our experiments and reaching general conclusions, we should like to indicate that the *extremities*, apart from their integration into the regulation of posture, can also be influenced independently. For example, elevation of the contralateral forelimb is elicited from a circumscribed area. It should be noted further that the effective stimulation may spread to the *face* where movements of the lips, the whiskers, the eyelids, and the ears are seen. These results, which are explained in detail elsewhere, may interest the clinician, particularly with respect to symptoms of the so-called extrapyramidal motor system. In fact our observations and their interpretation seem to provide the *physiological aspects* of motor innervation, which up to now has been mainly the domain of clinical investigation.

So that no misunderstanding may arise from the last remark, the following comment is given: in the widest sense the *extrapyramidal system* should be conceived as the entire motor innervation apparatus which has its influence other than over the pyramidal pathways. There are authors who reserve the designation mentioned for a neuromotor apparatus which, like the pyramidal system, *originates in the cerebral cortex* but makes contact with the effector system via other pathways. Whether one accepts this definition in this or another form, the term itself implies that morphological criteria are to be used. The question remains open whether the classification made according to this viewpoint corresponds to an arrangement on the functional level. In a certain respect, in fact, this does appear to be the case. The final evaluation will only be possible when a sector from the entire motor system can be isolated in terms of its differential performance. It is in this sense that the following presentation is to be understood if the *physiological aspects* of the extrapyramidal motor system are to be grasped.

Regulated movements of an individual are predisposed by a well-defined subjective orientation toward the environment, transmitted by the sensory organs. A corresponding organization is, of course, required which is responsible for the active interpretation or the moving subject into the spatial coordinates or the field of action. This is required for the baseline situation – static of dynamic – from which an action develops. This integration, furthermore, taking place for each successive phase in the flow of a movement, is what we have in mind if we consider the framework of the entire motor system as one which establishes the *dynamic foundation* for the other motor system which organizes goal-directed movements.

By contrasting in this way the extrapyramidal to the pyramidal system at the *functional level*, we emphasize that the physiological organization, according to our concept, is a dynamic expression of the corresponding arrangement of the morphological substrate. Yet, nothing is said of its precise topographical localization. Nevertheless, the experimental findings and also the evidence from pathology seem to confirm the presence of the close relationship between the two aspects. In this direction something still should be mentioned: the above-described findings concerning the site of stimulation and the type of effect prove a systematic relationship between function and substrate, whithout disclosing the exact morphological arrangement of the organizing apparatus. While we are inclined to assume that the motor stimulation symptoms involve responses of fiber tracts, we admit that their relation to morphologically defined cell areas is a question for which the technique and the differentiated knowledge of *brain histology* must be employed to obtain an answer.

A Theoretical Model

In the first part of this review an attempt was made to analyze causal relationships in the motor performance of animals and man. As a basis for such an undertaking the physical realities had to be clarified. The goal was to gain insight into the patterning of muscular forces which must fulfill a given *dynamic function*. The most instructive approaches consist in observing free nature, and partly also in carrying out studies under experimental conditions. In both instances, cinematography provides useful information about the sequential order of movement. The following comments are intended to provide insight into the problems of the organization of move-

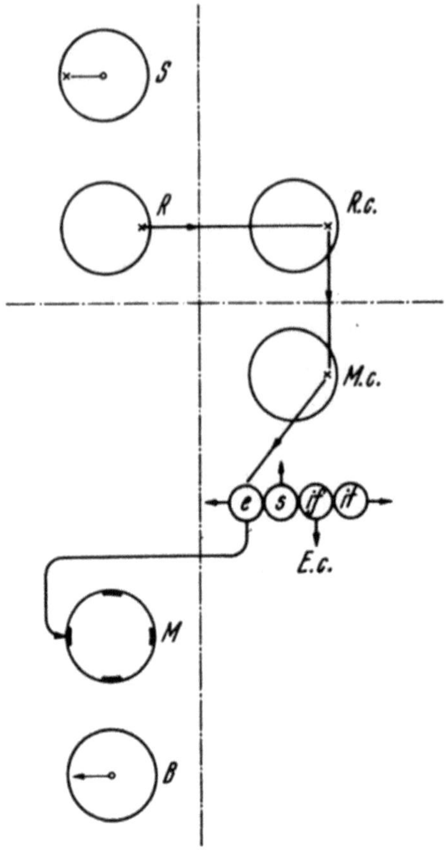

Fig. 15.9. Diagram of the transmission pathway for gaze movement: left eye to the left (oriented according to the direction of gaze!). At *X* a stimulus object appears suddenly in the lateral part of the visual fields *S*. It is projected (optically) to the nasal segment of the retina *R*. Through connections, a corresponding excitation focus in the central representation of the receptor system *R. c.* occurs. In the further process the analogously organized central motor field *M. c.* is activated with the corresponding connections to the effectors (*E. c.*, nucleus and subnuclei of the eye muscles). *E*, external rectus; *s*, superior rectus muscle; *if*, inferior muscle; *it*, rectus internus muscle. The forces of the external rectus are mobilized as effector link. The eye turns toward the left and fixates the stimulus object. This circuit operates with accuracy only from the primary eye position (gaze directly forward). Only the straight eye muscles have been considered here

ment in yet another way. We set ourselves the goal of understanding physiological performance on the basis of *a theoretical model* by which we seek to clearly reconstruct in detail relevant physical conditions for the biomotor system and to devise the scheme of an organization whose output is compatible with biological reality. Evidently, one can ask whether this technologically oriented approach has a place at all in biology. Let the outcome of this experiment decide. We are well aware of the dangers and yet conscious of the fact that nature itself uses this approach. If nothing else is gained, we shall at least obtain a deeper insight into the entire problem of the creation and evolution of motor systems.

Aside, perhaps, from the most primitive forms of movement, an organized pattern of muscle forces underlies the motor behavior of organisms. By means of combinations and temporal interlocking a spatially and temporally defined motor act is composed. These relationships are best illustrated by the gaze movement. The eyeball in its socket is movable in all planes of space. The moving forces are developed from three pairs of muscles which are oriented in all three planes. The biological performance consists in directing the optical axis onto an interesting object in the visual field.

In order to clarify the physical situation we shall give the following simplified account of the action of individual eye muscles.

A muscle attached to the eyeball on the temporal side and a muscle attached on the nasal side develop forces which turn the visual axis in the horizontal plane. The upper and the lower rectus muscles move the eye in the vertical plane, while the upper and the lower diagonal muscles rotate the eyeball about the axis of vision, and – with respect to the left eye and seen from *behind* (in the direction of the visual axis) – the first clockwise, the second counterclockwise. Through these three muscle pairs active movement in all three planes of space is possible. The release of the "potential" forces available in the muscles at rest occurs through nerve fibers that receive their innervation energy from cell groups which can be considered within the central nervous system (CNS) as "representatives" of these potential muscle forces.

The well-defined connections of the optic nerve between the circumscribed areas in the retina and the neural elements in the brain are also to be considered and can be thought of as a "projection" of the retina upon the CNS. The retinal elements themselves correspond, according to the laws of *optical physics*, with defined projections directed onto the visual field. The relationships described are presented in Fig. 15.9. In the upper left, the visual field (S) and the retina (R) are represented; a visually perceivable object (X) is depicted on the left in the former (S) and on the right in the latter (R). In the upper portion of Fig. 15.9 the central representation (Rc) is shown to be connected by a retinal projection pathway. In the lower half of the figure the effector apparatus is compared to the receptor apparatus in the upper half. The cell groups in contact with the four straight eye muscles (external, superior, inferior, and internal rectus muscles) are represented on the right. A stimulus object appearing in (X) of the peripheral visual field activates connections from the central motor field (Mc) to the cell group (e), which represents the peripheral motor apparatus (M), shown on the left. The consequence is an activation of the rectus external muscle and correspondingly the turning of gaze toward the left. Thus the stimulus

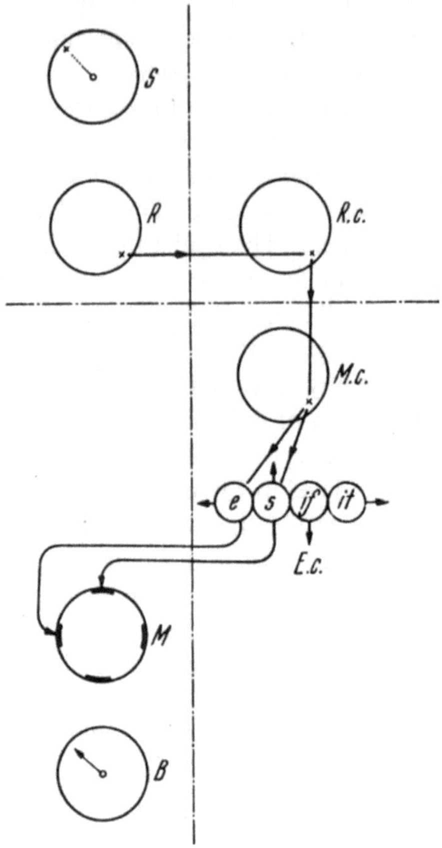

Fig. 15.10. Analogous to *Fig. 15.9.* A stimulus object appears in the lateral upper part of the visual field. The excitation flows through the geometrically and quantitatively structured transmission system into the external rectus muscle as well as into the superior rectus muscle. Muscle actions are combined according to the parallelogram of forces. The gaze turns out and up and fixates on the stimulus object

object (X) in the visual field is "visually grasped." It is important to note that this diagram is not hypothetical but represents the analysis of an actual functional situation which is valid for goal-directed visuomotor reflexes. Besides this, the diagram also has relevance for the design of a visually guided tracking system. In this context we should like to emphasize that the arrangement of all central connections, as indicated in the right half of the diagram, represents a functional and not an anatomical schema. So it is possible, for example, that the fields (Rc and Mc) overlap so that the elements of the receptor projection field (Rc) are in direct contact with the elements of the motor direction field (Mc).

A *rigid geometric arrangement*, on the other hand, must exist in the peripheral sector, namely between the position of receptors and effectors. On-

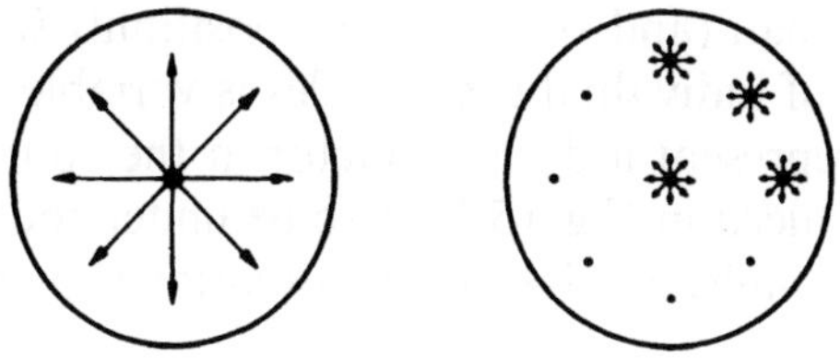

a b

Fig. 15.11. a Abstract representation of a field of gaze controlled only from the primary position of the eye; from this position objects in all possible directions can be sighted by the corresponding muscle combinations (=simple manifold). *b* Field of gaze in which the gaze can be turned in all possible directions from all possible initial positions (≕double manifold)

ly thus will the visual axis be guided with accuracy, especially when the stimulus object appears in the visual field somewhere between horizontal and vertical meridians (Fig. 15.10). Diagonal gaze movement requires, in this example, the proportional coactivation of two muscles, i.e., a change in the balance of forces to which all muscles continue to contribute. How the angle of deviation can be controlled will not be further described here [7]. However, another consideration is touched upon which further elaborates on the relationship between central organization and target-directed movements.

The functional scheme described in Fig. 15.9 is based on the assumption that the visual stimulus strikes the eye in its primary position (that is, forward gaze). In human and in visually more highly organized animals the eye muscle apparatus accomplishes more complicated tasks. An example is demonstrated in Fig. 15.11. The two disks represent fields of gaze. Figure 15.11a shows the organization of gaze *exclusively from the primary position*, the arrows corresponding to possible movements of gaze. Figure 15.11b represents the situation in which gaze can involve movements from any position of the eye in any possible direction. In order to understand what this greater complexity (double manifold) means for the central organization, we must recall the mechanics of the eye already discussed in this chapter. Figure 15.1 shows how the pull of a single muscle alters with changing eye position. Thus the external rectus muscle (Fig. 15.1a) develops an elevating component when the eye looks upward (Fig. 15.1b). Because the motor effect of each one of the six eye muscles is dependent on the eye position, the functional circuitry presented in Fig. 15.9, which is based on fixed relationships between the systems of receptors and effectors, no longer applies. For each baseline position of the eye, another muscle combination is called into action when a gaze movement in a given direction is required. The question arises as to how, in each case, the combination of muscle forces corresponding to the eye position may be selected. Here geometric reasons are decisive in that the *direction of pull* for each definite position of the eyeball (and only for this position!) *is clearly defined* although the action of individual eye muscles is variable. Thus the position of the eyes *must* represent a decisive factor in the coordinating function. The scheme reproduced in Fig. 15.12 is to be understood in this sense and represents a more detailed analysis of the situation shown in the right lower section of Fig. 15.9.

Instead of the simple relationship between the *central field* (Mc) and the individual muscles a *selecting* organ is introduced. Each functional unit in the central motor field maintains *a manifold of potential connections* with the representation of effector muscles. Whichever functional unit out of the multiplicity of possible impulse transmissions is to be activated in a concrete case is, according to what was said, dependent upon the momentary position of the eye, because the direction of pull of individual muscles is determined by the momentary position. In Fig. 15.12 it is assumed that a horizontal movement of the eye is induced by a visual stimulus when, by chance, it is looking directly upward. This position of the eye is represented centrally by a defined excitatory constellation, which is prerequisite to the activation of the appropriate effector pathway from the multiple possibilities. In the concrete case, the impulse is transmitted to the representations of the external rectus muscle (e), the lower oblique muscle (oi), and the rectus inferior muscle (if). Activation of the latter two muscles prevents that, with the eyes deviated upward, the elevating and rotating components developed by the external muscle come into play. For comparative purposes the connection by which the external rectus muscle would act, if the eye were in the primary position is designated by dotted lines in the model. We should like to emphasize once again that the *diagram* expresses geometrically conceived *functional relationships*, but it does not indicate how the connecting elements are topographically arranged in the brain. And again the diagram can serve as a model of a pertinent and, indeed, most simple technical device.

Other solutions are possible. *However, they must invariably contain the functional links described.* Biological research is committed to appreciating such insight as a challenge and as a guideline for continued physiological and morphological investigations on the mechanisms which organize motor performance.

These conclusions present us with still further problems. In man and in animals with a more highly organized visual system, the head is also involved in no small measure in the gaze movement. From the point of view of our organization scheme this means that impulses are transmitted to two effector systems arranged in parallel, i.e., the muscles of eyes and head. The fact that the topographical relations of the latter are not as simple and obvious as those of the former has little significance for the *structure of the central organizer*, since for the eye movements as well as for the head movements the requirements span and limit themselves to the motor control of the three planes of space. The functional diagram must include the splitting of the motor impulse deriving from a common motor center (Mc) into two parallel excitation patterns, which, according to the principle of the selecting switchboard, are separately transmitted to the two groups of effectors. A question in itself concerns the *reciprocal relationships* which may exist between the effector muscles of the two groups while the distribution of innervation energy is adjusted to changing circumstances (for example, resistance). We are satisfied to bring up these considerations without pursuing them here further, except for one point: the organization of target-directed head movements is more complex than that concerning the movements of the eye. While the eyeball can turn around its stabilized center of gravity, the center of gravity is as a rule displaced with movements of the head. In lowering the head, gravity adds itself with a positive sign while in raising the head it adds itself with a negative sign to the play of forces. The simple concept valid for the gravity-free field is further disturbed in that with changes of head position, displacements occur with respect to the eccentric support, and the muscle actions necessary for a *regulated positioning* of the head at rest or in movement differ according to the situation. In other words the innervation mechanism of gaze movement appears to be extraordinarily complicated. Yet, the diagram which we presented permits one to conceive only relatively simple situations. Basically the cooperation of two separately working organizational systems is involved, whereby the second reacts to the *stimulus of gravity* and is activated through the function of compensatory reflexes. In effect, this mechanism, by *neutralizing the target-alien forces*, sets the visually guided innervation mechanism into a gravity-free field. Owing to adequate design of the acceleration receptor mechanism, not only gravity but also an other kind of acceleratory force can be compensated for; as a result the accurate targeting of the visually guided system is guaranteed under all circumstances.

The question arises as to the functioning of two motor mechanisms which are controlled by receptor systems arranged in parallel. Various solutions may be considered. One is to have two separate organizations, each with its own muscle apparatus, whose actions would be superimposed; the forces brought into play would be combined independently at the effector level. A more interesting solution is the integration of the two systems at the central level, i.e., before the innervation is transformed into muscle action, but rather when motor impulses are processed by central cell groups whose excitations eventually activate the relevant muscles. This solution amounts to a superimposition of "*potential forces*," whereby the muscle apparatus carries into effect the centrally integrated innervation pattern derived from the two organizations. This second arrangement means simplification and therefore an advance. The performance of this organization is

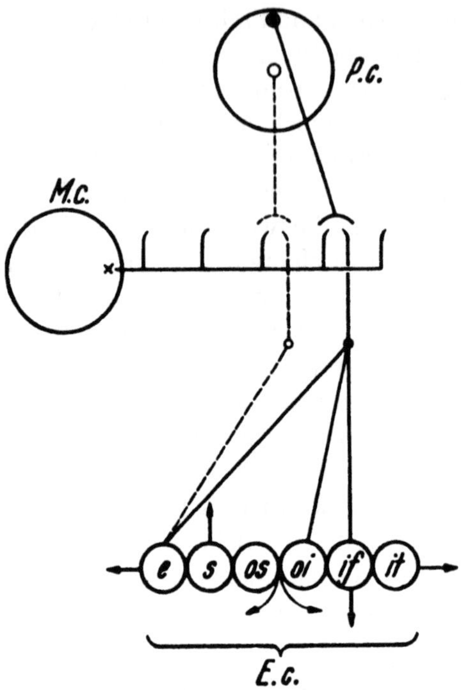

Fig. 15.12. Motor control of the field of gaze according to the principle shown in *Fig. 15.11b:* selective transmission of stimuli (by interposing a "selecting organ"). *M. c.*, central motor field, as in *Figs. 15.9* and *15.10*. The impulse arriving from the receptor system has potential connections to complex muscle combinations by means of branching of the excitatory transmission. The selective activation of a certain connection is determined by the position of the eyeball. Deviation toward the left is induced as in *Fig. 15.9*. The *dashed line* designates the path of the impulse with the eye in the primary position: only the external rectus muscle is mobilized. The *heavy line* designates the path of the impulse when the stimulus for turning appears with the eye looking upward: the impulse is distributed in conformity with the changed action of the external rectus muscle. Its elevating component *(Fig. 15.1)* is compensated for by the inferior rectus muscle *if*, the rotatory component is compensated for by the oblique inferior muscle *oi*. The resultant is purely leftward turning. In reality the respective antagonists are involved reciprocally; in each case the muscle apparatus acts as a whole in the sense of a dynamic unit. The effectors *e*, *s*, etc., are designated individually by *arrows*

perfect – if judged on the basis of the tracking ability – provided that the *summation of the innervation energies at the central point of convergence can be achieved quantitatively and in terms of vector analysis.* This model is of particular interest to us because its operation may well correspond in its manifestation with the motor performance of more highly organized animals and man.

Finally, a further thought is expressed in the sense of a *conclusion* and as a challenge for *further development*. Motor performance has attained its highest perfection in man and some animals and can be considered as an approach to understanding the organizing potentialities of the central nervous system. By clarifying the problems of motor innervation, we have attempted here to obtain an insight into the organization of the central nervous system. This approach appears to open a path for us which leads into still another field. The question is whether it is not the same laws which, in another context, control the *organization in the psychic domain*. Here as there, the functional factors are the same, namely the biological properties of neural elements arranged according to space and time. So it can be that for us the study of the biomotor system may also provide a valuable key to a successful scientific approach to psychophysiology.

Summary[1]

The prerequisite for a spatially accurate motor system is an organization of the movement forces which operates quantitatively in geometrical terms. The arrangements underlying the functional order may be best described by theoretical models.

The eye which is moved in all planes of space by three pairs of eye muscles is an ideal case for such a description. The organization is first demonstrated as a model which imparts to the eye freedom of movement in all directions from the primary position (single manifold). A second model describes the organization which provides freedom of movement for every possible initial position of the eye (double manifold). The organization demonstrated for the motor system of the eye is also valid for the motor activity of head and body. In the latter case the tonic innervation is transformed under some circumstances into a phasic mechansim of innervation (in the extremities). The various effector systems are arranged in parallel.

An organization has thus been conceived which, despite competition with gravity, leads the target-oriented forces to the goal without error. Disturbing factors – translatory acceleration forces of other origin and angular acceleration forces – are eliminated by control mechanisms derived from receptor systems which operate in parallel.

The synthesis achieved by the organization of goal-directed and posturally stabilized movements may be implemented at different levels; the forces arising from separate effector systems combine according to the parallelogram of vector forces. Integration of force representation at central levels leads to the same result as long as processing is based on the principle of vectorial analysis.

Organization at the level of the representation of forces opens a wide range of combinations. Only the biologically important ones, i.e., those act-

1 Taken from [7]

ing on the rotational planes of specific joints, have been developed by nature. This reduction means economy.

The anatomical arrangement of the representation of forces is irrelevant for the functional organization; the same is true of the connections. The geometric relationship between the direction-specifically oriented receptors and the direction-specifically performing effectors remains decisive.

Automatic postural regulations are integrated with volitional movements so that the course of the voluntary performance is stabilized against competing forces and impeding reflex mechanisms. The role of recoil forces is clarified; their position in the organization process is discussed and the relationship to the development of individual movements (learning process) is formulated. Space perception is organized in parallel with motor performance.

Results of previously reported experimental findings are in clear agreement with the model treatment of motor organization. The "functional model" is expounded as an aid in dealing with biological order and exemplified by specific prototypes of motor organization.

References

1. Hess, W.R.: Beiträge zur Physiologie des Hirnstammes. I. Die Methodik der lokalisierten Reizung und Ausschaltung subkortikaler Hirnabschnitte. Leipzig: Thieme 1932
2. Hess, W.R.: Zwischenhirn und Motorik. Pflügers Arch. Ges. Physiol. Menschen Tiere *243*, 634–650 (1940)
3. Hess, W.R.: Motorische Symptome im Gesicht bei Reizung im Zwischenhirn. Pflügers Arch. Ges. Physiol. Menschen Tiere *243*, 678–685 (1940)
4. Hess, W.R.: Diencephale Reizsymptome am Körperstamm; Beziehungen derselben zu Bewegungseffekten im Gesicht und an den Extremitäten. Pflügers Arch. Ges. Physiol. Menschen Tiere *243*, 741–747 (1940)
5. Hess, W.R.: Charakter der im Zwischenhirn ausgelösten Bewegungseffekte. Pflügers Arch. Ges. Physiol. Menschen Tiere *244*, 767–786 (1941)
6. Hess, W.R.: Experimenteller Beitrag zur Frage der extrapyramidalen Motorik. Z. Ges. Neurol. Psychiatr. *172*, 639–642 (1941)
7. Hess, W.R.: Die Motorik als Organisationsproblem. Biol. Zentralbl. *61*, 545–572 (1941)
8. Hess, W.R., Bartorelli, C., Bucher, V.: Diencephale Motorik und Marchi-Degeneration. Z. Ges. Neurol. Psychiatr. *174*, 340–352 (1942)

Chapter 16

Teleokinetic and Ereismatic Mechanisms* and Biomotor Functions

W.R. Hess

In connection with the analysis of the "extrapyramidal" responses to electrical stimulation of the brain stem, the need was felt to study the physical aspects of biomotor functions, because it was thought that the central organization can only be properly understood if the system of forces underlying any movement is correctly grasped.

The following is a brief contribution to the subject:

1. Goal-directed movement develops by virtue of the forces of propulsion. The activation of the proper muscle groups corresponds to coordinated action organized in terms of a *teleokinetic* innervation.
2. Particular circumstances are attendant upon the fact that we are subject to the laws of gravity. The course of movement is deformed, which can be avoided by the use of an appropriate group of forces resulting from antigravitational innervation (Sherrington, Magnus, and Hess).
3. Every goal-directed action elicits a reactive force (same intensity, opposite direction). Examples: throwing a stone or a ball or a gesture with the hand.

Teleokinetic innervation is successful only when the recoil forces are intercepted. A particular system of forces is in operation, providing support. It is not rigid but flexible; i.e., it adapts itself to the teleokinetic system of forces according to the situation.

Without this supportive group of forces, any directed movement would break down. As this does not happen, it is obvious that these forces intervene. The reactive mobilization of this dynamic system requires a particular central organization. In view of this supportive function we propose the term *"ereismatic"* innervation system (*ereisma* = support). It is an integral part (defined by the fulfilment of a particular task) of a functionally nonuniform extrapyramidal innervation.

---

* Hess, W.R.: Teleokinetische und ereismatische Kräftesysteme in der Biomotorik. Helv. Physiol. Acta *1*, C62–C63 (1943)

a d
b e
c f

Fig. 16.1a–f. Goal-directed movement and its dynamic foundation. At the *left*, *a–c*, jump from the shoulder; success with dynamic support; at the right *d–f*, failure without dynamic support (original still pictures from a slow motion film). *a–c* Preprogrammed jump. The supporting man is informed of the imminent jump. In anticipation of the jump, he leans forward and counters the recoil. The jumper reaches the expected goal. Teleokinetic and ereismatic innervation are coordinated. *d–f* Goal-directed jumping without dynamic support. The supporting man is *not* informed of the imminent jump. Because of the unforeseen recoil he tumbles backward and the jumper falls short of the attempted goal. Analysis of the mechanical forces. *Left series:* repulsive shock is absorbed, i.e., compensated for. Compare the forward leaning of the supporter in *b* and *c* with *a*. The posture of the jumper in *b* and *c* is goal directed. The pole in *a* points to the target. *Right series:* in *d* the assistant fails to counteract the repulsive shock in time. The jumper falls short of the target *e,f*. Compare the jumper in *b* (correct performance) and in *e* (failure). (Editor's note: The man pointing to the target with the pole is W.R. Hess.)

a b

Fig. 16.2a, b. When throwing the mass of a large stone the inertial compensation forces are transmitted over shoulder, back, hip joints, and legs to the platform. The action causes a recoil of the support. Shortly after onset, the reaction forces are counteracted by an assistant so that the projectile almost reaches the goal (original from the slow-motion film). *a, b* Throwing a projectile forward to the ground. *a* The muscles of the right arm and the legs (with a stiffened vertebral column) are tensed for the throw; *b* pattern of a goal-directed synergistic coordination of almost the entire body musculature. The entire body is thrown forward together with the arm whose mass is accelerated by the momentum of the arm musculature. The supporting mechanisms add to the accelerating forces: by being collaterally activated they contribute to the system of forces released in the goal-directed performance. Note the sliding of the mobile platform. The recoil is transmitted to the platform. The left leg of the assistant actively compensates for the recoil which is transmitted via the body of the stone thrower to the platform

A slow-motion film [1] has been prepared showing the two systems and the two types of motor forces which are produced by means of two examples: goal-directed jumping and stone-throwing. Still pictures taken from the slow motion film are shown in Figs. 16.1 and 16.2.

Goal-directed jumping: (a) One man stands on the shoulders of a second. He attempts to jump on a given mark (target). The man underneath is supported by a man standing behind him, who holds and supports him so that he is now able to withstand the recoil. The man on top can unfold his propulsive force without hindrance and lands on the target (Fig. 16.1 a–c). (b) The man underneath falls backward and the man on top misses the target (his jump is too short) (Fig. 16.1 d–f).

Stone throwing represents, by physical necessity, an analogous situation. The recoil forces are shown by using a mobile platform on which the goal-directed movement is to be performed (Fig. 16.2).

1 Schematic representations of this movie appeared in later publications [1, 2]

References

1. Hess, W.R.: Cerebrale Organisation somatomotorischer Leistungen. I. Physikalische Vorbemerkungen und Analyse konkreter Beispiele. Arch. Psychiatr. Nervenkr. *207*, 33–44 (1965)
2. Jung, R., Hassler, R.: The extrapyramidal motor system. In: Handbook of physiology, Sect. 1: Neurophysiology, Vol. II. Field, J., Magoun, H.W., Hall, V.E. (eds.), pp. 863–927. Washington: American Physiological Society 1960

Chapter 17

Motor Functions of Tectal and Tegmental Areas*

W. R. HESS, S. BÜRGI, and V. BUCHER

Contents

Introduction

Man and higher mammals have the ability to carry out astonishingly differentiated motor acts. To the fore stands locomotion, a seemingly relatively simple performance, in reality, however, the outcome of a complex

* Hess, W.R., Bürgi, S., Bucher, V.: Motorische Funktionen des Tektal- und Tegmentalgebietes. Monatsschr. Psychiatr. Neurol. *112*, 1–52 (1946)

and finely regulated interplay of forces. As a prerequisite to locomotion, but pertaining to it in a dynamic context, is the active starting position. The initiation and maintenance of this spatially oriented posture set relatively high demands on motor innervation, for we are here again dealing with the delicately balanced action of a great number of muscles, whereby the interplay and counterplay of directional forces ensure fixation of the various joints. Regulated posture and movement of the eyes serve to maintain visual contact with the environment, and in connection with this function, the postural mechanisms governing head and trunk as well as the support provided by the limbs should be mentioned. Conditions shift and change as locomotion sets in, and muscle activity begins to alternate in cycles. Propulsion then is the integrated result of all the forces deriving from a sequential pattern.

Maintenance of an active starting position or during locomotion increases in complexity as soon as disturbing forces from without exert their action on the body. If the posture adopted or locomotion nevertheless continues, we have proof that the balance of forces does not merely rest on a simple, rigid innervation scheme, but is capable of continuous adjustment to changing conditions. Examples of this are walking on an unstable surface, standing or walking against gusts of wind, or readjustment of poise when the body is exposed to more or less vigorous blows.

Having thus penetrated somewhat into the physiology of the motor system we are now faced with an *innervation problem* as theoretically interesting as it is of practical importance. For the muscles are merely acting as a means of supplying dynamic balance with mechanical forces, a balance which finds its expression in active posture or locomotion. These forces come into coordinated action under the command of a pattern of nervous impulses representing the centrifugal projection of a well-ordered centrally operating scheme of excitations. The latter stands under the shaping influence of afferents arising from peripheral pressure and tension receptors. The task and mode of operation of the nervous "steering" apparatus directing skeletal muscle activity have thus been briefly outlined.

Analysis of experimental results by eminent workers has revealed that simple and ever more complexly organized reflex mechanisms form a hierarchical innervation system. Numerous experimental investigations have been carried out in recent years, the results of which have already been reported [23–26, 34] (see translation of [26] in this volume, p. 223) with the idea of extending our knowledge of this organization and its functional implications to higher levels, namely the diencephalon. The present paper completes and extends these studies. However, before broaching this subject it may be advisable to present a *short survey of pertinent experimental results.*

We begin with the well-known fact that even the simplest responses to a stimulation depend on the presence of a specialized receptor, conduction by at least two nerves (one sensory, the other motor), and an effector (the muscle). If we first examine the effector elements, we see a single nerve strand derived from a motoneuron, each strand supplying 30–150 muscle fibers. This *basic motor unit* obeys the so-called all-or-none law by which a stimulus, according to the prevailing situation, evokes a maximal contraction or no response at all, exactly like the nematocyst of *Hydra*, which responds with maximal explosion or remains closed as soon as its sensory hair is touched. In other words, the power of quantitative adaptation does not exist in the basic motor element.

However, adaptation of a dead-certain response to stimulus strength is of vital necessity in higher organisms. This fundamental property is indeed a physiological fact. Its organization is guaranteed by the specific conditions which prevail at the site where excitation is transmitted from one conducting element to the other, i.e., the *synapse*, whose properties have been made known to us by Sherrington [51, 52]. The gradation of responses rests on the fact that, depending upon stimulus strength, a varying number of basic motor elements are excited, that numerous subthreshold stimuli become effective by summation at the synapse, and finally, that competition between synchronous responses results in mutual facilitation or inhibition. Synaptic organization even at the spinal level plays a predominant role in every posture, in the form of the tonus at rest, a phenomenon which is due to excitations arising in the muscles themselves and alternately transmitted to various motor units. This weak tetanic contraction manifests itself electrophysiologically by the appearance of sharp asynchronous "platoon-firing" discharges [56].

It is this sustained tension which occurs under optimal, i.e., most economical, conditions when agonists are activated and antagonists are simultaneously inhibited. This mechanism obeying the general principle known as *reciprocal innervation* was discovered and analyzed by Sherrington [51, 52]. The inhibitory effect on those muscles which disturb the economy of motor reactions is moreover found not only in spinal reflexes, but also, for instance, during stimulation of the prerolandic area [16, 51]. The *innervation of the agonists* should be viewed in the light of *functional* rather than anatomical criteria. Sherrington early pointed out that the sum of the nerve fibers found in the ventral root in no way corresponds to a "reflex figure." In a later work [52] he specifically studied the problem in connection with the flexor reflex and found that, under certain circumstances, an anatomical antagonist is also innervated when contraction thereof essentially sustains the action of the agonist. Reciprocal innervation considered in the broader context of coordinated achievement does not always represent an optimum, e.g., when an extremity should provide support during a motor performance. *The various innervation patterns*, depending on the type and aim of the movement, have been the object of electrophysiological studies by Foerster and his school in Breslau [2, 56–58]. To mention only the following results: only the agonists are innervated for maintaining posture against sustained forces; in almost all phasic movements, agonists and antagonists work alternatingly except when the need to protect a joint arises. Reciprocal innervation is best seen in casual, to-and-fro movements carried out at moderate speed, during which elastic forces and inertia are fully exploited.

In this connection, we should like to mention the electrophysiological analysis of movement by Scherb and Arienti [50], which has led to significant applications in the clinical field.

If we have so far considered motor manifestations under the aspect of stimulation from without, we should, to be complete, add that the *impulses* also come *from within*. Notable preliminary studies in teleosts have been carried out by von Holst [36], who pointed out that the hypothesis of "chain reflexes" does not suffice to explain the periodicity of locomotion (i.e., the coordinated phase-shift between fin movements, relative coordination of fin movements); instead he postulated a central automatic control which generates a *genuine* rhythm (besides reflex control responding to external stimuli).

Motor performances, considered as a whole, normally do not evolve under steady-state conditions, for they are constantly under the influence of *more or less considerable disturbances*. In order to ensure their well-aimed and smooth execution, the central nervous system disposes of a series of additional mechanisms. In the first place come the *Eigenreflexe* studied by Hoffmann [35], through which a sudden increase in resistance is parried by a corresponding reflex increase in tension. All muscles exposed to the possibility of such exogenous disturbances have this mechanism. However, according to Hoffmann, it seems to be lacking in the eye muscles and in the diaphragm. A similar device, predominantly but not exclusively existing in the so-called antigravity muscles, cancels out reflexly any gradual, i.e., not too sudden, increase of resistance. These reflexes are the "myotatic reflexes" demonstrated experimentally by Liddell and Sherrington [38, 39] (stretch reflexes [17, 18]). In contrast to the *Eigenreflexe*, we are here dealing with tetanic contractions which are so finely graded that only the muscle segment concerned may contract during partial stretch.

To the mechanisms subserving compensation of external disturbance belong also all those concerned with an adequate starting position of body and limbs. It is mainly to Magnus and his co-workers that we owe our understanding of these mechanisms which depend on the integrity of the medulla oblongata, the mesencephalon and, to a certain extent, the thalamus [40–42, 47]. Magnus (as a pupil of Sherrington), proceeding from observations made on spinal and decerebrate animals, asked himself what additional performances appear when the brain stem levels remain in connection with the spinal cord. His experiments revealed a group of "reactions," i.e., reflex mechanisms, responsible for stance and maintenance of the upright position *(Halte- und Stellreflexe)*. The suspended spinal animal can carry out coordinated running movements, but cannot stand; the medullary animal steadily maintains its exaggerated extensor tonus in all positions and cannot get back on its feet alone; in contrast, normal distribution of muscle tone is found in thalamic and mesencephalic animals during standing, and these preparations are able to right themselves without help. These functions are mediated on the one hand by the *vestibular system* and, on the other, by *proprioceptive afferents* from the limbs and body, in particular the neck region. The corresponding central organization reaches from the medulla oblongata toward thalamic levels. In higher animals visual influences mediated by the cerebral cortex play a decisive role.

Depending on the primary (starting) position the movement is carried out by quite different types of innervation patterns. To cope with the situation is the main task of *proprioception* and its central representation, which must comprise a complex *switchboard system* whose significance was already underlined by Magnus and has been discussed by us elsewhere [27, 29, 30] (see translation of [29–30] in this volume, p. 243). Several components of the proprioceptive regulation derived from muscles and joints have already been mentioned. They adjust normal muscle tone, trigger the *Eigenreflexe*, and are in great measure responsible for righting reflexes. Here too belong the "autogenctic inhibition" of Fulton et al. [19], i.e., the muscular self-inhibition arising during movement, which ensures that reflex contraction does not ensue too abruptly, a reflex control characteristically suppressed by posterior root section [2]. As far as circuits arc concerned, von Uexküll [55], Sherrington [51, 52], and Magnus [40–42] and others have shown that the specific responses largely depend on the momentary situation of the effectors, which means that the condition of the periphery of the body, in one way or another, is constantly reflected in the central organization and controls the corresponding switchboard operations.

All these types of proprioceptive and vestibular influences must undergo central processing for coordinated performances to occur. One of the fundamental tasks of the cerebellum, the head ganglion of the proprioceptive system, in Sherrington's own terms, is to ensure this equilibrium. It is significant that Rademaker [47] noted that all standing and righting reflexes still existed in a dog deprived of the cerebellum, although they were modified in form (and were hence not organized in this part of the brain). The same remark holds for the *Hinkebein*, *Schunkel*, and *Stemmbein* reactions described by these authors, which involve reflex adaptation of the paws to changing inclination of the underlying surface. Thus, passive abduction of the supporting limb induces active stretching and abduction of the free leg (*Schunkel* reaction). An attempt to push the animal to one side induces an increase in muscle tone of the corresponding limbs (*Stemmbein* reaction). Displacement of one paw placed in midposition results in a decrease of supporting tone, and if the abduction is pronounced, even in sagging of the limb concerned and vice versa (*Hinkebein* reaction). It is clear that these reflex mechanisms must be coordinated in degree and time if a sure footing on an unstable base is to be achieved. These reactions, as already pointed out, still exist after cerebellar ablation, but they are now carried out abruptly, hypermetrically, and with delay.

In our analysis of reflex performances, we have now reached the point where we can direct our attention to the data assembled by means of electrical stimulation [21–26] of the *diencephalon.* These studies were carried out in the *freely moving* animal with *intact central nervous system.* To the fore stand deviations of head and body, out of the primary, i.e., normal posture. Especially, rotation of the head, upward or downward movements, and turning to the right or to the left were obtained during stimulation. Histological examination revealed that the type of response *depends on the site of the stimulating electrode.* What is more, *definite zones* within the motor response area stand in systematic relation to movements carried out in the various *planes in space.* Stimulation in the fields of Forel, for example, evoked a movement either in the transverse or in the sagittal plane, according to whether the electrodes lay in the medial or in the lateral parts of the field; intermediate sites yield a vectorial combination of the responses [26]. Rotations in the transverse and sagittal planes are strictly bound to onset and cessation of stimulation, while turning the head in the horizontal plane – the response may culminate in circling movements – appears with latencies ranging from one to several seconds and may outlast the stimulation. In view of cortical motor innervation, it is of interest that the head and body are set into movement from *one and the same site.* The extent of the stimulation effect, i.e., the spread of the response from head to the forepart of the body and finally the whole body, depends merely on *stimulus strength.* Whereas the cortical motor zones are arranged according to muscle groups, the morphological ordering here corresponds to *spatial coordinates* of movements. *Rotation* around the sagittal axis (transverse plane), in its pure form, is symmetrically represented and consistently elicitable from the fields of Forel, and even better from the intermediate portions of the ventral thalamic nucleus. During rotation the head is inclined to the ipsilateral side.

Upward movement of the head, in some instances combined with raising of the body, is in its purest form evoked from ventral parts of the prerubral field (capsule) and posterior hypothalamus (see Figs. 17.1 and 17.2). *Downward movement* of the body in the sagittal plane is produced from a field comprising the posterior commissure and adjacent areas. This effect was never evoked from more rostrally lying levels. *Turning movements* evoked

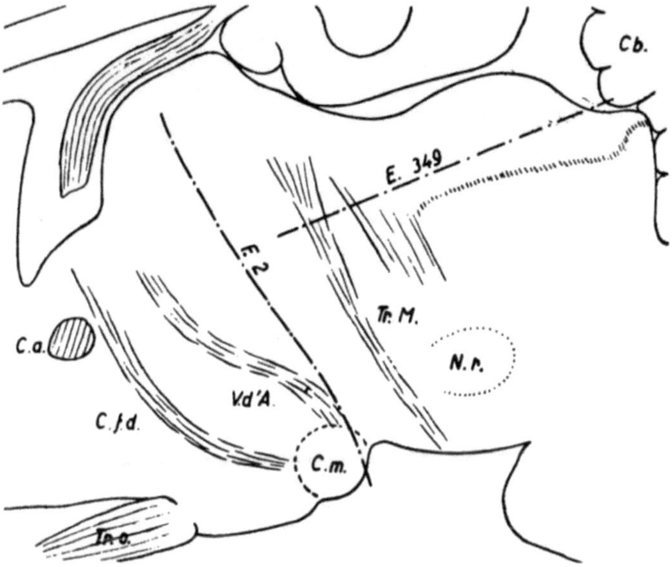

Fig. 17.1. Schematic drawing of a sagittal section through the di- and mesencephalon showing planes of sections F. 2 and E. 349 in *Figs. 17.2 and 17.4*, respectively. *C. a.*, anterior commissure; *Cb.*, cerebellum; *C.f.d.*, fornix; *C.m.*, mammillary body; *Tr. M.*, habenulo-interpeduncular tract (Meynert); *Tr.o.*, optic tract; *V.d'A.*, mammillothalamic tract (Vicq d'Azyr); *N.r.*, red nucleus

from the diencephalon fall into two groups: ipsilateral and contralateral deviations. If the stimulus strength is increased, the movement spreads to the forepart and then to the whole body, ending in circling movements. The direction, however, is not due to chance, but depends on the stimulation site. In this connection, the following should be pointed out. Deviation to the *side of stimulation* was evoked from the tegmental reticular formation dorsolateral to the red nucleus and could be followed into caudo-intermediate parts of the thalamus. Deviation to the side *opposite stimulation* was evoked from the zona incerta, but also sporadically from the tegmental reticular formation.

The histological evaluation of a great number of experiments has revealed that we are dealing, as far as turning to the left and right is concerned, with fiber tracts which connect infra- with supradiencephalic parts of the whole motor coordinating apparatus. This conclusion is based on the observation that turning movements – in contradistinction to rotations about the longitudinal and the transverse axis – arise with some delay and continue for a few seconds after cessation of stimulation. The typical reflex character is hence missing. A further difference noted between the deviations carried out in the sagittal and the transverse planes and those executed in the horizontal plane is that, in the former, a series of low-

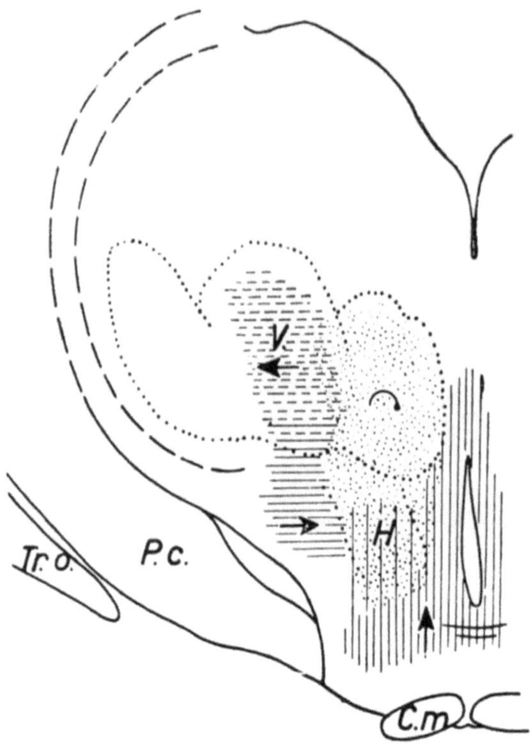

Fig. 17.2. Diencephalic areas from which stimulation evoked upward movements *(vertical lines)*, rotation *(dotted)*, turning in the horizontal plane to the side of stimulation *(broken lines)*, and turning to the side opposite stimulation *(horizontal lines)*. The drawing represents a frontal plane corresponding to F. 2 in *Fig. 17.1*. *V*, ventral nucleus of the thalamus; *H*, H field of Forel; *P.c.*, cerebral peduncle. *Tr.o.*, optic tract; *C.m.*, mammillary body

frequency stimuli (eight per second) give rise to saccades whereby the separate responses add up to give way to a continuous movement (summation effect). In contrast, separate movements are barely noticeable at the onset of turning in the horizontal plane (i.e., about the vertical axis) and soon give way to a smooth motion. This difference in character is worthy of attention; whereas the driving of the response by the stimulus rhythm is more pronounced the closer the electrical stimuli are to the effector, the merging into a smooth movement points to a "precentral" site of the substrate. Striking examples to support this interpretation have been found in our experimental material. As reported elsewhere [22], on several occasions long electrodes were used and the tips sometimes lay close to the ophthalmic branch of the trigeminal nerve. The reflex response to a series of eight impulses per second was then conspicuously smooth. In contrast, only minimal summation (probably mechanically induced) was seen in the eyes when the electrode tips lay more rostrally and laterally and thus not far from the oculomotor nerve fibers. Analogous movements in saccades (reflecting the stimulus frequency) were also observed at more caudal levels when stimulation involved the trochlear nerve, eliciting rotatory movements of the eyes. If we apply these criteria to the data produced by electrical stimulation of the diencephalon, we may conclude that deviations in the sagittal and transverse planes represent a more direct motor effect than does turning in the horizontal plane. We can complete the relation between localization and the stimulation response by additional observations which throw light on the dynamics of center function. To these for example belongs the observation that simultaneous stimulation of symmetrically lying sites abolishes the response produced by stimulation of one side alone, not through opposing muscle forces, but exclusively through *intracentral processing*. This conclusion is based on the fact that no bilateral increase of muscular tension could be observed (which would have fixed a mechanically induced equilibrium). Instructive, too, were the experiments in which stimulation was followed by electrocoagulation in the same area (i.e., by passing a high-frequency current). The immediate response to destruction was deflection of the head from its normal position, the direction being the reverse of the one produced by stimulation. The deviation resulting from circumscribed destruction proves that the elements mobilized by stimulation were in a *normal state of tension*. The experiments demonstrate at the same time that the arrangement of the diencephalic elements governing the motor performance of the axial musculature organizes the innervation of the effectors with regard to the spatial coordinates.

The survey just presented shows that a gap exists between the *results of diencephalic stimulation* and the concepts of Magnus and co-workers [40–42]. Above all, more knowledge is required about the tectal and tegmental fields. In order to reach this goal it is necessary to study these regions by near-threshold stimulation and circumscribed destruction, as well as by means of an analysis of the fiber connections arising from, ending in, or traversing the area of stimulation. This prompted us to extend our exploration of the diencephalon to include the more caudally lying structures of the brain and to complete the study by systematically applying the Marchi method. The following pages will deal with some of the results obtained.

Results

1. Turning in the Horizontal Plane to the Side Opposite Stimulation. Stimulation of the tectal area in many experiments produces a characteristic stereotyped response beginning with turning of the eyes *to the side opposite stimulation* (where they then remain) as soon as the stimulus is applied. The effect spreads almost immediately to the head, which turns to the

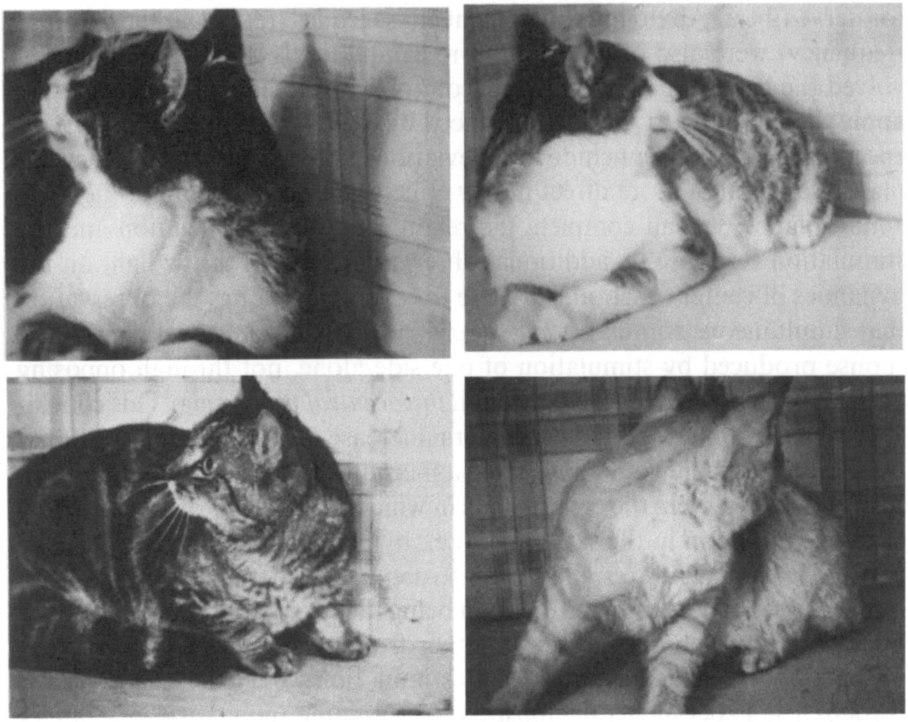

Fig. 17.3a–d. Turning in the horizontal plane to the side opposite stimulation evoked from the tectum opticum. *a* Stimulation on the left side elicits deviation to the right side (experiment 317, 1 V); *b* stimulation on the right side elicits deviation to the left (experiment 317, 1 V). Note the symmetry of stimulation effects elicited from symmetrical sites. Further examples of turning to the side opposite stimulation are shown in *c* (experiment 339, 1.5 V) and *d* (experiment 338, 2 V) with marked backward turning of head

same side as the eyes and even backward. The eyes, in the meanwhile having returned to their primary position, gaze in the direction determined by the head (Fig. 17.3). Cessation of stimulation does not, however, produce instantaneous return of the head to its initial position, for it "gets caught" (as we say) for a short time in its artificially induced pose. Turning of the eyes and head carried out *exclusively* in the horizontal plane is rarely seen (experiment 338, 339). The movement is in most cases associated with a deviation in the sagittal plane, i.e., an upward movement of the head, and less often, a downward movement. Variations will be dealt with later. It should, however, be pointed out that a more or less pronounced rotation of the head (in the transverse plane) sometimes evoked during tectal stimulation only arises as a weak *secondary*, i.e., mechanically induced accompanying symptom and hence not as a direct stimulation effect (which will be dealt with more fully later). In its typical form following weak stimulation (maximum 2 V), we were able to elicit a contralateral gaze response from over 20 stimulation points in nine experimental animals.

Further symptoms *accompanying deviation of the head* are motor responses affecting face or forelegs. Typical for this is a vibrissae response, the ipsilateral vibrissae moving forward, the contralateral ones backward during turning, so that the plane passing through the vibrissae is turned with the direction of gaze. Even the ears can move correspondingly, often only the contralateral ear. Lid responses are easily evoked: lid closure is obtained from many points in the brain stem; *lid opening*, however, is usually elicited from the tectum, and this as soon as stimulation sets in. Other effects, mainly concerning the contralateral half of the face (lips, cheeks), may be present but not as a general rule. Often seen, in contrast, are accompanying effects of the forelimb, especially the contralateral one. As has been said, all the motor effects associated with turning of the head to the side opposite stimulation can, however, arise as single responses (experiment 316, 317, and 338) – above all those of the vibrissae. Latency in general is short and the threshold many times is low.

The responses begin with eye movements, spread to the forepart of the body and, if stimulation is increased or lengthened, may finally lead to a true circling movement (experiment 316, 319, 321, and 339). The head is held strikingly high and the eyes at the same time are opened wide. It should be pointed out that the initial posture of the animal has no decisive influence on the response; eye and head deviations in principle are executed in exactly the same way, whether the animal is standing, sitting, or even lying on its side. During experimentation and analysis of the motion pictures, one gets the impression of a synergically coordinated "compact" performance. Most interesting was the observation that spontaneously induced movements may interrupt the motor effects evoked from the tectum. We have observed several times that the animal at first carried out the turning induced by stimulation, but that the response, probably owing to the tendency of the alert animal to fix its gaze, may then be modified because the eyes happen to remain fixed on some outstanding object in the direction of gaze. If the head is immobilized during stimulation, only conjugate deviation of the eyes takes place, and this in augmented form. The turning can, however, fall out, for example, when the animal fixes its gaze on the investigator. It is significant that the eyes turn to the ipsilateral side the moment stimulation is interrupted, which shows that the tectally induced response was contending with antagonistic innervation.

Finally, it should be pointed out that the response at onset is usually carried out in saccades, i.e., that the response to each stimulus impulse is a slight jerk. These saccades are not followed by a return phase, for the deviation is maintained until the next motor impulse arrives. Thus, marked summation of the stimulus effects prevails. If stimulation is continued, deviation to the side increases and may culminate in a backward turning of the head carried out in saccades or smoothly. The animal finally remains in this stimulus-induced posture, and only minimal twitches allow one to

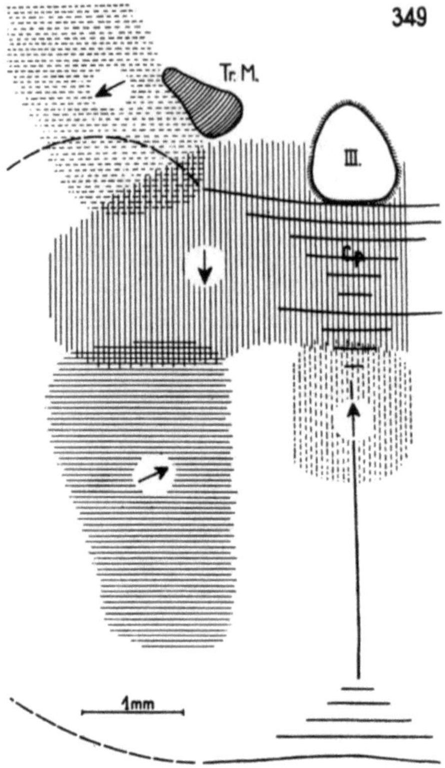

Fig. 17.4. Tectal responses represented in horizontal section through the superior colliculus (plane E. 349 in *Fig. 17.1*). *Horizontal lines* indicate turning to the side opposite stimulation, sometimes associated with upward movement. *Vertical broken lines* represent pure upward movement. *Vertical lines* show downward movement (in the region of the posterior commissure). *Horizontal broken lines* indicate turning to the side of stimulation associated with downward component, the former predominantly produced from the thalamus. *Blank parts* of the drawing represent regions not explored. *C.p.*, posterior commissure; *Tr. M.*, tract of Meynert; *III*, third ventricle

conclude that they have been evoked by direct current impulses at eight per second. Histological analysis shows that the reactive points lay at dorsal levels when the movements were executed in saccades (experiment 316, 317, 319, 320, and 340).

The localization of the responsible sites of stimulation was carried out in two stages. Position of electrode and stimulus responses were entered in the protocol. Specific effects were then studied independently in a motion picture taken during the experiment, and the sites from which the effects were evoked were finally plotted in a specially prepared horizontal atlas series [21]. Figure 17.4 shows the results for contralateral deviation just described. The active field lies in the *optic tectum*. The rostral boundary is sharp; no boundary, however, could be determined laterally or caudally, as exploration was not extended far enough in these directions. Marked effects were evoked from the level of section 324 passing through the stratum opticum, but also from plane 362, and even 387 extending into the base of the optic tectum and into the dorsal part of the central gray.

Stimulation points yielding marked conjugate deviation of the eyes to the side opposite stimulation are found in the *brachium colliculi superioris* and its fan-like projection into the *stratum opticum* and also the *stratum lemnisci*. Predominantly saccadic responses are observed when the electrode tip reaches the deeper levels of the optic tectum from which the *tecto-spinal tract* emerges, but sometimes only when stronger currents were applied, e.g., 3 V as compared with 1 V at a 2-mm more dorsal level. Vegetative effects, about which more is said later, are evoked from the electrode tips lying partially in the *central gray substance*. These vegetative effects in no way interfere with the turning responses which appear simultaneously.

2. Upward Movements. We have already drawn attention to the fact that upward movements of the eyes and head often accompany deviation to the side, the transition from one movement to the other occurring in a swift, smooth manner with accent on the upward component. This response is most pronounced in experiment 338, with electrode tips at the base of the optic stratum, approximately 2 mm from the midline (Fig. 17.5).

Furthermore, pure upward movement of the head in stimulus rhythm is seen in experiment 321, associated with marked excitement and pupillodilatation. The eyes, however, are not raised. The electrode lies in the optic tectum close to the midline, with its tip in the central gray. We are unable to state whether the effect is specific or of mixed character (cf. Fig. 17.4).

3. Downward Movements. Pure or almost pure downward movement of the head is more often obtained than the reverse effect and involves the forepart of the body (Fig. 17.6) if the animal is upright at stimulation onset (experiments 265, 316, 317, 319, 320, 321, 324, 338, and 339). Attention should be drawn to the fact that the hindpart of the body, although participating in the event, is not raised; at any rate no saccades corresponding to stimulus rhythm occur. The direct response is therefore limited to the forepart of the body. Symmetrical foreward movements of the pinnae or the vibrissae are occasionally seen; the protocol in most cases, however, underlines that no facial expression effects are observed. In contrast, stimulation frequently affects the forelegs, especially the contralateral one. In two experiments (338 and 339) the cat is crouched with eyes closed. They are opened the moment stimulation sets in, and the head and forebody swing forcefully down in stimulation rhythm. In one case the animal begins to sniff rhythmically (synchronously with the frequency of stimulation) the moment the head nears the floor. In the second case, the cat gazes down. A similar experiment

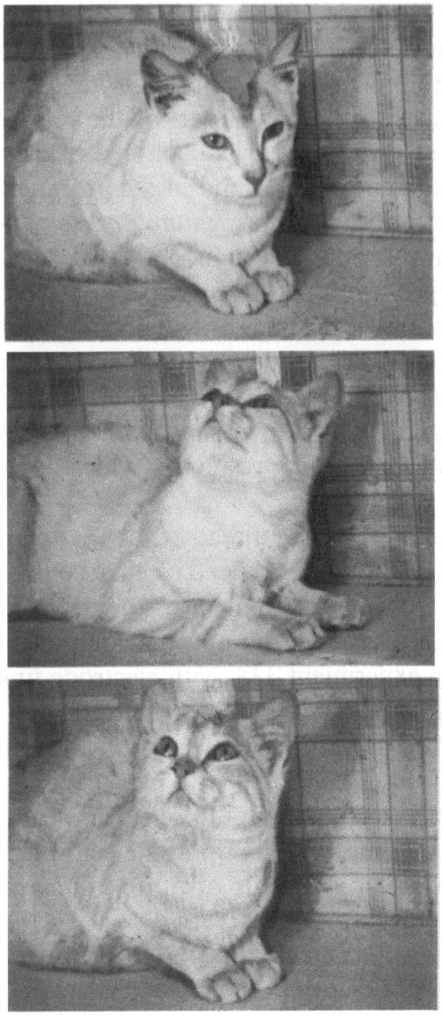

a b c

Fig. 17.5a–c. Upward movement of head with upward gaze evoked from the optic tectum (experiment 338, 2 V). *a* Before stimulation; *b and c* stimulation responses

(265) yields *deviation of the eyeballs down* and slightly to the contralateral side. Verification of eye position during saccades is not easy, because the head is lowered with each stimulus, and often so much so that the head touches the floor. The head is raised the moment stimulation ceases. As regards localization, we discovered that dorsal levels yield *smooth movements* (experiments 319, 320, and 339), whereas *saccade movements* are produced from levels lying 2-mm deeper.

Two experiments throw light on the *results of tissue destruction.* At the end of the experiment, electrocoagulation was carried out at the site from which a marked response had been obtained. Stimulation was then once more applied to make sure that the site was "out of action." Two stimula-

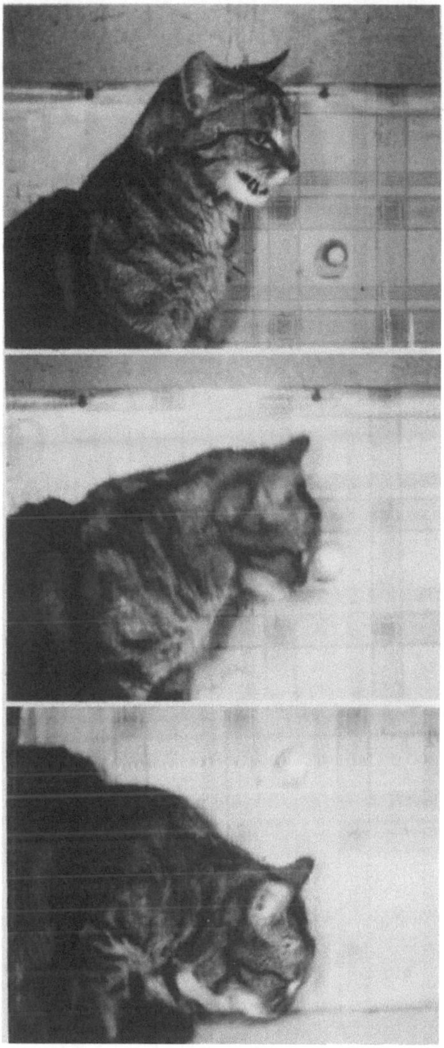
a
b
c

Fig. 17.6a–c. Downward movement of head and forepart of the body (experiment 339, 1.5 V). *a* Before stimulation; *b* downward movement; *c* position reached 0.5 s after onset of stimulation. Note that only 1.5 mm more caudally the electrical stimulation elicited turning in the horizontal plane to the side opposite stimulation (as shown in *Fig. 17.3c*)

tion sites (experiments 320 and 339) that had yielded downward movements no longer responded to stimulation. For several days, however, the cat held its head high, the pose sometimes resembling an opisthotonus (Fig. 17.7). In one case, we observed that a moving object is followed by raising the eyes or by gazing to the right, or to the left, but *never downward.* Abolition of the downward and release of the upward component was so complete that the animal could no longer move its head downward during feeding, an act normally mediated by volitional innervation. Solid food had to be fed

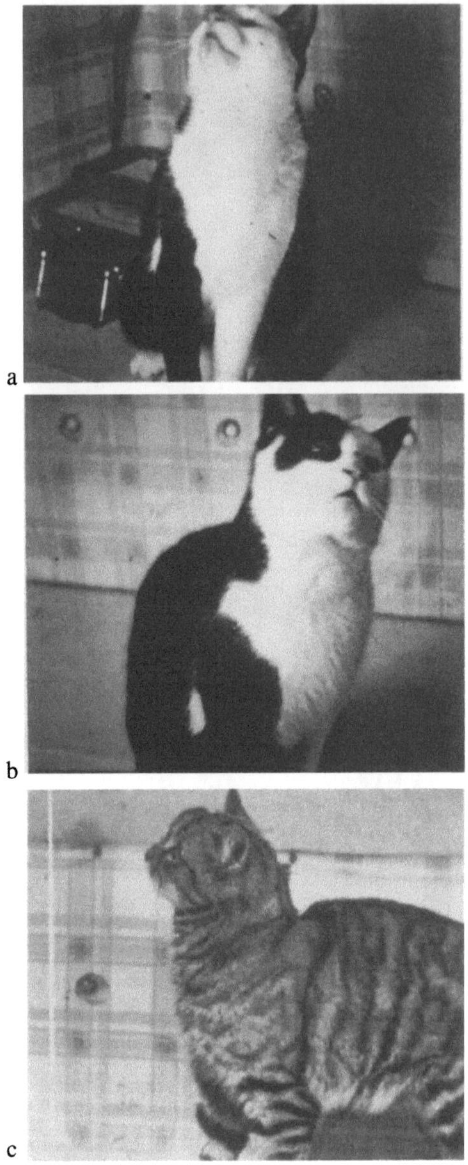

Fig. 17.7a–c. Effect following destruction of the posterior commissure region. *a* Shortly after coagulation (experiment 320); *b* the same animal 1 day later, partial restitution; *c* another (experiment 339) with focus in the same area, 24 h after coagulation. Note similarity of this effect with response to stimulation of tegmentodiencephalic transition area, i.e., upward movement of head and forepart of body (see *Fig. 17.8d*)

"from above" and liquid food given by pressing a sponge soaked in milk. The forelegs were stretched in exaggerated fashion and the cat, head held high, advanced "stork fashion."

Histological examination of the points yielding downward movements showed that the effect was obtained from a region comprising the *posterior commissure* (Fig. 17.4), reaching laterally to the side of the nucleus of the posterior commissure [54]. Exploration was not carried out farther laterally, but rostrally to the Tr. Meynert, and caudally into the rostral border of the tectum opticum. The area thus lies – roughly – in a pretectocommissural field. The response was obtained from level 324 (plane through the brachium colliculi superioris and the stratum opticum), well from level 349 (plane through posterior commissure and stratum lemniscale pretectale), possibly still from level 387 (plane through descending fibers of posterior commissure, aqueduct wall and basal layers of the colliculus superior). The electrode tips therefore lie in the various layers of the pretectum and tectum, but usually also somewhat more ventrally. The fibers of the posterior commissure were not always directly touched by the electrode tips but lay within reach of stimulation (assumed to be approximately 0.5 mm).

4. Turning in the Horizontal Plane to the Side of Stimulation. Ipsilateral turning is often observed; the response, however, is quite different in character from that contralateral to the stimulated side. Coordination of head and eyes, it is true, does occur; eye movements, however, do not appear first, and glancing does not hence "lead" the response, which may be carried out with closed or half closed eyes. It is striking that downward movement components often accompany the deviation; the head is turned to the side but also somewhat down, and when, because of increased or prolonged stimulation, circling movements appear, the latter are executed in a crouched posture. The effect involves the head, and then spreads to the body, first to the fore- and then to the hindparts. However, the response appears to be neither synergic nor coordinated. One is left with the impression that the body is forced – head first – around an axis passing through its hindquarters. In contrast to contralateral turning, ipsilateral deviation is almost consistently accompanied by contralateral facial motor effects (including lid closure). The deviation itself is usually in saccades at onset, which then soon give way to smooth movements often outlasting stimulation. The facial response, however, always follows the stimulus rhythm.

Ipsilateral deviation at *dorsal levels* (sections 324–362) was always evoked from the caudal thalamus (region of the center median of Luys) and the rostral pretectum (see Fig. 17.4). The response, however, could also be obtained when the electrode was lowered (experiments 316, 339, and 340) into the rostral *tegmentum (mesencephalic reticular formation)* (experiments 318, 320, 321, and 322) (see Fig. 17.8a, b), but at this ventral level, overlapping occurred of points yielding ipsilateral deviation and sites yielding rotatory reactions. At the level of the red nucleus the active points for ipsilateral deviation lay dorsal and lateral to this structure.

In this connection a few *special observations concerning contraversion* should be added. Of particular interest are three experiments in which deviation culminating in circling movements to the side opposite stimulation

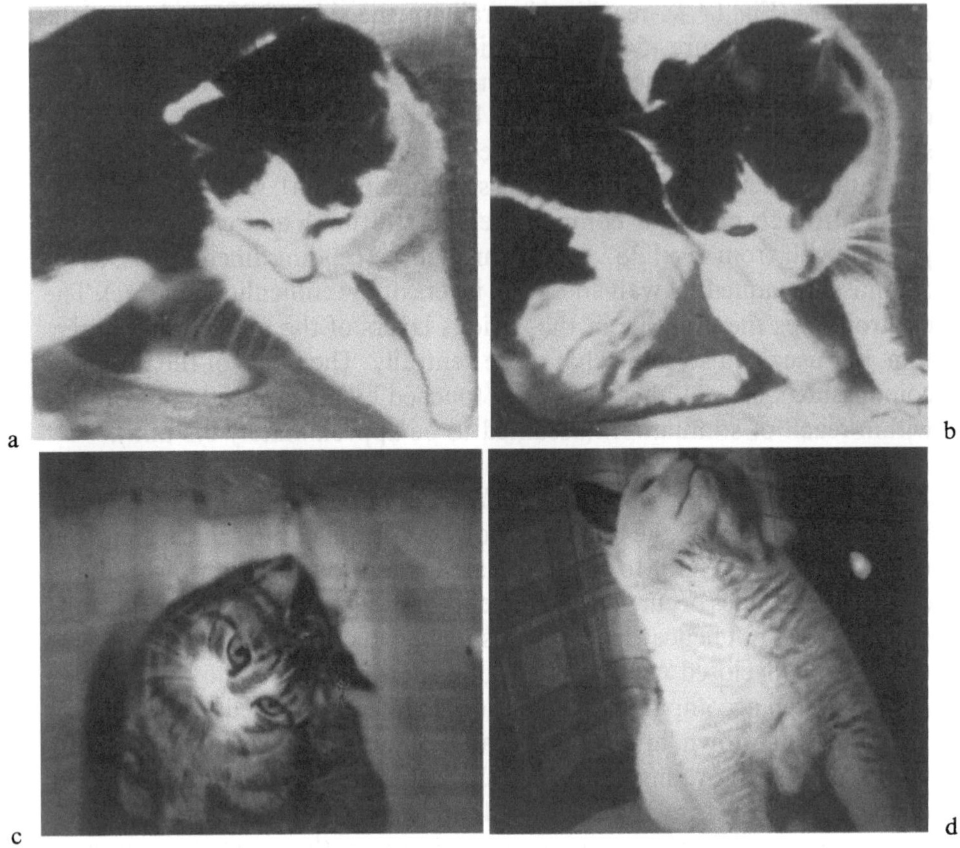

Fig. 17.8a–d. Deviations of head and body. *a* Turning in the horizontal plane to the side of stimulation. Localization: caudal thalamus just in front of pretectum (experiment 318, 1 V); *b* the same response in another animal following stimulation of the mesencephalic tegmentum (experiment 322, 0.5 V). Note similarity of stimulation effects. *c* Rotation in the transverse plane (about sagittal axis) with contralateral half of face raised by stimulation in the mesencephalic tegmentum (experiment 265, 0.5 V); *d* upward movement by stimulation in the mesencephalic tegmentum (experiment 323, 1 V). Compare with similar effect produced by destruction in the region of the posterior commissure *(Fig. 17.7)*

was observed with electrode tips lying in the tegmentum. In experiment 321 the rostral electrodes yielded ipsilateral turning, and the caudal electrode evoked contralateral circling movements, with the head high and the eyes wide open. Histological analysis showed that the caudal electrode had produced degeneration of fibers of the tectospinal tract descending to their decussation (the fountain decussation of Meynert). Similar findings were made in experiment 320. In a third experiment (324, 1 V), contralateral circling movements were obtained with the forepart of the body raised: the electrode tips at both levels were located entirely in the tegmentum, somewhat lateral to the red nucleus so that stimulation of the optic tectum was out of the question.

5. *Rotation.* Rotation (in the transverse plane) toward the side of stimulus (involving the head and forepart of body), in confirmation of earlier results [6], was evoked from 20 reactive sites in the region of mesodiencephalic transition (experiments 318, 320, 322–324). The response is similar to rotation elicited from the diencephalon [26] and occurs predominantly in stimulus-dependent saccades, associated moreover with summation of the motor effects. The response is seldom limited to the head, but usually includes the forepart of the body so that the contralateral forelimb is raised (flexed or extended) in abduction (Fig. 17.8c). Thus arises a flowing transition to a stage in which the animal rolls over completely. In confirmation of previous experiments, we repeatedly observed that whenever the reactive points lie *close to the midline*, an ipsilateral component appears which culminates eventually in pure raising of the forepart of the body with extension of the forelimbs (experiments 323, 324) (Fig. 17.8 d). The accompanying motor effects of the face – to be dealt with later – are, however, usually confined to the contralateral side. It should be added that the upward movements sometimes evoked from the tegmentum are produced from points lying lateral to the red nucleus.

The reactive sites extend from the rubral region, rostrally into the H field of Forel and the ventrally lying field H_2, and supramammillary region bounded medially by a sagittal plane passing through the tractus retroflexus (Meynert), thus comprising medial and medioventral portions of the tegmentum and caudal diencephalon. It has already been pointed out that a clear separation between structures responsible for ipsilateral turning, rotation, and upward movements cannot be established at this level; the fiber tracts mediating these responses appear to lie too close to one another.

6. *Contralateral Motor Effects in Face and Body.* Almost all cases of rotation, upward movements, and most cases of ipsilateral turning are associated with motor effects in the contralateral face and foreleg, less often the contralateral hindleg. The effects are always in saccades. The upper part of the face (forehead, upper eyelid) moves down and the lower part (lips, cheeks, lower eyelid) moves up. The forelimbs show marked abduction of the shoulder, or proximal flexion and distal extension, or, finally, abduction of the upper arm associated with raised shoulder.

These responses were evoked from the tegmental areas yielding rotation and turning, particularly the sectors through which pass the brachium conjunctivum and the medial lemniscus. Correlations between these stimulation effects and localization have been dealt with in another paper [13], and a separate communication is being prepared concerning the physiological interpretation of motor effects of face and forelimb [33].

7. Vegetative Effects. Although the emphasis of the present paper is on motor effects, we will touch upon a few typical observations which clearly show that certain areas are *not responsible for motor effects.* We regularly observed changes of the animal's activity, namely, a decrease associated with pupilloconstriction and prolapse of the nictitating membranes (experiments 316, 319, 321) or excitement and arousal (experiments 317, 319, 320, 321) associated with pupillodilatation. These effects arose when the electrodes lay, at least partially, in the central gray matter. The dorsal longitudinal fascicle (Schütz's bundle) was seen to have degenerated in our Marchi sections; it could be followed a short distance caudalward, most of the fibers running rostralward, some to the massa intermedia, most to the caudal part of the hypothalamus. In this connection it should be mentioned that whereas stimulation near the massa intermedia was associated with somnolence, stimulation of the hypothalamus led to general arousal in the form of ergotropic prevalence [31, 32] (see translation of [31] in this volume, p. 131). A second field from which vegetative responses were evoked also yielded rotation and contralateral motor effects of the face. The vegetative responses were characterized by decreased vigilance (experiments 321, 323, 324).

8. Anatomical Observations. The Marchi method proved to be a useful tool when carrying out our histological analysis, for the electrodes only damaged relatively few fibers and the electrocoagulation of the response sites was limited to small foci. Here we shall discuss observations relating to functional aspects; the details of the histological examination have been published elsewhere [8]. Degenerating fibers in the *brachium colliculi superioris* and the *stratum opticum* run caudalward and could not be followed into deeper layers (Fig. 17.9a). A small number of medially lying fibers, after having reached caudal levels, sweep around laterally and can be followed to the lateral border of the colliculus; some appear to cross the midline before running laterally. These decussating fibers can be seen in sagittal and horizontal sections as a distinct bundle lying in front of the commissure of the inferior colliculus.[1] The *stratum lemnisci,* in which lemniscal contributions could not be identified in our preparations, conveyed few caudally running fibers; most run rostrally through the deeper layers of the pretectum and then sweep into the region of the center median of Luys, and that of the suprageniculate nucleus *(fasciculus tectothalamicus)* (Fig. 17.9b).

1 A sharp differentiation between stratum zonale and stratum opticum could not be established at midline levels in our sagitally and horizontally cut Marchi sections

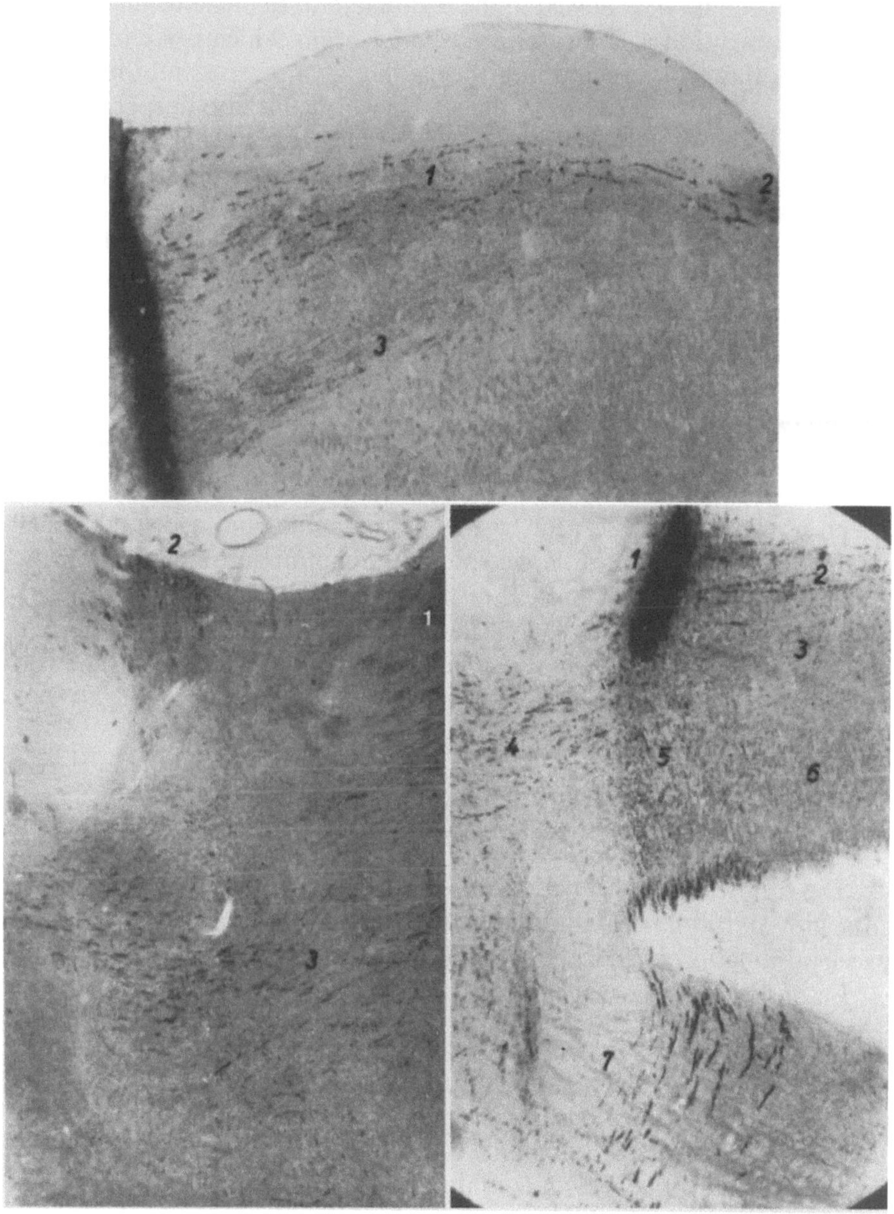

Fig. 17.9a–c. Fiber degeneration as revealed by the Marchi stain following small lesions. *a* Optic tectum, sagittal section. Note the horizontal course of the degenerating fibers. *1*, stratum opticum; *2*, partial crossing of fibers from optic layers; *3*, stratum lemnisci (experiment 318, section 513). *b* Stratum lemnisci of pretectal area as shown in a sagittal section. *1*, coagulation site; *2*, brachium colliculi superioris; *3*, stratum lemnisci of pretectal area (experiment 319, section 395). *c* Sagittal section of optic tectum, degenerating fibers. *1*, focus; *2*, stratum opticum; *3*, stratum lemnisci; *4*, stratum lemniscale pretectale; *5*, commissura colliculi superioris; *6*, intermediate and basal layers; *7*, tectospinal tract descending to Meynert's decussation (experiment 319, section 399)

A few fibers could be followed to the region lateral to the habenula. Finally we should mention a *fasciculus pretecto-incertalis* which descends behind the subparafascicular region into the caudal zona incerta and the H field. No fibers could be traced from the stratum opticum into deeper layers in our Marchi preparations. Fibers of the *commissura colliculi superioris*, whose origin could not be determined, run lateralward and these could be then followed into the pons *(tectopontine tract)*; other elements could be followed into ventral portions of the suprageniculate region. The *tecto-spinal tract* leaves the intermediate and deep layers of the colliculus, crosses the midline in the fountain decussation of Meynert and then runs caudalward in the contralateral predorsal fascicle (Fig. 17.9 a). Before reaching the decussation one finds degenerating fibers (perhaps collaterals), which turn off and can be followed into the H field and zona incerta. As regards the *posterior commissure*, we agree with Spatz [54] that this structure predominantly represents an interstitio-interstitial connection. A few commissural fibers, however, are also sent to the subthalamus, and a distinct, finely myelinated bundle can be followed out of the commissure alongside the medial longitudinal fasiculus of either side. We could not establish any connection with the nucleus of Darkschewitsch. Further observations made in connection with the *tegmental region* have no direct bearing on the problems dealt with in the present communication [8].

Discussion

I. Symptom and Substrate

The following section will be an attempt to correlate the physiological data with morphological findings in order to establish the relationships between stimulation effects and circumscribed substrates.

1. Contralateral Turning Movements. These effects are initiated by conjugate deviation of the eyes in the same direction and appear to be a gaze response. As the active points lay in the *optic tectum*, predominantly in the upper layers, we shall first examine what is known about this structure. There is evidence, at any rate in the cat, that the brachium colliculi superioris and the stratum opticum contain crossed and (fewer) uncrossed fibers from the retina. Comparative anatomical studies have been carried out on the subject by Minkowski [43, 44], and similar results have been obtained in cat by Barris [4] and Barris et al. [5]. It is therefore permissible to conclude that the response to stimulation of the stratum opticum represents a *visually induced response*. In addition it should be remarked that the fibers seen degenerating in the upper layers of the tectum (the latter, like the cortex, is built up in layers) in our Marchi sections could not be followed into deeper layers; this is true for the stratum opticum, less so for the stratum lemnisci. As cell layers alternate with fiber layers, we may safely assume that we are dealing with a system of synapses in which direct retinal impulses are processed and, by stages, conveyed to lower layers from which they are transmitted to the effectors via the tectospinal tract. The significance of fibers distributing to the thalamus and subthalamus and the role played by the projections will be examined later. At present we shall consider the reason why the tectospinal tract may act as an effector bundle for the responses evoked from the superior colliculus.

We have seen that stimulation elicits coordinated deviation of eyes, head, and the forepart of the body and that the hindpart rarely participates in the response (and then only as a secondary reaction). The ipsilateral vibrissae are moved forward during deviation of the head, and the contralateral ones, backward; vibrissae point forward bilaterally during upward movements. The ears show the same tendency. The opening of the eyelids is consistent with the interpretation of the response as a gaze mechanism. Involvement of the forelimb, particularly the contralateral one, can be explained by the fact that tonic innervation of the forelimb is synergically altered when the head is turned to the side and backward. One cannot of course expect that artificial stimulation necessarily excites all the elements ensuring coordinated cooperation of synergists and only these. When the accompanying symptoms appear as single factors, or when only a few of these accompany turning of the head, we may conclude that the artificial stimulation has failed to excite a single focus of retinal representation. Spread of the response to the hindpart of the body, i.e., a circling movement, only occurs after prolonged stimulation or when using stronger voltages, and should be considered a secondary effect, i.e., a proprioceptively mediated "giving in of the whole body" to marked tension in muscles of the cervical vertebrae evoked by the stimulation. The physiological data are consistent with the striking fact that the *tectospinal tract* in cat, like the pyramidal and the rubrospinal tracts, carries strongly myelinated fibers and, as anatomical studies have shown, influences the motor activity in the upper half of the body. The bundle arising in intermediate and deeper layers of the tectum, according to Rasmussen [48], crosses completely in Meynert's decussation, runs caudalward in the predorsal bundle, and terminates in the motor nuclei of cranial and spinal nerves. It can, however, only be followed to the C_7 level. In addition, fine fibers have been described as running from the superior colliculus to the oculomotor nucleus. This means that we are dealing with a tectofugal system transmitting motor impulses to cells that innervate the eye, face, head, and forepart of the body. Two experimental observations are hence of special interest. In both cases, stimulation yielded contralateral deviation, and the electrodes were placed predominantly at tegmental levels in the fibers of the still uncrossed tectospinal tract.

It should also be pointed out that the responses evoked from the more dorsally lying electrodes are on the whole smooth, while those evoked from the ventral layers occur in saccades, i.e., following the rhythm of the stimulation. This difference is important because numerous studies carried out in this laboratory have shown that pulsative excitations, after passing several synapses, generally give way to smooth movements. Stimulation of efferent bundles or of the afferent limb of a short reflex, however, produces a saccadic response to each stimulus pulse associated with minimal summation or no summation at all.

Correlation of physiological and functional data shows that contralateral deviation represents a visually induced movement that occurs when the periphery of the retina is physiologically stimulated and the excitation is directly transmitted to the optic tectum and, having undergone functional processing, from there to the motor apparatus. The innervation pattern organized in the tectum and representing a regulated vectorial system of forces, is transmitted to the corresponding effectors by the tectospinal tract to motor nuclei subserving eye and neck muscles.

The physiological significance of these observations will be dealt with later. A peculiar experimental observation should, however, be mentioned now in relation to morphological aspects. As pointed out above, turning elicited by stimulation often appeared to be subject to modification from the voluntary sphere. Barris et al. [5] have shown that the optic tectum, besides receiving direct retinal afferents, also receives fibers from the visual cortex, which, as is well known, receives impulses originating in the retina and relayed in the lateral geniculate body. These anatomical conditions therefore signify that optic impulses in the cat excite the primitive optic tectum and, in parallel, the visual cortex from which impulses are sent to the superior colliculus and the pretectal area. There is thus striking agreement between the anatomical data and the physiological convergence of two mechanisms.

2. Upward Movements. We shall be brief with regard to upward movement combined with turning and preceded by a corresponding movement of the eyes. The active points also lay in the optic tectum, i.e., the region from which contralateral deviation was evoked. In one experiment in which pure upward movements were elicited, without, however, any ocular response, the electrode tips lay close to the midline, suggesting a correspondence between functional as well as morphological organization with regard to gaze. The question, however, requires further experimental research. As concerns the afferent (retinal) and the efferent (tectospinal) pathways we refer the reader to what has been said in the preceding section.

3. Downward Movements. These effects of the head and upper trunk are evoked by stimulation from a region clearly grouped around the *posterior commissure* and from the adjacent *pretectal area.* The latter, like the optic tectum, comprises an optic and a lower-lying lemniscal layer. The response evoked was essentially limited to the upper part of the body. Electrocoagulation has shown that direction of gaze is a decisive factor in the reaction, for after destruction of the responsive field the cat can no longer follow a low-lying object and has to be fed from above.

The fact that the hindpart of the body is not raised during downward movements of the forepart, shows in addition that this stimulation response does not correspond to a vestibular compensation [7], since vestibular reflexes entail raising of the hindpart of the body when the forepart is lowered; the animal thus appears to be turning about its transverse axis There is therefore a characteristic difference between this mechanism and the stimulation responses which we shall now examine. The same arguments presented for turning in the horizontal plane and upward movements might suffice, were it not for the remarkable fact that downward movements were produced by stimulation of the posterior commissure and its surroundings. The commissure essentially interconnects the region of the interstitial nuclei of Cajal, concerned with vestibular influences and closely connected with the oculomotor nuclei. Whether Cajal's nuclei are concerned with downward movements of the eyes is not known, but attention should be drawn to pathological and anatomical observations, i.e. ocular crisis and vertical gaze paralysis (*"plafonnement,"* Parinaud's syndrome) resulting from stimulation and destruction, respectively, of the region of the posterior commissure. We shall see that downward movements have another meaning in the cat than does glancing to the side and up. It should also be pointed out that an abnormal posture ensues after destruction in the region of the posterior commissure, but not after lesions in the optic tectum. All these considerations point to the possibility of the posterior commissure being a constituent of a specific organization subserving downward movements. A final conclusion is, however, at present not possible. Neither can we define the fiber tracts proceeding to the effector nuclei. Damage of the interstitial nucleus of Cajal itself leads to degeneration of strongly myelinated fibers which descend to spinal levels.

4. *Results of Other Investigators.* We shall close the discussion about the results obtained from the tectal and pretectal fields with a brief survey of the results of other workers. Ferrier [15], as early as in 1877, noted a deviation of the eyes and head to the contralateral side after stimulation of the superior colliculus in monkeys, observations subsequently often confirmed (for survey see [46]). The authors without exception based their observations on animals in which major surgical interventions had been carried out; hence all were anesthetized or hampered in their movements, so that observations that can only be made on freely moving animals escaped attention. Monnier, in his detailed survey, which also included his own observations, says

with regard to contralateral deviation of head and forepart of the body: „Cette réaction semblable à la réaction tegmentale *n'en diffère que par le sens de la déviation.*" Based on our own observations and conclusions, we must, however, point out that the "tectal" (i.e., visually induced) response in no way resembles the "tegmental" effect. Deviation may be dissimilar even if the direction of the turning movement is the same. We shall return to this subject later. It is furthermore necessary to underline that the simple topographical concept of "tectal" reactions based on earlier literature should be viewed with caution. This concept ran as follows: Upward movements follow stimulation of dorsomedial parts of the tectum, deviation that of lateral parts, and downwards movements that of ventral parts. However, according to our numerous results, organization of the latter response points to the *region of the posterior commissure and adjacent pretectum*, whereas *upward movements are not only* evoked from medial parts of the optic tectum.

An interesting hypothesis of Monnier should be mentioned. Based on his extensive review of the pertinent literature, he concludes that a crossing of the fibers concerned with contraversive deviation takes place at isthmus levels, for the reverse effect arises more caudally. We not only can confirm this,[2] but can also add that the fibers mediating visually induced contralateral deviation do indeed cross in the fountain decussation of Meynert (tectospinal crossing). At more caudal levels the reaction can occur toward the ipsilateral side only by stimulation of the predorsal bundle.

5. Ipsilateral Turning in the Horizontal Plane. We have mentioned that this effect is obtained from a region of transition thalamus/pretectum; it must be clearly differentiated from the contralaterally directed "tectal" response. With regard to the former, as an identical response was evoked from the tegmentum and could be traced dorsally and laterally to the red nucleus and into the caudal thalamus, we have come to the conclusion that ipsilateral turning is not related to the optic tectum, but rather, is similar to or identical with what has been called the "tegmental reaction" [37, 45, 46]. A connection seems to have been established with the mesencephalic reticular formation, a widespread structure carrying caudally but also (according to our data) rostrally running fibers that all require further study. So, to sum up, we conclude that ipsilateral deviation was predominantly obtained from the dorsolateral portion of the rostral tegmental reticular substance and stands in connection with similar symptoms elicited by stimulation of circumscribed portions of the diencephalon, effects which may correspond to vestibular manifestations [26].

2 The exception is the contralateral deviation evoked occasionally from the lateroventral tegmentum, a response that we would like to bring into connection with stimulation of the zona incerta

Finally, attention should be drawn to a response related neither to the tectally induced gaze response nor to the tegmental reaction. It concerns a contralateral turning associated with upward movement of the forebody and was obtained as an isolated effect from the tegmental reticular substance. A similar, perhaps identical effect was evoked from the zona incerta [26]. Further typical findings have been described by Bürgi [9]. Detailed analysis of this effect and the structures mediating it has still to be undertaken.

6. *Rotation.* As regards rotation evoked from the region of mesodiencephalic transition, we are dealing with the same effect that Hess [26] evoked from the diencephalon itself, namely, from the H field as well as from the intermediate thalamus. A mirror-image effect, i.e., upward turning of the *ipsilateral* half of the face, was evoked by stimulation of the medial portion of the brachium conjunctivum as well as the lateral reticular formation at the isthmic and pontine levels [10–12]. As the crossed components of the brachium run in part through the red nucleus and H fields to distribute mainly to more rostral regions of the ventral thalamic nucleus, it may be possible that these fibers are responsible for the effect. The question is at present still open and has been discussed by Bürgi and Müller [13].

II. The Dynamic Foundation of Motor Innervation

The foregoing section dealt with the relation between symptom and stimulation of definite structures. The next problem to be examined concerns the functional interpretation of the responses produced by stimulation. What could be directly observed were stimulus effects, and as such, elements (i.e., phases) which, under physiological conditions, in one way or another form parts of an integrated motor performance. Study of the pathology of the central nervous system does not usually go beyond establishing the relation between structure and symptom, and this is particularly true for the localization of motor disturbances. For the physiologist, however, the goal is only reached when motor effects are brought into the general concept of a meaningful function. It is permissible to hope that this extended analysis in the field of empirical symptomatology may serve the clinician. Certain conditions should, however, be fulfilled, and these as a rule do not receive the attention they deserve. Explanations are all too lightly advanced, although the dynamic aspects of the *whole biomotor problem* and the manifold factors which come into play have not been assessed. We shall therefore attempt to outline briefly the base upon which a consistent interpretation should be built [27–30].

Motor activity considered as a whole enables the organism to assert itself in space and to assume an active role toward the environment. Two categories of motor functions should be held apart: one concerns the basic postural mechanisms; the other consists of goal-directed movements. The first, in conformity with its supporting character, is called *"ereismatic"* (Greek *ereisma* = support), an expression based on the fact that it is concerned with subcortically organized reflexes governed by converging vestibular and proprioceptive excitations combined with unconscious tactile afferents. This system provides the essential dynamic foundation on which goal-directed movements can be properly carried out. The second system, the *"teleokinetic"* (Greek *telos* = goal, aim) is characteristically dominated by the will and primarily organized at the cortical level under the guidance mainly of the eye and the modulating influence of proprioception and touch.

The following aspect of the relation between teleokinetic and ereismatic innervation requires special consideration. When a stabilized posture is attained against disturbance, we might expect that the same mechanisms ensuring this stability must – unless precautionary measures are taken – at the same time prevent the release of a goal-directed action. Head turning, for instance, would be hindered by the automatic intervention of vestibular reflexes the moment the movement is started. Corresponding counterforces are indeed set into action, but this only occurs in the case of *passive* changes of posture. This difference in the responses to active versus passive movements means that a qualitative and quantitative inhibition of the vestibular reflexes arises from volitional impulses. It could also mean that the stabilizing equilibrium of forces is displaced through corollary impulses to the ereismatic innervation system.

It is well known that a distinction between two forms of motor function, i.e., pyramidal and extrapyramidal, has been made by clinicians – a distinction based on symptomatology and pathological anatomy. There is good agreement between our two functionally defined systems and the two divisions based on clinical and morphological arguments. The agreement, nevertheless, is far from complete because subcortically organized functions are so dissimilar in nature that they cannot be classed under one heading (as is the case with the terms "pyramidal" and "extrapyramidal" reactions). It should also be said that morphological criteria are basically an inadequate means of characterizing a functional system. A striking example of this will be shown later.

Having pointed out how a goal-directed movement depends upon a stable start position we shall now examine the factors underlying the latter. An important role is played by compensation of the force of gravity on the basis of opposing muscle tension, quantitatively and reflexly mobilized, an operation by which the organism is transferred so to speak into a gravity-free field because the volitional forces can now be carried out undisturbed. Activation of antigravity forces is not only responsible for the starting position, but it also accompanies teleokinetic innervation during the various phases of the evolving movement. Ereismatic innervation, besides opposing the ever-present forces of gravity, also cancels out other factors interfering with motor coordination, for example, unforeseen forces arising by chance, such as blows or gusts of wind.

Every goal-directed *propulsion*, in physical terms, necessarily releases a quantitatively equal *counterforce*, a phenomenon corresponding to what is observed when a missile is fired off. The projectile shoots forward and the barrel is thrown back, but gets caught by a particular device to avoid dislocation of the firearms. In like fashion reflex tension, adjusted in quantity and direction, must counteract the forces arising from voluntary motor activity. Conditions are indeed so that effective goal-directed movements could not be properly carried out without this continuous play of postural reflex mechanisms. We can thus see that manifold and high demands are set on ereismatic innervation. A final observation should be added: in the course of a movement, a smooth displacement occurs between origin and insertion of each particular muscle, which results in constant change as regards *direction and degree of traction*. Muscles, therefore, do not represent *forces;* rather they stand in relation to a *constellation of forces* which undergoes dynamic change with the change in the spatial situation. The force of a muscle as an integral part of a well-ordered dynamic whole can only be released in connection with the *momentary posture*, i.e., the dynamic range of the moving limb. Applied to the notion of an organized center, it is clear that motor impulses do not reach the muscle directly but over an organ which acts as a "switchboard," which, by drawing upon a given set of proprioceptive afferents, selects the appropriate combination of muscles. This "switchboard" is particularly complex when movements involving a ball-and-socket joint with its three-dimensional freedom of motion have to be organized. The "switchboard" function is unnecessary for mechanisms involving joints with only one-dimensional freedom.

III. Physiological Aspects of the Experimental Findings

We may finally ask ourselves to what extent stimulation effects evoked from the tectum and tegmentum represent motor functions which arise under normal conditions. The question presupposes some acquaintance with the normal motor behavior of the cat. Because we are briefly considering this subject, some physiologists and even neuroanatomists may think we are straying beyond the limits set by laboratory explorations. In answer to this criticism it is important to state that an exhaustive answer to every problem concerned with morphological and physiological organization can only be obtained by bringing the problem into relation with the motivational forces essential in the struggle for survival. In the cat we can observe that visual orientation is highly developed, but also that the reaction of this animal to acoustic stimuli is prompt and vigorous. The cat, however, can only follow a scent for a short distance, for example, when it approaches a (visually) striking object and examines the latter by sniffing. A reaction to temperature may then arise, for even coveted food is at once rejected if too warm or too cold. Touch plays a decisive role as soon as the snout or vibrissae come into contact with an object. Depending on the situation, the cat responds by moving the head away or retracting only the vibrissae on one side. The contact can, however, be intensified; i.e., the motion toward increases. When an edible object touches the lips, the tongue and inner surface of the mouth, licking up, chewing and, finally, swallowing set in, under the control of touch and taste. That a subcortical coordination mechanism mediates these acts has been shown by Goltz [20] and Rothmann [49] on decorticated dogs. An analogous subcortical organization appears to be probable, at least partially so for the cat. A special role is played by the nociceptive receptors in the snout and face, by eliciting withdrawal of the head and sometimes the body. Suprathreshold stimulation when the head is held fast gives rise to *secondary* defensive movements of foreleg and jaws. All these observations show that the receptors mediating the above-mentioned sensory modalities are primarily responsible for the release of specific motor mechanisms and for their continuous guidance. As regards motor coordination of single functions, it is noteworthy that sniffing and chewing are both normally carried out with lowered head and forepart of the body. When analyzing these reactions it should be remembered, particularly as regards visually induced movements, that a cat prefers an elevated spot from which it may easily survey the surroundings, and this includes looking *down*. Physiologically, fields of vision and gaze have the same biological value in all directions. All this is the case when an object (a mouse scuttling past, a bird taking off) enters the visual field from the periphery. As quick as lightning the cat responds by turning the head and, just before, the eyes. The adjustment brings the blurred image of the object (due to blurred peripheral vision) to those parts of the retina with the greatest resolving power; i. e., the target is "grasped" so that the features of the image are detectable under optimal conditions. It should be added that, in the cat, perception of a moving object in the lower visual field, besides eliciting fixation of the eyes, often leads to immediate seizure of the object. An unwary mouse is not only seen but also grasped. This may be due to the fact that the area responsible for downward movements is extensive and morphologically specialized. Lesions placed in the region of the posterior commissure give rise to an abnormal posture not seen after tectal lesions.

These experimental results, in agreement with the proof that there is a direct fiber connection between the retina and the superior colliculus, allow us to conclude that visual fixation is organized in the optic tectum and that it can be conceived as a "visual grasp reflex." The result is an intensified visual contact brought about by the activation of a highly differentiated central mechanism such as is found in the occipital cortex. This brings up the problem of the relationship between primitive and more highly developed visual centers. In teleosts, the most highly developed analysis of visual stimuli as well as the organization of goal-directed behavior, e.g., striking at prey, is carried out at the level of the midbrain tectum. It cannot be said that this coordinating function loses importance with the evolution of a highly differentiated cerebral cortex. The cat, being an animal of prey, takes aim by moving its eyes and head, but at the same time consigns the target to a higher organization capable of more elaborate functions. The primitive – in our case, tectal – influence is not, as one says, deprived by a higher center, but is preserved, the tectum becoming an associate of a *partner* with greater competence. The question arises whether the processes mediated by primitive visual mechanisms reach *conscious awareness*. Only conjectures are possible, although not quite negligible from a psychological point of view. Self-observation tends to give a positive answer to the question. The unexpected appearance of a moving object in the outer peripheral visual field leads to simultaneous involuntary turning of the eyes and head toward the object. The situation induced by the ocular movement, kept in mind for a moment, is then replaced by what is perceived by fixation of the eyes. There is reason to believe that species controlled only by the tectum do not "see" *more* than what is conveyed by subcortical pathways to the tectum in the cat: the voracious fish darts toward every glittering object, e.g. a fisherman's "spoon," even when the latter does not resemble the normal prey in shape. Differentiation is beyond the creature's range of knowledge, often with fatal outcome.

Among the problems deserving attention, besides those related to psychology, should be mentioned the fact that rotation of the eyes and head was never evoked from the tectum – a reaction, it is true, not seen operating during normal gaze. It is remarkable too that conjugate deviations to the side opposite stimulation, but also downward and to some extent upward, are all localized within limited tectal fields. There exists, therefore, a *systematic relation* between topographically defined portions of the tectal visual center and the retina, on the one hand, and tectal and pretectal areas and defined muscle groups on the other. We would like to raise the question whether the fibers running directly (i.e., without synapse in the lateral geniculate body) from the retina to the optic tectum originate from receptors lying in the periphery and, by working as a collective, respond to moving stimuli. Small electrolytic lesions or electrocoagulations of the corresponding retinal areas would suffice to establish the topographical organization of retinotectal projections. Postmortem examination of the brains of patients with minute pathological damage of the peripheral retina would also throw light on the problem. A better understanding of pathways connecting olfactory, tactile, and proprioceptive receptors in the nose, snout, and mouth to the optic tectum and the pretectocommissural region (as regards downward movements) would also be of value.

The connection of the latter field with the cranial motor nerves or supranuclear organization is documented by material provided by human pathology. Conjugate vertical paralysis (Parinaud's syndrome) is often brought into connection with damage to the superior colliculus and in particular the region of the posterior commissure. It is true that the differentiation between upward and downward paralysis is not always carried out consistently. It seems important to us that oculomotor reflexes are still retained in patients with paralysis of voluntary eye movements in the sagittal plane resulting from cortical or subcortical lesions [1]. Also, the often-seen oculogyric crisis associated with Parkinsonism (*Plafonnement* of French authors) is attributed generally to irritation produced by lesions in the quadrigeminal plate. Our data suggest that damage to the pretectocommissural field and consequent *abolition* of downward gaze may also play a role. The limitation of paralysis and cramps to eye movements in human patients is consistent with the fact that physiological turning in man is often carried out by the eye muscles alone. The cat when looking around uses its exquisitely mobile head much more. The question whether the tectal "visual grasp reflex" also exists in man can be answered in the affirmative, and again it seems to be limited to the eye muscles. Most optic fibers are relayed in the lateral geniculate body, but even in man direct fibers run to the optic tectum, according to most authors [3, 53]. On the effector side, the tectospinal tract shows a regressive trend in evolution. An optic-acoustic defense and flight reflex organized from the quadrigeminal plate has been said to innervate the eye muscle nuclei directly. Whether the heterogeneous medial longitudinal fasiculus also carries direct tectospinal fibers [14] appears to be questionable. Such a direct connection could not be established in our Marchi material [8].

With regard to the responses evoked by stimulation of the *tegmentum*, the following can be said: Active posturing of the individual in space depends on the delicate balance of influences arising from proprioceptive and tactile receptors as well as from the vestibular apparatus. Direct and indirect afferents (via the cerebellum) are known to reach specific portions of the diencephalon from which statokinetic control movements [26] are evoked. That these effects were produced by central stimulation of the medial lemniscus and the composite bundle making up the brachium conjunctivum [13] has been ascertained by stimulation of the tegmentum. Taking into account that they all seem to lack coordination and often appear meaningless and forced in character, we may conclude that we are predominantly dealing with stimulation of fibers pertaining to the *ereismatic inner-*

vation system. A systematic relationship between stimulation site and tegmental response is as a rule only possible to a limited degree (in contradiction to what holds true for the diencephalon). The former structure usually yields more or less complex effects, probably due to the fact that the electrode tips measuring 1–1.5 mm force their way between cells and fiber systems pertaining to different types of reaction. Some pathways descending from diencephalic and supradiencephalic levels to the motor and "premotor" nuclei may be functionally related and run in close proximity to one another; others may possibly be more dispersed among functionally unrelated bundles. Of particular interest is the observation that ipsilateral turning evoked from the midbrain differs not only with respect to direction but also quantitatively from the contralateral deviation elicited from the tectal region.

These observations confirm the view that differentiation between pyramidal and extrapyramidal systems is not an adequate basis for carrying out a functional analysis. Tectal gaze is a teleokinetic act, directed toward a goal and, in spite of its "extrapyramidal" innervation, should be strictly held apart from the tegmental motor response, which is also extrapyramidally innervated but plays a totally different role. Because gaze-directed movements depend, however, on concurrent innervations (excitatory or inhibitory) by postural mechanisms, these requirements, in the case of tectal gaze, are fulfilled by the cooperative action of extratectal organizations connected with the tectum by means of various pathways (such as the tectothalamic tract, fasciculus pretecto-incertalis, and the tectopontine tract).

Summary

1. The experiments on the organization of the diencephalon, dealt with elsewhere, have been extended to the rostral portion of the mesencephalon. Electric stimulation was combined with cinematographic recordings of the effects obtained, and the Marchi method was used for the study of the secondary degeneration of those fibers which had been stimulated electrically.

2. The most striking effects obtained from the tectum and adjoining pretectocommissural region (elicited from definite areas described in the text) were typical changes in posture of the eyes and head: turning in the horizontal plane to the side opposite stimulation and upward and downward movements. Rotation of the head or eyes was never observed. These symptoms are interpreted as special reflexes ("visual grasp reflexes") for the perception of moving objects.

3. Histological analysis to the following results: Direct visual projections from the retina to the superior colliculus reach a system of synapses in the tectum. An adequate reflex response is organized at these levels and

mediated by way of the tectospinal tract to the motor nuclei connected with the eye, neck, and upper trunk.

4. The tegmentum as a rule yields complex movements which appear to be forced – at any rate, unnatural movements – in striking contrast to the reactions obtained from the tectum. In several cases ipsilateral deviation of head and trunk was evoked. These responses were obtained from artificially isolated components of a motor system whose role is to assure the body basic posture in movement or repose.

5. The comparison of motor reactions elicited from the tectum and tegmentum tends to show that the anatomical distinction made between pyramidal and extrapyramidal motor systems falls short when a functional interpretation is attempted. We suggest a distinction that takes into consideration the effect to be realized. The teleokinetic system mobilizes the forces that direct the eyes, ears, head, and body toward the goal aimed at. The ereismatic system ensures basic posture and counterbalances the forces which might impede faultless execution of directed movements.

6. Finally, the significance of the relation between tectally induced eye movements and cortical vision is discussed.

References

1. Alajouanine, T., Thurel, R.: Révision des paralysies des mouvements associés des globes oculaires. Rev. Neurol. *1*, 125–169 (1931)
2. Altenburger, H.: Untersuchungen zur Physiologie und Pathophysiologie der Koordination – Adiadochokinesis. Z. Ges. Neurol. Psychiatr. *142*, 373–412 (1932)
3. Ariens Kappers, A.C.U., Huber, G.C., Crosby, E.C.: The comparative anatomy of the nervous system of vertebrates, including man. New York: MacMillan 1936
4. Barris, R.W.: A pupillo-constrictor area in the cerebral cortex of the cat and its relationship to the pretectal area. J. Comp. Neurol. *63*, 353–368 (1936)
5. Barris, R.W., Ingram, W.R., Ranson, S.W.: Optic connections of the diencephalon and midbrain of the cat. J. Comp. Neurol. *62*, 117–153 (1935)
6. Bartorelli, C.: Esperienze di stimolazione elettrica della regione mesencefalica. Arch. Fisiol. *42*, 384–415 (1941)
7. Bartorelli, C., Wyss, O.A.M.: Kinematographische Analyse der labyrinthären postrotatorischen Körperdrehreaktionen. Pflügers Arch. Ges. Physiol. Menschen Tiere *245*, 511–523 (1942)
8. Bucher, V., Bürgi, S.: Untersuchungen über die Faserverbindungen im Zwischen- und Mittelhirn der Katze. Confin. Neurol. *6*, 317–340 (1945)
9. Bürgi, S.: Die „tegmentale Reaktion" und ihre physiologische Bedeutung. Helv. Physiol. Acta *1*, 3–22 (1943)
10. Bürgi, S.: Reizung und Ausschaltung des Brachium conjunctivum. I. Experimenteller Teil. Helv. Physiol. Acta *1*, 359–380 (1943)
11. Bürgi, S.: Reizung und Ausschaltung des Brachium conjunctivum. II. Bindearme und Kleinhirnfunktion. Helv. Physiol. Acta *1*, 467–487 (1943)
12. Bürgi, S., Monnier, M.: Motorische Erscheinungen bei Reizung und Ausschaltung der Substantia reticularis pontis. Helv. Physiol. Acta *1*, 489–510 (1943)
13. Bürgi, S., Müller, H.-R.: Proprioceptive Einflüsse auf den Ventralkern des Thalamus. Helv. Physiol. Acta *3*, 497–518 (1945)

14. Clara, M.: Das Nervensystem des Menschen. Ein Lehrbuch für Studierende und Ärzte. Leipzig: Barth 1942
15. Ferrier, D.: The functions of the brain. London: Smith Elder 1876
16. Foerster, O.: Symptomatologie der Erkrankungen des Großhirns. Motorische Felder und Bahnen. In: Handbuch der Neurologie, Vol. 6. Bumke, O., Foerster, O. (eds.), pp. 1–357. Berlin: Springer 1936
17. Foerster, O., Altenburger, H.: Zur Physiologie und Pathophysiologie der Sehnen- und Knochenphänomene und der Dehnungsreflexe. I. Zur elektrophysiologischen Analyse der Sehnen- und Knochenphänomene bei Gesunden. Z. Ges. Neurol. Psychiatr. *146*, 641–660 (1933)
18. Foerster, O., Altenburger, H.: Zur Physiologie und Pathophysiologie der Sehnen- und Knochenphänomene und der Dehnungsreflexe. II. Dehnungsreflexe bei Gesunden. Z. Ges. Neurol. Psychiatr. *147*, 169–183 (1933)
19. Fulton, J.F., Scholar, Ch.W., Liddell, E.G.T.: Observations on ipsilateral contraction and "inhibitory" rhythm. Proc. R. Soc. London Ser. B *98*, 214–226 (1925)
20. Goltz, F.: Der Hund ohne Großhirn. 7. Abhandlung über die Verrichtungen des Großhirns. Arch. Ges. Physiol. Menschen Tiere *51*, 570–614 (1892)
21. Hess, W.R.: Beiträge zur Physiologie des Hirnstammes. I. Die Methodik der lokalisierten Reizung und Ausschaltung subkortikaler Hirnabschnitte. Leipzig: Thieme 1932
22. Hess, W.R.: Ergebnisse von Reizversuchen in Zwischenhirn und Nachbargebiet. Zentrale Erregung von Niesen, Schnuppern und Erbrechen. Pflügers Arch. Ges. Physiol. Menschen Tiere *243*, 409–430 (1940)
23. Hess, W.R.: Beitrag zur Technik des zentralen Reizversuches. Pflügers Arch. Ges. Physiol. Menschen Tiere *243*, 431–438 (1940)
24. Hess, W.R.: Zwischenhirn und Motorik. Pflügers Arch. Ges. Physiol. Menschen Tiere *243*, 634–650 (1940)
25. Hess, W.R.: Motorische Symptome im Gesicht bei Reizung im Zwischenhirn. Pflügers Arch. Ges. Physiol. Menschen Tiere *243*, 678–685 (1940)
26. Hess, W.R.: Charakter der im Zwischenhirn ausgelösten Bewegungseffekte. Ein Beitrag zur extrapyramidal gesteuerten Motorik. Pflügers Arch. Ges. Physiol. Menschen Tiere *244*, 767–786 (1941)
27. Hess, W.R.: Die Motorik als Organisationsproblem. Biol. Zentralbl. *61*, 545–572 (1941)
28. Hess, W.R.: Physiologische Aspekte der extrapyramidalen Motorik. Nervenarzt *15*, 457–466 (1942)
29. Hess, W.R.: Biomotorik als Organisationsproblem I. Naturwissenschaften *30*, 441–448 (1942)
30. Hess, W.R.: Biomotorik als Organisationsproblem II. Naturwissenschaften *30*, 537–541 (1942)
31. Hess, W.R.: Das Schlafsyndrom als Folge diencephaler Reizung. Helv. Physiol. Acta *2*, 305–344 (1944)
32. Hess, W.R.: Von den höheren Zentren des vegetativen Funktionssystemes. Bull. Schweiz. Akad. Med. Wiss. *1*, 138–166 (1945)
33. Hess, W.R.: Zwischenhirn und Motorik. Helv. Physiol. Acta [Suppl.] *V*, 1–55 (1948)
34. Hess, W.R., Bartorelli, C., Bucher, V.: Diencephale Motorik und Marchi-Degeneration. Z. Ges. Neurol. Psychiatr. *174*, 340–352 (1942)
35. Hoffmann, P.: Die physiologischen Eigenschaften der Eigenreflexe. Ergeb. Physiol. *36*, 15–108 (1934)
36. Holst, E. v.: Die relative Koordination als Phänomen und als Methode zentralnervöser Funktionsanalyse. Ergeb. Physiol. *42*, 228–306 (1939)
37. Ingram, W.R., Ranson, S.W., Hannett, F.I., Zeiss, F.R., Terwilliger, E.H.: Results of stimulation of tegmentum with Horsley-Clarke stereotaxic apparatus. Arch. Neurol. Psychiatr. *28*, 513–541 (1932)

38. Liddell, E.G.T., Sherrington, C.S.: Reflexes in response to stretch (myotatic reflexes). Proc. R. Soc. London Ser. *96*, 212–242 (1924)
39. Liddell, E.G.T., Sherrington, C.S.: Further observations on myotatic reflexes. Proc. R. Soc. London Ser. B *97*, 267–283 (1924)
40. Magnus, R.: Körperstellung. Berlin: Springer 1924
41. Magnus, R., De Kleijn, A.: Körperstellung, Gleichgewicht und Bewegung bei Säugern. In: Handbuch der normalen und pathologischen Physiologie, Vol. 15/I/1. Bethe, A. (ed.), pp. 29–54. Berlin: Springer 1930
42. Magnus, R., De Kleijn, A.: Haltung und Stellung bei Säugern. In: Handbuch der normalen und pathologischen Physiologie, Vol. 15/I/1. Bethe, A. (ed.), pp. 55–87. Berlin: Springer 1930
43. Minkowski, M.: 2. Über den Verlauf, die Endigung und die zentrale Repräsentation von gekreuzten und ungekreuzten Sehnervenfasern bei einigen Säugetieren und beim Menschen. Schweiz. Arch. Neurol. Psychiatr. *6*, 201–252 (1920)
44. Minkowski, M.: 4. Über den Verlauf, die Endigung und die zentrale Repräsentation von gekreuzten und ungekreuzten Sehnervenfasern bei einigen Säugetieren und beim Menschen. Schweiz. Arch. Neurol. Psychiatr. *7*, 268–303 (1920)
45. Monnier, M.: Syndromes déviationnels provoqués par l'excitation et la destruction du système reticulaire bulbo-protubérantiel chez le chat; (manège et réaction tegmentale). Monatsschr. Psychiatr. Neurol. *107*, 84–102 (1943)
46. Monnier, M.: Physiologie du tronc cérébral; le rôle du système réticulaire dans l'organisation de la motricité extra-pyramidale. Ergeb. Physiol. *45*, 321–422 (1944)
47. Rademaker, G.G.J.: Das Stehen. Berlin: Springer 1931
48. Rasmussen, A.T.: Tractus tecto-spinalis in the cat. J. Comp. Neurol. *63*, 501–525 (1936)
49. Rothmann, H.: Zusammenfassender Bericht über den Rothmannschen großhirnlosen Hund nach klinischer und anatomischer Untersuchung. Z. Ges. Neurol. Psychiat. *87*, 247–313 (1923)
50. Scherb, R., Arienti, A.: Ist die Myokinesigraphie als Untersuchungsmethode objektiv zuverlässig? Schweiz. Med. Wochenschr. *75*, 1077–1079 (1945)
51. Sherrington, C.S.: The integrative action of the nervous system. New Haven: Yale University Press 1906
52. Sherrington, C.S.: Flexion-reflex of the limb, crossed extension-reflex, and stepping and standing. J. Physiol. (London) 28–121 (1910)
53. Spalteholz, W.: Handatlas der Anatomie des Menschen, Vol. 40/3, 9th ed. Leipzig: Hirzel 1920
54. Spatz, H.: Anatomie des Mittelhirns. In: Handbuch der Neurologie, Vol. I. Bumke, O., Foerster, O. (eds.), pp. 474–540. Berlin: Springer 1935
55. Uexküll, J. v.: Die ersten Ursachen des Rhythmus in der Tierreihe. Ergeb. Physiol. *3/II*, 1–11 (1904)
56. Wachholder, K.: Willkürliche Haltung und Bewegung insbesondere im Lichte elektrophysiologischer Untersuchungen. Ergeb. Physiol. *26*, 568–775 (1928)
57. Wachholder, K., Altenburger, H.: Beiträge zur Physiologie der willkürlichen Bewegung. IX. Fortlaufende Hin- und Herbewegungen. Pflügers Arch. Ges. Physiol. Menschen Tiere *214*, 625–641 (1926)
58. Wachholder, K., Altenburger, H.: Beiträge zur Physiologie der willkürlichen Bewegung. X. Einzelbewegungen. Pflügers Arch. Ges. Physiol. Menschen Tiere *214*, 642–661 (1926)

Section V

Biography and Bibliography

Chapter 18

Biographical Data: Walter Rudolf Hess

1881	Born March 17 in Frauenfeld, Switzerland
1900–1905	Studied medicine in Lausanne, Bern, Zurich, Berlin, and Kiel
1906	Graduated as M.D. in Zurich
1906	Assistant in Surgery, Kantonsspital Münsterlingen (Conrad Brunner)
1907–1908	Assistant in Ophthalmology, University of Zurich (Otto Haab)
1908	Married Louise Sandmeier; children: Gertrud (1910) and Rudolf Max (1913)
1908–1912	Practiced Ophthalmology in Rapperswil
1912–1913	Assistant in Physiology, University of Zurich (Justus Gaule)
1913–1917	Privatdozent, University of Zurich
1915	Assistant in Physiology with Max Verworn in Bonn
1917–1951	Professor of Physiology and Director of the Physiology Department in Zurich
1931–1937	Cofounder (1931) and Chairman of the Board of the International High Altitude Research Laboratory at Jungfraujoch
1932	Prix Marcel Benoist (Switzerland)
1934	Doctor honoris causa, University of Berne
1938	President of the Sixteenth International Physiological Congress held in Zurich
1938	Carl Ludwig Medal of the German Society of Circulation Research
1944	Doctor honoris causa, University of Geneva
1949	Nobel Prize
1953	Doctor honoris causa, McGill University, Montreal
1959	Doctor honoris causa, University of Freiburg im Breisgau
1967	Moved to Ascona, Switzerland
1973	Died August 12, in Locarno, Switzerland

Chapter 19

From Medical Practice to Theoretical Medicine: An Autobiographic Sketch*[1]

W. R. Hess

Ancestry

I should like to begin this short sketch of my life with a few comments concerning my forebears. This will come as no surprise to readers interested in biology, who are well aware of the decisive influences one's inheritance has in the shaping of one's life. In my own case it is rather interesting, as two different folk-groups are represented in my pedigree. My father's ancestors belonged to the Alemanni, a Germanic tribe of the upper Rhine valley formerly under Roman control. In several expansive movements these peoples drove the Romans southward and over the Alps; they then occupied the now German-speaking region of northeastern Switzerland and are considered the principal ancestors of its present population. My mother's people came from the region of the former kingdom of Saxony, where East European types were to be found along with the Nordic. This divergence between the maternal and paternal inheritance generally is of a certain genetic interest.

A rather broad range of trades and professions is to be found in my family tree. It includes farmers, craftsmen, and merchants as well as more distantly related clergymen and politicians who were rather gifted orators.

My father, a college teacher in physics, combined a remarkable sense of order with a teacher's talent. Although fundamentally a very serious man, he always retained a good sense of humor. My mother had an altogether happy and pleasant disposition, which stood in striking contrast to the rather sober mood and reserved attitude so characteristic of the inhabitants of eastern Switzerland at that time. It is no wonder that we had little contact with our neighbors, who looked upon us almost as strangers with different customs and social standards. My mother also had an unusually

---

* Hess, W.R.: From medical practice to theoretical medicine. An autobiographic sketch. Perspect. Biol. Med. *6*, 400–423 (1963)

1 This autobiographic sketch was translated from the original German manuscript by Dr. E. Castagnoli, an American physician, presently in Zürich

lively disposition. She bubbled over with energy – so much so, in fact, that she often had a hard time falling asleep. I, too, was to suffer from this difficulty throughout my life. I have always had to struggle to shut out the impressions and experiences of the day.

Early Environment

The editors who invited me to write a short autobiography encouraged me to include a psychological background to my life story. This is in keeping with modern autobiographic writing and means that the role played by the external environment in developing, shaping, and limiting an individual's inherited capabilities must not be overlooked.

My own childhood was provided with ample opportunities to become aware of nature at firsthand. My early curiosity about the world around me was stimulated and supported by my father. Even at the age of five, I used to explore the fields and meadows to collect plants, and every new specimen meant an exciting experience, for it was brought home and carefully classified with father's help. Soon I had by herbarium and a collection of butterflies. As time went on, I became aware of the significance of the ecological setting, that is to say, of the specific interrelationship between flora and fauna. I could see, for example, that a particular species of caterpillar was always to be found on a certain plant. Such observations were unforgettable. Even much later I remembered the exact place where they had occurred. More and more it became clear that functional manifestations, such as the germination of a seed or the rapid sprouting of a shoot from a willow, were more apt to capture my mind than purely morphological features. Perhaps most fascinating to me was the observation of the metamorphosis from caterpillar to chrysalis to butterfly. I just couldn't get enough of nature. How beautiful were those solitary walks through fields and forests and along streams and rivers. By the time I entered elementary school I had thoroughly discovered the countryside surrounding my village.

Student Years

The phase in my life that did much to form my personality began with my entrance into the Gymnasium at the age of twelve. My father soon allowed me to visit his physics laboratory and to help out in setting up the experiments for his classes. This early contact with classical physics was of great value to me in later life. It was during this time that I saw how the seemingly stationary processes in so-called static systems were in reality a system of antagonistic forces resulting in a dynamic equilibrium.

During my free time I used to make toys such as bows and arrows, sail boats, and airplanes from improvised materials to be found in and around the house. This did much to develop not only manual skills but also a certain practical sense and inventiveness. On the other hand, I had to overcome many difficulties with my studies. It was hard for me to remember isolated data which could not be associated with significant events. I also had trouble memorizing the vocabularies in Greek, Latin, and French. In short, in the linguistic subjects I was not a good student; I was poor also in composition. Throughout my life I have had difficulties in expressing my thoughts in writing. On the other hand, I have always been able to express myself clearly and convincingly in speaking. When involved in an interesting debate or discussion, the words flow freely. But when I try to put these same ideas on paper, I run into trouble, the sentences usually being too long and complicated. I have so often envied the writing skills of others.

The question naturally arises how it came about that someone like myself, whose true interests lie in natural science, should come to study medicine. I think that this can be attributed to the influence of our family doctor who treated me for pleurisy. This bout with tuberculosis cast the only shadow on my otherwise happy youth. Being very lively by temperament, it was all the harder for me to heed the warnings against overexerting myself.

I decided during the last semesters in the Gymnasium on the path I would take following graduation. The medical profession seemed to offer a course of training which would not be too severe in its demands. In addition, both seeing the application of science for the benefit of people and the doctor's role in the community appealed to me.

University life opened up new dimensions for an essentially small-town boy. First of all, I had to decide at which university I would study. My first choice was the University of Lausanne, located in the French-speaking area of Switzerland. There I would also have the opportunity of getting to know the essentially different traditions and customs of my fellow citizens who were of Roman descent. Following the first year, I took advantage of an opportunity to study for one semester in the capital city of Bern, where I was exposed to the well-functioning democracy of my country. I then went to Zürich, where I completed the preclinical courses in anatomy, histology, biochemistry and physiology. The urge to travel – inherent in all youth – overcame me and I went to Berlin, at that time still under the Kaiser's rule. In addition to its famous clinics, Berlin had many interesting things to offer, such as museums, a zoological garden, an oceanarium, and an astronomical observatory. The countryside of Berlin, with its wide plains and open spaces, was especially interesting and novel for a Swiss accustomed to the hills and mountains of his country. Everything in the fields of art and theater was also at one's disposal. Despite all these varied possibilities, I remained faithful to my love for nature.

In 1903 I had an excellent opportunity to experience the sea in the German city of Kiel, whose harbor also served as a base for the German navy. How longingly my eyes followed the ships as they disappeared over the horizon into distant lands.

For me, the most unforgettable personality on the medical faculty of Kiel was the internist Quincke. He had the talent of presenting his cases in a very interesting way; I missed hardly any of his lectures. I also learned a great deal of surgery under Helferich, who was an excellent teacher, although to be sure an air of cynicism wafted through his lectures or demonstrations. The lectures in dermatology and venerology were excellent; the clinical case presentations were especially impressive, as more than adequate material was available from the port district.

Following the completion of my studies in Berlin, and after having devoted much time to personal scientific interests during the summer semester spent in Kiel, I returned to Switzerland. At the turn of the century, the length of study for doctors was ten semesters – that is, for an industrious student. Because I had devoted much time to my side interests, I felt that my prospects of passing the final examinations were rather poor. Nevertheless, I made up the classes I had missed in Berlin and Kiel by adhering to a rigid work schedule during the last two semesters in Zürich. I took no notes whatsoever, but gave my full attention to the lectures and demonstrations. For reading material, I consulted textbooks and atlases. This study procedure was supplemented with periodic discussions with my fellow students who were also preparing for final exams and who were informed about the "specialties" of the examiners. I successfully passed the examinations in 1905 and received the diploma bestowing the right to practice medicine in Switzerland.

Inklings of Research

Looking back on my days in medical school, I can still remember one incident which is worthy of mention. During a dissecting session in the Anatomical Institute in Zürich the instructor called our attention to an anomaly in the arterial blood supply of the foot. The arteria dorsalis pedis was not formed as a direct continuation of the arteria tibialis anterior as it usually is. Instead, a well-developed anastomosis with the arteria tibialis posterior was present. This fact, pointed out without further elaboration by the anatomist, interested me and I asked for an explanation. The teacher's answer that the anomaly was not at all infrequent did not satisfy my scientific curiosity. I finally gathered up enough courage and presented my own interpretation. It seemed clear to me that the differential resistance between the arteria tibialis anterior and posterior must be of considerable significance. The path with the least resistive drop in pressure carries the blood more effectively to the peripheral segments since it maintains the greater pressure reserves. It is this pressure reserve which procures blood for the vascular system of the foot via the anastomosis between two arterial branches. Thus, the above mentioned anomaly is a direct example of the relationship between hemodynamics and morphogenesis in the arterial system.

I can no longer remember the content of the discussion which followed in the dissecting room. In any event, the question remained open and arose again in Berlin in a different context. There I met W. Roux, the anatomist at the nearby University of Halle, who was the founder of the science of developmental mechanics. Because I was interested in the question, I was keen to learn the opinion of such a distinguished scientist, who had also worked on the problem of the relationship between hemodynamics and the morphological formation of the arterial tree with special reference to the angle of its branches. I therefore presented my ideas to him in the form of a letter and was very happy to receive a reply inviting me to Halle for a personal discussion. As a result, Roux suggested that the theory be published in the *Archiv für Entwicklungsmechanik*. It is understandable that my interest in scientific work received a strong stimulus. It led to a student's paper entitled „Eine mechanisch bedingte Gesetzmässigkeit im Bau des Blutgefässystemes" ("A Mechanically Induced Conformity in the Structure of the Vascular System"). My concern with this problem led to further consequences, as the reader will discover later.

Choice of the place for my first internship was dictated by financial circumstances, which did not permit me to pursue a course of study in theoretical medicine, as I would have liked, following the completion of my training in 1905. From then on I had to earn my own living. My first assignment was a residency in the state hospital of my home canton rather than an unpaid position in a university clinic.

The department of surgery at the hospital was under the direction of Dr. Conrad Brunner, whose reputation extended beyond the borders of Switzerland. Retrospectively, I can see how much I profited from having worked in the well-disciplined team under Brunner. In addition to the routine daily chores, I came to understand and appreciate the skilled technique and high sense of responsibility of this great surgeon. Furthermore, I had the opportunity to see parts of the vascular system in vivo, which intensified my early interests in hemodynamics.

The relationship between the morphological organization of the vascular system and the flow characteristics of blood factors became more apparent in determining circulatory phenomena. I could also see that in practical medicine more attention was paid to the dynamic action of the heart pump than to peripheral factors such as the resistance and blood viscosity. The role of the latter was almost completely ignored, and therefore I decided to construct a convenient apparatus to measure blood viscosity for clinical usage. The problem was intriguing because the existing viscosimetric apparatus (e.g., Ostwald) applicable to so-called idealized fluids could not be used for measuring blood viscosity. Due to the coagulability of blood, the measurements would have to be completed within 2 to 3 minutes, and, for practical purposes, 1 to 2 drops of blood – as used in determining hemoglobin values – would have to suffice. In addition, it was desirable to develop a procedure which would allow the results to be read off directly, thereby avoiding calculations. How to construct an apparatus which could fulfil all these requirements occupied almost all my free time. I would concentrate on the problem especially during walks. After I had probed many varied constructions in my mind, a possible solution all at once occurred to me. It had the character of a sudden inspiration whereby the latent psychological tension becomes discharged. Unforeseen difficulties had to be overcome in order to actualize the mental concept into a practical, usable instrument. Much patience was required and modifications had to be made often to arrive at the goal.

At last the plan was fulfilled and measurements were made. The results and conclusions were published in a small treatise entitled "Viscosität des Blutes und Herzarbeit" ("Viscosity of the Blood and the Work of the Heart"). Dr. Conrad Brunner, my chief, who freely admitted that he had an inadequate knowledge of the problem, left the future of the work entirely in my hands. I therefore personally submitted the paper for publication in *Pflüger's Archiv für Physiologie.* I had to undergo a rather painful experience because of the authoritarian attitude then prevalent toward youthful scientists. Today, of course, the necessity and duty of promoting and encouraging young people in scientific research are well established. Pflüger wanted to know under whose direction the work had been carried out. Upon explaining to him that it had been done entirely on my own, he rejected the paper without comment. Fortunately for me, a member of the Zürich medical faculty who heard of the situation interested himself on my behalf. It was Professor Zangger who took over the printing of the work in *Vierteljahresschrift der Zürcher Naturforschenden Gesellschaft.*

The main finding in the paper was the demonstration that the viscosity of the blood varies with the number of circulating erythrocytes, which is one of the factors that may influence peripheral resistance. It further suggested that decreased viscosity of the blood can lead to turbulence and may be responsible for the so-called anemic heart murmur.

In the following years the science of viscosimetry of blood plasma and serum expanded considerably and, in combination with refractometry, it became possible to differentiate the albumin and globulin fractions in plasma or serum. An increasing clinical interest in blood viscosity manifested itself between the years of 1908 and 1920. Thereafter interest declined, probably because more convenient and practical methods (e.g., measurement of the sedimentation rate) had been developed. Nevertheless, for special cases, interest in the viscosimetry of blood and other body fluids has remained until the present time.

During my first year of internship I was naturally concerned about my future, so I set a firm goal for myself. I wanted to work in some branch of medicine which would give me an adequate income while at the same time allowing me sufficient time to devote myself to problems of basic research. A practice in ophthalmology seemed ideal to meet these two requirements. I therefore seized the first opportunity to enter the department of ophthalmology at the University of Zürich as a resident early in 1906. My new chief, Professor O. Haab, was an outstanding clinician and skilled surgeon. The somewhat one-sided morphological orientation prevalent in the clinic did not quite suit my views; however, I had ample opportunities to make diagnostic observations of a more dynamic character – for example, in analyzing oculomotor disturbances.

According to the technique employed at that time, the examiner relied on the patient's report as to the site of the double image when a moving light with changing directions was flashed before his eyes. With such a method the examiner was, of course, too dependent on the patient's intelligence, and quite often an exact diagnosis had to be foregone. I decided, therefore, to develop a method which would give more objective results concerning the co-ordination of eye muscles. I shall not describe the procedure in detail. Suffice it to recall that by employing an apparatus called "co-ordimeter," it was possible to determine the motor defect quantitatively. Furthermore, it was possible to evaluate with satisfactory accuracy the effects of treatment through repeated controls. The apparatus has undergone improvements and modifications since its inception, but the novel principle has held its own till now. More important to me personally was the clear insight I gained into the problem of motor co-ordination which – many years later – added greatly to my understanding of the analysis of motor responses during stimulation in the region of the thalamus and subthalamus.

The same was true of a thorough investigation of stereoscopic vision, which presents an instructive example of the integrative function of the central nervous system. In conjunction with the above interests, I tried to develop a procedure for making stereoscopic photographs, with the hope that I might capitalize on it and obtain thereby a certain degree of financial freedom. In fact, I succeeded in making three-dimensional transparencies which gave a real view into the depths of the object. The pictures, however, lacked certain desirable qualities which I was trying to incorporate when World War I halted my attempts. At the end of the war my position was more secure and other problems were occupying my attention.

Before the war I had completed my residency in ophthalmology and had furthered my education through a brief period of study in Paris. Soon thereafter I took over the practice of an ophthalmologist in an attractive city and important railroad intersection in the vicinity of Zürich. By the spring of 1908 I was living rather comfortably as a practicing eye specialist and was, therefore, in a position to marry my fiancée, Louise Sandmeier, who had worked in the outpatient department of the Zürich Eye Clinic. As a former doctor's aide, she was well prepared to offer efficient help in her husband's office, and thus we were fortunate enough to overcome our financial difficulties for the time being.

Leap Into the Unknown – New Years of Training

The security and compensation derived from an extensive practice had negative aspects. I had set up a laboratory in which I spent my free time working on problems of hemodynamics, particularly in relation to blood viscosimetry, and on problems of motor co-ordination. The ever increasing demands of the practice, and especially the many professional consultations and opinions, left me with less time to pursue my scientific interests. I would come home from stimulating scientific meetings in Zürich and feel downright unhappy and dissatisfied because my hands were tied. A deep conflict resulted then between my sense of duty to provide for my family and my longings for pure science. Finally, a critical decision had to be made, as a position became available for me at the Institute of Physiology in Zürich. I can remember so well the exact spot in my garden where I decided in favor of science, despite the fact that I now also had a child to support, and that the decision entailed a considerable reduction in income. Of course I discussed the problem beforehand with my wife, who agreed to the step which was so significant for my later life.

The move was made in 1912. It wrought mental tensions which were not always easy to bear. My new teaching assignment made it necessary for me to review scientific material – old and new – and much was actually learned while teaching. Because of the poor health of my chief, Professor Gaule, a pupil of Carl Ludwig, personal contact with him for purposes of scientific discussion was rather limited, and I was more or less on my own.

I found no lack of pressing problems to tackle. Predominant among these was the economy of energy in the mechanics of circulation, which presents a dynamic parallel, for example, to the architectural patterns of spongeous bone, known to be built so as to insure optimal bearing strength.

Take the head of the femur as an example. While the conditions obtaining in hemodynamics are not so obvious as in static systems, they are, nevertheless, of fundamental significance. In order to understand properly the hemodynamic situation, we must realize that the energy output of the heart is greatly dependent upon the configuration of the vascular system, especially of the arteries. Significant criteria are (*a*) the relation between the cross-sectional area of a vessel and the sum of the areas of its branches; (*b*) the angle between the branching vessel and the axis of its trunk; and (*c*) the distance to be covered between the heart and the capillary bed of an organ.

Given the above parameters, it was possible to set up an idealized model for an optimal vascular energy system for purposes of comparison with the performance of conditions prevailing in a living circulatory system. The formulation of such an ideal model may be regarded as an example of theoretical biology. The second procedure, testing the theory in vivo, represents, on the other hand, unequivocal experimental biology. I presented the above work for my *Habilitationsschrift*, which was accepted by the medical faculty in 1913. My presentation was not as convincing as the evidence found in osteology in regard to the adaptive structuralization of spongeous bone, partly because the former requires mathematical treatment before it becomes obvious and the latter presents itself directly to the eyes of a scientifically naïve observer. As I glance back after fifty years of experience, I would now tackle the problem somewhat differently; nevertheless, my basic attitude of laying a theoretical foundation before setting up an experimental procedure would remain unaltered.

In my position as *Privatdozent* I was authorized to give laboratory courses, which I organized according to my own inclinations – i.e., biophysically, but not without clinical perspectives. This ideal situation was, unfortunately, terminated by the outbreak of World War I. In the general mobilization of the Swiss army which followed, I, along with everyone else fit for military service, was called to active duty. My work at the physiology institute was completely interrupted. As the war went on, and the increasing superiority of the Allies became apparent, it was sometimes possible to return to work for a few months. I even contemplated continuing my scientific training abroad. Above all, I wanted contact with general physiology.

I found it in the laboratory of Max Verworn, Pflüger's successor in Bonn. Verworn had extensive experience with animal physiology, acquired as a pupil of Haeckel. He was especially familiar with so-called elementary organisms and the basic requirements for life. A widely read book by Verworn entitled *Allgemeine Physiologie* (1922) contains more information concerning his work. Along with the formal lectures and laboratory work, the almost daily private contact at tea time with Verworn, when we discussed teaching methods in general, the work of our colleagues, and even topics in art, literature, and philosophy, did much to broaden me and prepare me for the future. I will be forever thankful for the kindness shown to me by this man. He never let me be aware of the vast distance which separated him – a famous, experienced, and wise man – from myself, a young and inquisitive scientist. I was allowed complete freedom in experimental work and was able, therefore, to concentrate further on problems of hemodynamics.

As a ready supply of dogs was available for experimental purposes, I was able to investigate in living animals the regulation of regional blood flow, i.e., its adaptation to the changing needs of specific organs. At that time the concept of the "peripheral heart" was a topic of much discussion, especially among clinicians who understood thereby the total muscular component of the arterial system. The pulse was ascribed partially to an active force, i.e., the centrifugal wave of arterial contraction released by the contraction of the heart and resulting in the forward propulsion of the blood.

The nonbranching sector of the common carotid artery seemed to me to be well suited for experimentally testing the above concept. The technique used was as follows. Through a small longitudinal incision in the proximal part of the common carotid artery a glass stopper was so inserted as to stop the flow of blood. To prevent any backflow into the area under investigation, the system was also closed off cranially with a stopper provided with a system for optically registering pressure changes in the thus isolated arterial section. The systemic blood pressure was occasionally monitored in the femoral artery, particularly in connection with artificial pressure increases in the carotid artery. All attempts to alter the blood pressure, either in the isolated section of the carotid or in the peripheral arterial system, ended in failure. At the end of my year's leave of absence I could only report negative results.

Soon thereafter I learned by reading the papers of Hering, Jr., and Heymans that I had just missed an opportunity to make an important discovery. My own procedure involved the common carotid up to the point where it bifurcates, thus excluding the carotid sinus. Had I included the sinus, I could not have failed to observe a massive fall in pressure due to the now well-known reflex control of blood pressure. This oversight really irked me, but I consoled myself with the thought that similar failures had certainly happened to many other investigators. In any case, I didn't allow myself to be detracted from the study of hemodynamics.

Upon returning to Zürich, it was not possible for me to continue the experimental work aimed at clarifying the adaptation of blood supply to the changing requirements of individual organs. I was called upon for further military service in addition to my teaching schedule, which took up all my time. Further research, then, had to wait until the end of World War I.

An Unexpected Turn of Events

As mentioned previously, my chief, Professor Justus Gaule, head of the department of physiology in Zürich, had suffered from poor health for many years, which hindered him from being very active in teaching and prevented him completely from doing research. Conditions finally became so critical that Gaule announced his resignation just before the winter semester of 1916. I was asked to take charge of the lectures and laboratories temporarily until the appointment of the permanent holder of the chair. I would have personally preferred to spend several more years as a *Privatdozent*, with adequate time for research, rather than to be acting chairman with all the administrative and teaching duties. Nevertheless, with healthy optimism and steadfast determination I began the work ahead of me. In relying on the experience gained during my years with the handling of mechanical and optical equipment, I planned to use such extensively in the laboratory experiments. I was also counting on drawing from my internship experience, and particularly on the operative skills I had acquired as an assistant surgeon and as an ophthalmologist. Finally, conceptual references between normal and pathological physiology as well as the insights gained into general physiology under Verworn turned out to be most helpful in the new situation.

Zürich Faculty

It should be noted that in the German-speaking region of Switzerland appointments to professorial chairs were at that time rather complicated and entailed far greater attention to the so-called "school" of the prospective candidate than, for example, in the United States, where much more liberal principles for selection operate. Additional difficulties arose in general from the fact that the relatively young University of Zürich lacked a corps of graduates trained in science. Therefore, German candidates usually had an advantage from the beginning, and many positions were occupied by Germans, some of whom were really outstanding. This method for gathering academic personnel was in vogue during the time of the expansive tendencies of the Wilhelmian epoch. In any event, the German professors took it for granted that a Swiss person could not be considered seriously for an academic position in a theoretical medical discipline; a Swiss scientist was more likely to be intrusted with a practical assignment.

Under these circumstances, then, a proposal was prepared at the end of the winter semester which foresaw calling a German professor to the chair in physiology. With due consideration of "his seniority and much larger training," he had produced far more scientific papers than I and had undeniably good qualities, especially in the field of physiological chemistry. In spite of this, he had not yet been called to a leading position in Germany and was still in line for one.

Since, however, the administration and management of the institute during the winter semester had gone along smoothly under my acting chairmanship, and because the lectures and laboratories had been well attended, the university authorities decided to re-examine the entire situation more thoroughly. An invited opinion written by Verworn which had been withheld by the faculty till then was also brought to my support. Finally, the summer semester turned out to be another success, to which two very capable assistants generously contributed, E. Rothlin and A. Fleisch – subsequently professors, respectively, of pharmacology at Basle and of physiology at Lausanne. The decision of the nominating committee came out in my favor, and in the fall of 1917 the chair for physiology was officially intrusted to me.

Development of a Program for Instruction and Research

Now that I was responsible for the development of physiology, it was mandatory for me to become familiar with the management and setup in other universities. I had the great fortune of being able to get to know and to observe the English physiologists of the classical period, as they were still working in their laboratories. Starling gave me complete freedom to gain an insight into his programming and methodology for lecture sessions and lab work. Through Langley I came to understand how the institute in Cambridge was run. I saw Sherrington at work only briefly, as he was very busy and implied that he did not want to be bothered. With the help and kindness of one of his assistants, however, I was allowed to find out about everything of interest to me. Many years later I met this assistant – who now holds a leading position in the United States – at an international congress (I believe in Montreal). Naturally the topic of conversation turned to Professor Sherrington. Because of my own later experience, I had long since come to understand Sherrington's reserve and aloofness.

In addition to this visit to England and later travels, I also participated in various congresses to which I owe a great deal for the broadening of my knowledge. I can still vividly remember the demonstration of dogs suffering from rickets presented by E. Mellanby, later Lord Mellanby, with the help of his wife, later Lady Mellanby, at the International Physiological Congress in Paris in 1920. At that time Mellanby demonstrated the effects of malnutrition – the concept of vitamins was still controversial. One could hear the differences of opinion being argued more critically in the corridors than in the demonstration hall in the presence of the speakers themselves.

A great scientific event also occurred at the congress in Edinburgh in 1923, where, under the leadership of J. J. R. MacLeod, a series of lectures included those by Banting and Best in which the chain of events which led to the discovery of insulin was unfolded. I returned from still other scientific meetings enriched in experiences and knowledge which could be incorporated into my lectures by way of illustrating them with experimental material. The meetings of the *Deutsche Gesellschaft für Naturforscher und Ärzte* were particularly stimulating and important for me. At these sessions I attended primarily the general lectures. Otherwise, I sat in on those given by the internists. As a teacher, I was much concerned with the advances in internal medicine, as my students, upon entering the clinical semesters, should be able to correlate clinical material with their knowledge of physiology.

A special feature of these meetings was the chance to become personally acquainted with other scientists in the more relaxed atmosphere outside the lecture halls. Apart from shop talk, these personal contacts provided important insights for evaluating, to some extent, the publications of the various institutions. Such first-hand impressions saved me a lot of time in selecting literature for study.

Research Projects

For the time being hemodynamics remained the chief subject of research in my institute. It included research work concerning the autonomic properties and the reactivity of isolated arteries. It was clear to me that information obtained on an isolated arterial section could not by itself be decisive for the intact living arterial apparatus. On the other hand, I was convinced that the inherent functional properties of the arterial tube, even though normally subject to nervous influences, would be of great significance in themselves. In line with these thoughts, arterial rings were removed from various parts of the arterial system and tested for their spontaneous behavior while submersed in physiological salt solution. In addition, the rings were stimulated mechanically and electrically with varying parameters as well as chemically with biologically active substances – among others, adrenalin and acetylcholine. These experiments were extended over a period of years and yielded the material which, with inclusion of the reflex mechanisms, provided a general insight subsequently elaborated in 1930 in the monograph *Regulation of the Circulatory System*. From the close relationship and synergic functioning of the circulatory and respiratory apparatus, it followed automatically that *The Regulation of Respiration* be handled in an analogous manner (1931).

From discussions with a few colleagues, I occasionally heard that reference to the concept of a meaningful organization of physiological systems was here and there regarded as an expression of teleological tendencies. In replying to such comments, I have always maintained that living matter can only exist in the form of organization oriented toward success; indeed, that the investigation of such organization poses the central problem of biology. The knowledge and insight gained by experimental methods and the collected pool of facts allowing a synthesis between organization and performance truly constitute the grand aim of an *integral physiology*, which can never be fulfilled with the mere collecting of facts. The deleterious effects of disturbances, as in sickness, confirm the above viewpoint. The development of this theme was worked out and later formally expressed in the monograph (1948) *Die funktionelle Organisation des vegetativen Nervensystems (Functional Organization of the Vegetative Nervous System).*

The shift to a seemingly new topic – namely, examination of active regulatory dispositives at the higher level of the diencephalon – will readily be understood as a logical consequence of a long-range project. From an earlier programmatic work entitled "Correlations between Psychic and Vegetative Functions" (1924), in conjunction with and from knowledge of certain clinical experiences and from exploratory experimental investigations of Karplus and Kreidl, certain conclusions could be drawn. It became apparent that before further progress could be expected, a new experimental technique which could adequately fulfil the more differentiated demands of the modern investigation would first have to be developed. For it was my goal to test the behavior of experimental animals (cats) subjected to electrical stimulation and diathermic lesions without impairing the freedom of movement of the animal by immobilization or narcosis.

For this purpose, the very finest electrodes were needed which could be inserted into the hypothalamus without flexing and bending. In addition, to use the induction coil as a stimulating device was out of the question, for the relatively slowly arising autonomic effects in the face of the rapidly occurring potential deflections produced by induction could be all too easily overshadowed by reactions from somatic sensorimotor systems and could, under certain conditions, not even have time to be formed. I was more hopeful for a form of stimulation with pulsating direct current, provided that the cathodic electrode points could be depolarized between the single impulses, thus maintaining a consistent stimulus intensity also during prolonged series of stimulation. The decisive advantage therein proved to lie in the possibility of adapting duration and shape of the stimulating pulses to the time properties of the nervous elements under examination.

Apart from these precautionary measures concerning stimulation technique, the problem of anatomical identification of the stimulated structure had to be solved so as to allow the closest possible correlation between symptom and reacting substrate. Since required accuracy can never be achieved on the basis of electrode placements alone, microscopic control of serial histological sections of each experimental brain is necessary. Three atlases composed of photomicrographic reproduction of serial sections cut perpendicularly to each other gave invaluable assistance in the topographical analysis of stimulation sites. In addition, these atlases served to register in each individual case the histological finding as well as to prepare cumulative map files of all points stimulated during the many years of investigative work. Cross-references among these three standard atlases allowed localization of each site of stimulation in all three planes, thereby providing three-dimensional registration, which in many cases was indispensable to demonstrate the relation between topographical and functional organization. Cinematography was used extensively from 1927 on as a further aid in recording observations. It enabled us to make detailed comparisons between symptoms of experiments separated from one another by years. This would have been impossible by reliance on verbal notes alone. Finally, a filing system was developed which contained references to the symptoms produced by stimulation; this enabled one in the train of data integration to obtain full particulars of a desired case together with the verbal protocol, photomicrographic localization, and cinematographic documents.

The technical aspects of this entire program to investigate the functional organization of the diencephalon were reported in detail in a monograph (1932) entitled *Die Methode der lokalisierten Reizung und Ausschaltung subkortikaler Hirnabschnitte* ("The Methodology of Localized Stimulation and Destruction of Subcortical Brain Areas"). This as well as the second part, published 1938, concerning blood pressure and respiratory effects elicited by localized diencephalic stimulation did not become too well known because of World War II. The group of competent workers in this field was small at that time, and all the type of this edition was destroyed in Leipzig during World War II.

At the very beginning of the exploratory experiments (1925) I had reckoned with from two to three years for gathering data sufficient to obtain a certain insight. But the mere manual work involved in the technique described above used up this much and even more of my time. Also, a heavy teaching schedule did not make things go any faster. Nevertheless, in 1927, the first results were made known at a meeting of the Deutsche Physiologische Gesellschaft in Frankfurt. Of course, following the conclusion of the first series of about fifty cases, no insight into general laws governing the functional organization was obtained; the symptoms were too inconsistent. Therefore, a second experimental series was carried out – which did not, however, do much to clarify the situation.

Only little by little and ever so slowly did the veil lift a bit here and there so that twenty-five years had passed before I could finally think of putting together all the many single research publications that had appeared over the years which had been concerned with individual symptoms. The vast number of experiments turned out to be decisive; for generalization concerning symptoms, syndromes, and localizations could be supported only by such a large body of data. An integrative review of all these efforts was published in two monographs (1947, 1948). They were well received, and in 1949, "for your discovery of the functional organization of the diencephalon and its role in the coordination of the functions of the inner organs," the Nobel Prize was bestowed upon me. The reward was shared with Egaz Moniz, who had developed the prefrontal lobotomy as a therapeutic method in the management of certain psychoses.

The long way which culminated in this high distinction has been presented here in a rather detailed manner because during the war Switzerland was totally closed off and precious little information concerning my investigative methods and results had reached readers abroad. Even with the war's end the fact that all my publications were written in German, the accepted written language of central and eastern Switzerland, proved to be for many years an almost insurmountable barrier in re-establishing a close scientific contact with the Anglo-Saxon countries.

Apart from behavioral and autonomic symptoms, interesting somatic motor effects came to light during the course of the experiments. Through these the relationships of certain formations of the thalamus as well as of the mesencephalon with the so-called extrapyramidal motor system were established. Last but not least, our attention was drawn to psychomotor phenomena on which I reported in a recently published treatise (1962) entitled *Psychologie in biologischer Sicht* ("Psychology from the Standpoint of a Biologist").

Incidents, Personalities, and Anecdotes

During my active years as professor of physiology on the medical faculty at Zürich, the above described work was often interspersed with noteworthy incidents and problems which have remained vividly in my memory. One of these could be cited as an example of "fortune in misfórtune." When I first came to the physiological institute of our university, there was a noticeable shortage of both equipment and space, without which a successful work program was rather difficult – or, more plainly, impossible. Significant improvement was out of the question for the time being. The basement and first floor of the four-storied building which had been constructed during the time of the famous physiologist Hermann, the predecessor of Gaule, were occupied by the physics department, which was also cramped for space. As a result of these unsatisfactory conditions, the department of theoretical physics and a succession of its famous staff members, such as Einstein, von Laue, and Schrödinger, had to work in the smallest space of all. A "higher power" intervened in this precarious situation. On the evening before New Year's Eve in 1923, just before retiring, I received a telephone call informing me that our institute was going up in flames. As I arrived on the spot, the entire top floor was already burning. The fire department went to work immediately. The captain of the fire squad was assured that it would not be necessary to evacuate material from the lower floors. As though to spite this comforting prognosis, however, the fire suddenly broke through the ceiling and threatened to destroy the main portions of our institute. It was high time, then, to get books – and anything else of value – out of the building as quickly as possible and to take them through the deep snow to neighboring institutes. During this "operation evacuation" our basic sense of democracy, valid for all social levels, was dramatically illustrated. For, in addition to the staff members of the physics department and other institutes, members of the executive branch of our state government were also on hand doing their very best to help out. Since only my chief assistant and myself were informed about the value of the inventory, we became responsible for giving directions as to the priority for objects to be saved. Thus it arose that the shoe came to be on the other foot, in that our employers, the political authorities, had to take orders from their employees, i.e., from the professor and his assistants. On the following day, after the fire was finally extinguished, we had to collect our possessions from the neighboring buildings. Of course, everything stored in the top floor was lost. In addition, the building itself had been seriously damaged. This state of affairs gave the cue to the second, happy, part of my story; for, during the repair work, an additional floor was added and thus our need for space alleviated. We moved into the newly repaired and expanded institute with the sincere hope of being able to devote ourselves to our work without disturbance.

This expectation, however, all too soon proved to be illusory. In our direct democracy a constitutional law exists whereby action can be initiated by the people provided, that a sufficient number of signatures are submitted. A movement of the community touching at the very heart of medicine and biochemistry arose banning all animal experiments in the canton of Zürich. Members of the physiology institute in Zürich were called upon by the medical faculty to combat this attack, which was aimed at destroying a technique indispensible to scientific progress. We had to prepare ourselves for the battle, which, unfortunately, would take up a lot of time. In place of scientific work, it now became necessary to write popular newspaper articles and to speak at public gatherings in order to present the true interest of the people. The problem of financing the campaign, in the face of the opposite community drive on behalf of the protection of animals, was not a simple matter. When the polls opened, our side was understandably on pins and needles. Members of our campaign headquarters and many interested colleagues had assembled in a hotel room to keep informed on the incoming results from the city and the rural areas. Happily, our efforts had not been in vain. The popular vote resulted in a two-thirds majority in favor of freedom of research with reasonable restrictions, It is impossible to imagine what the fate of our faculty might have been had animal experimentation been declared a criminal offense.

Another equally laborious but more constructive undertaking into which I was drawn by chance turned out to be more than I had bargained for. The Naturforschende Gesellschaft (Swiss Society for Scientific Research) had taken upon itself, due to the promptings of one of its members, a competent meteorologist, to set up a committee for the purpose of constructing a research station at the highest point in Switzerland accessible by railroad – namely, on the Jungfraujoch, over 11,000 feet above sea level. The scientific promoter of the plan died before he could carry it out. Therefore, a new presidium had to be selected for the commission, and it came about that the job fell to me. I had suspected that acceptance would entail no small sacrifice, but in various respects this was a great underestimate.

Originally a Swiss undertaking was envisioned. I myself tried to have the project internationalized. One of the main reasons was the thought that thereby new contacts between representatives of countries which had been enemies during World War I could arise, which could help promote understanding among peoples of different nations. A second point in my program was to include all natural sciences, and particularly biology. One source of resistance was laid down in the initial formulation of objectives of the research station, according to which it was to be used primarily for meteorological studies and for work on geophysical problems. Fortunately the branch of the federal government charged with the main authority was farsighted enough to interpret the formulation of the objectives of the plan so that all disciplines of scientific research having an established interest in work in the high Alpine region should receive equal consideration.

With agreements on internationalization and nonrestriction as to field of research secured, it was time to look around for partners who were prepared to make a financial contribution for the project. We had the satisfaction of receiving positive responses from many countries, among which may be mentioned the then Kaiser Wilhelm Society in Berlin (presently Max-Planck-Gesellschaft), University of Paris, Royal Society of London, Academy of Sciences in Vienna, National Foundation for Scientific Research in Brussels, and, finally, the Rockefeller Foundation in New York, which, however, declined to participate on the international committee. Naturally, it was of prime importance that the Jungfrau Railway Company reaffirmed its active and financial co-operation, which it had promised before the grant for construction of the railway had been made, as well as its readiness to help out in any other problems concerning construction and operation.

No effort could be spared in overcoming the conditions which were to be expected in working on the rugged, fissured precipices. One had to reckon with unexpected stone slides and water breakthroughs as well as unseasonal heavy snowfall. Another problem was the extreme diurnal variation in temperature. The stone material to be used for the building was carefully selected to withstand the intensive heating by high-altitude solar radiation during the day and exposure to freezing temperatures during the night. A completely unexpected event made a joke of all our careful calculations – for in an area which had been stable and inactive for seventy years, an avalanche suddenly broke lose, interrupting the flow of supplies for valuable weeks.

But at long last, in the summer of 1931, the research station was dedicated in the presence of prominent representatives of the participating countries and institutions as well as Swiss government authorities of the highest rank. A special volume prepared for the memorable occasion included contributions from W. Nernst and v. Ficker (Berlin), Ch. Fabry and L. Lapicque (Paris), Sir F. G. Hopkins and G. M. B. Dobson (Oxford), A. Durig, and W. Schmidt (Vienna), E. de Wildeman, Gand, and J. Jaumotte (Liège), as well as from many others. Every donor country was represented by two delegates. With the reading of the declaration statement "For the Scientific Brotherhood" which had been prepared by the physiologist Lapicque, we fraternally joined hands.

After being responsible for directing the commission through five years of development and construction in addition to a further five years of service as director of the newborn institute, I felt relieved of a great burden when I finally stepped down. A. v. Muralt, from Bern, proved himself more than adequate and an excellent choice for the succession.

I should mention another event which brought additional honors and duties to our institute – the Sixteenth International Congress of Physiology in Zürich (1938). It should not be forgotten that we stood at that time at the brink of World War II. The annexation of Austria by Hitler was then an accomplished fact. At the last minute, partly out of political, partly out of racist motives, we were exposed to a considerable change in attitude from foreign colleagues and officials. Also, personal tensions between participants of the congress could not be ignored. It was not even certain whether the congress would be held at all. Nevertheless, it must be credited to those attending this memorable meeting that proper discipline was maintained and unpleasant incidents were avoided by suppressing any allusions of a political nature in the lectures and in the other events of the congress.

In connection with the deterioration of the warlike situation, the Swiss physiologists were put into an unusual position. Till then they had been associated partly with the German Physiological Society and partly with its French-speaking counterpart. A new way had to be found in loosening our ties from both of the warring factions. With the support of K. Spiro, the biochemist on the Basle medical faculty, I made a motion for the founding of a Swiss Physiological Society.

Inevitably, the next step was the founding in 1943 of a periodical, *Helvetica Physiologica et Pharmacologica Acta.* It is understandable that at first the circulation of the new periodical was rather limited. Everyone's interest was completely directed toward the war. In addition, the passage of printed matter out of Switzerland into, for instance, the United States was barred. Even after the war, the demand for the new journal was small. Yet we had to be satisfied with what had been achieved. Today we note that the journal is enjoying an increase in circulation.

To end this sketch, I have the satisfaction of announcing a gratifying undertaking which is intimately connected to my life's work. As early as 1940 I had assumed the initiative for founding a study group devoted to brain research. Periodically, physiologists, psychiatrists, brain surgeons, internists, pharmacologists, neuro-anatomists, psychoanalysts, and psychotherapists would come together at the physiological institute. At first, we wanted to get to know one another's points of view and to find a common language for the various terms and concepts.

Thanks to the financial support of the Rockefeller Foundation and later to that of the Swiss Academy of Medical Sciences, through its Commission for Brain Research, this study group achieved its goal with ever expanding participation until 1950. Today the original plan has culminated in an institute for brain research, to be opened in 1963.

If one considers my own interest in the functional organization of the brain along with its psychic manifestation – an interest which was expressed as far back as 1924 and which was never lost sight of – one will well understand that the decision of the faculty and the responsible authorities to create a chair for brain research has filled my young old heart with the very deepest satisfaction. Equally gratifying is the fact that I can regard the future of the institute with confidence, as one of my earlier co-workers, K. Akert, presently on the medical faculty, has been named its director.

A last point remains to be mentioned – research costs money and I am glad to acknowledge the help I received. While the small communities which support our Swiss universities were not able to provide adequate means, financial assistance by the Rockefeller Foundation and by a personal benefactor, Benjamin Stern, was of decisive value. In the last few years the Swiss National Foundation for Scientific Research furthered our endeavors.

In anticipation of a possible question from my readers, especially those in America, I must answer whether or not during my active period I found time for a hobby. Through fortunate circumstances, in view of my inclination to financially unpromising basic research, I was awarded a cash prize from the Swiss Marcel Benoist Foundation during the relatively early years of my work. This money was invested in a simple summer house south of the Alps in the region of the so-called "sun terrace of Switzerland." There I tended my vines and cared for my garden whenever I found it necessary or possible to relax a bit. My reward was not only a harvest of grapes and the pleasure given by the southern plants, but also acquisition of a variety of bits of information as well as the manual skills practiced by the farmers there since, presumably, as far back as Roman times. And the Italian attitude and language contributed much to my relaxation and enjoyment. I would quickly recover from overwork and in the presence of lizards, scorpions, Aesculapius snakes, and other crawling beasts so alien to eastern Switzerland, feel myself a new man. In this rustic setting, my mind would often turn to contemplating questions of prehistory, and I would become mentally refreshed and rested.

And now, after sixty years of service on behalf of science, I contentedly put down my pen with the hope that perhaps my work has contributed to scientific progress in my own field, especially toward the understanding of regulatory functions at the integrative level of the organism.

Chapter 20

Bibliography of Writings*

1903–1973

1903 Hess, W.R.: Eine mechanisch bedingte Gesetzmäßigkeit im Bau des Blutgefäßsystems. Arch. Entwicklungsmech. Org. *16*, 632–641

1906 Hess, W.R.: Zum Thema: Viskosität des Blutes und Herzarbeit. M.D. Dissertation Vierteljahrsschr. Naturforsch. Ges. Zürich *51*, 236–251

1907 Hess, W.R.: Ein neuer Apparat zur Bestimmung der Viskosität des Blutes. Münch. Med. Wochenschr. *54*, 1590–1591
Hess, W.R.: Die Bestimmung der Viskosität des Blutes. Münch. Med. Wochenschr. *54*, 2225–2229
Hess, W.R.: Ein neuer Apparat zur Bestimmung der Viskosität des Blutes. Corresp.-Bl. Schweiz. Ärzte *37*, 73–75

1908 Hess, W.R.: Un nuovo apparecchio per la determinazione della viscosità del sangue. Policlinico *15*, 4–6
*Hess, W.R.: Eine neue Untersuchungsmethode bei Doppelbildern. Arch. Augenheilkd. *62*, 233–238 [p. 55]
Hess, W.R.: Die Viskosität des Blutes bei Gesunden. Dtsch. Arch. Klin. Med. *94*, 404–408
Hess, W.R.: Der Einfluß warmer Bäder auf die Viskosität des Blutes. Wien. Klin. Rundschau *32*, 595–597

1909 Hess, W.R.: Die Viskosimetrie des Blutes (Beitrag zur Apparatenfrage). Med. Klin. *5*, 1397–1400

1910 Hess, W.R.: Reibungswiderstand des Blutes und Poiseuillesches Gesetz. Z. Klin. Med. *71*, 421–427

1911 Hess, W.R.: Blutviskosität und Blutkörperchen. Pflügers Arch. Ges. Physiol. Menschen Tiere *140*, 354–362
Hess, W.R.: Zur Kritik der Viskosimetrie. Dtsch. Med. Wochenschr. *37*, 1854–1855
Hess, W.R.: Die graphische Darstellung von Bewegungsstörungen der Augen mit Beispieltafeln zur Diagnose von Augenmuskellähmungen (+14 Erläuterungstafeln für den Text). Arch. Augenheilkd. *70*, 10–16

1912 Hess, W.R.: Über Relationen zwischen Blutbeschaffenheit und Hämodynamik. In: Festschrift zur Eröffnung des gerichtlich-medizinischen Institutes der Universität

* References marked with asterisks are translated into English in this volume, see page numbers in brackets

Zürich, pp. 91–108. Berlin: Hirschwald 1912. Vierteljahrsschr. Gerichtl. Med. 3. Fol. *44*, 93–108
Hess, W.R.: Reibungswiderstand des Blutes und Poiseuillesches Gesetz. Z. Klin. Med. *74*, 428–432
Hess, W.R.: Aufzeichnungs-Formulare für die graphische Darstellung von Bewegungsstörungen der Augen nach Dr. W. R. Hess. Fünfzig Blatt. Wiesbaden. Bergmann
Hess, W.R.: Herstellung plastischer Photographien (mit Demonstrationen). Vierteljahrsschr. Naturforsch. Ges. Zürich *57*, XXXIV–XXXVII
Hess, W.R.: Der Strömungswiderstand des Blutes gegenüber kleinen Druckwerten. Arch. Anat. Physiol. (Abt. Physiol.) *1912*, 197–214

1913 Hess, W.R.: Der Einfluß des Druckes auf den Koeffizienten der Blutviscosität. Berl. Klin. Wochenschr. *50*, 197
Hess, W.R.: *Das Prinzip des kleinsten Kraftverbrauches im Dienste hämodynamischer Forschung*. Master's thesis. Leipzig: Veit. Arch. Anat. Physiol. (Abt. Physiol.) *1914*, 1–62

1914 *Hess, W.R.: Direkt wirkende Stereoskopbilder. Z. Wiss. Photogr. Photophys. Photochem. *14*, 33–38 [p. 61]
Hess, W.R.: Unmittelbar wirkende Stereoskopbilder. Vierteljahrsschr. Naturforsch. Ges. Zürich
Hess, W.R.: Über das stereoskopische Bild. Wissen und Leben 424–430
Hess, W.R.: Über die funktionelle Bedeutung der Arterienmuskulatur. Corresp.-Bl. Schweiz. Ärzte *44*, 993–1000
Hess, W.: Die Wärme im Energiehaushalt des Organismus. Corresp.-Bl. Schweiz. Ärzte *44*, 1281–1292

1915 Hess, W.R.: Gehorcht das Blut dem allgemeinen Strömungsgesetz der Flüssigkeiten? Pflügers Arch. Ges. Physiol. Menschen Tiere *162*, 187–224

1916 Hess, W.R.: Die Arterienmuskulatur als „peripheres Herz“? Pflügers Arch. Ges. Physiol. Menschen Tiere *163*, 555–593
Hess, W.R.: Ein einfaches messendes Verfahren zur Motilitätsprüfung der Augen. Z. Augenheilk. *35*, 201–219

1917 Hess, W.R.: Über die periphere Regulierung der Blutzirkulation. Pflügers Arch. Ges. Physiol. Menschen Tiere *168*, 439–490
Hess, W.R.: Die Kinematographie in Unterricht und Forschung. Verh. Schweiz. Naturforsch. Ges. Zürich 99. Vers., 307–308

1918 *Hess, W.R.: *Die Zweckmäßigkeit im Blutkreislauf.* Akademische Antrittsvorlesung, gehalten am 15. Juni 1918. Basel: Schwabe [p. 35]

1919 Hess, W.R.: Der erste Herzton. Verh. Schweiz. Naturforsch. Ges. Lugano 100. Vers., 143–144
Hess, W.R.: Untersuchungen über den Antrieb des Blutstromes durch aktive Gefäßpulsationen. Pflügers Arch. Ges. Physiol. Menschen Tiere *173*, 243–264

1920 Hess, W.R.: Viscosimeter mit Temperaturregulierung. Pflügers Arch. Ges. Physiol. Menschen Tiere *180*, 61–67
Hess, W.R.: Beitrag zur Theorie der Viskosität heterogener Systeme. Kolloid Z. *27*, 1–11
Hess, W.R.: Die innere Reibung gelatinierender Lösungen. Kolloid Z. *27*, 154–163

Hess, W.R.: Die graphische Aufzeichnung der Herztöne nach neuer Methode. Pflügers Arch. Ges. Physiol. Menschen Tiere *180*, 35–60
Hess, W.R.: Die Entstehung des ersten Herztones. Dtsch. Arch. Klin. Med. *132*, 69–95
Hess, W.R., Gundlach, R.: Der Einfluß des Adrenalins auf die Sekretion des Magensaftes. Pflügers Arch. Ges. Physiol. Menschen Tiere *185*, 122–136
Hess, W.R., Gundlach, R.: Der Einfluß von Hypophysenextrakt auf die Magensaftsekretion. Pflügers Arch. Ges. Physiol. Menschen Tiere *185*, 137–140

1921 Hess, W.R.: Die physiologischen Grundlagen für die Entstehung der reaktiven Hyperämie und des Kollateralkreislaufes. Beitr. Klin. Chir. *122*, 1–19
Hess, W.R.: Die Rolle der Vitamine im Zellchemismus. Hoppe Seylers Z. Physiol. Chem. *117*, 284–308
Hess, W.R.: Die Sensibilität der Kreislaufregulierung. Verh. Schweiz. Naturforsch. Ges. Schaffhausen 102. Vers., 160–161
Hess, W.R., Takahashi, N.: Nachweis eines stofflichen Defizites im Gewebe an Avitaminose erkrankter Tiere. Biochem. Z. *122*, 193–203

1922 Hess, W.R.: Photographische Konzentrationsbestimmung einer Farbstofflösung. Hoppe Seylers Z. Physiol. Chem. *119*, 172–175
Hess, W.R.: Die Rolle der Vitamine im Zellchemismus. Bemerkungen zu der „Ergänzung" von Emil Abderhalden. Hoppe Seylers Z. Physiol. Chem. *120*, 277–280
Hess, W.R., Messerle, N.: Untersuchungen über die Gewebeatmung bei Avitaminose. Hoppe Seylers Z. Physiol. Chem. *119*, 176–189
Hess, W.R., Wyss, W.H. v.: Beitrag zur Kenntnis der Eingeweidesensibilitäten. Pflügers Arch. Ges. Physiol. Menschen Tiere *194*, 195–205

1923 Hess, W.R.: Viscosimetrische Untersuchungen an lyophilen Kolloiden. Rec. Trav. Chim. Pays Bas *42*, 1097–1103
Hess, W.R.: Eine Station für Höhenforschung an der Jungfrau. Q. J. Exp. Physiol. Suppl. Vol.: Proc. XIth Int. Physiol. Congress, Edinburgh 143–144
*Hess, W.R.: Über die Wirkung von Acethylcholin auf den Skelettmuskel. Q. J. Exp. Physiol. Suppl. Vol.: Proc. XIth Int. Physiol. Congress, Edinburgh 144–146 [p. 67]
Hess, W.R.: Die Regulierung des peripheren Blutkreislaufes. Ergeb. Inn. Med. Kinderheilkd. *23*, 1–65
Hess, W.R.: Die physiologischen Grundlagen der pathologischen Blutdrucksteigerung. Schweiz. Med. Wochenschr. *53*, 1077–1082
Hess, W.R.: Die Blausäurevergiftung als Methode der Avitaminoseforschung. Pflügers Arch. Ges. Physiol. Menschen Tiere *198*, 483–489
Hess, W.R.: „Die Rolle der Vitamine im Zellchemismus". Erwiderung auf die Antwort Emil Abderhaldens. Hoppe Seylers Z. Physiol. Chem. *127*, 196–198
Hess, W.R.: Unbewußte Sinnesleistungen. Vierteljahrschr. Naturforsch. Ges. Zürich *68*, III–IV
Hess, W.R.: Metodo grafico per lo studio dei disturbi della motilità oculare. Boll. Oculistica *2*, 614–616
Hess, W.R., Rohr, K.: Über den Einfluß thermischer Vorbehandlung von Trockenhefe auf ihre Reduktionsleistung und ihre Vitaminwertigkeit, mit einem Beitrag zur Kenntnis der Vogelberiberi. Hoppe-Seylers Z. Physiol. Chem. *129*, 268–283
Hess, W.R., Ryffel, W.: Atmungsversuche an überlebender Fischmuskulatur. Ein Beitrag zur Biologie des Muskels. Pflügers Arch. Ges. Physiol. Menschen Tiere *200*, 124–131
Hess, W.R., Zimmermann, A.: La viscosité du suc pancréatique. Arch. Intern. Physiol. *21*, 191–203

1924 Hess, W.R.: Über die Wirkung der Vitamine. Dtsch. Med. Wochenschr. *50*, 163–164
Hess, W.R.: Toxikologische Untersuchungen im Dienste der Avitaminoseforschung. Arch. Exp. Pathol. Pharmakol. *103*, 366–379
Hess, W.R.: Zur Physiologie der Vasomotoren. Schweiz. Arch. Neurol. Psychiatr. *14*, 20–29
Hess, W.R.: Die reflektorische Ruhigstellung schmerzender Körperteile. Pflügers Arch. Ges. Physiol. Menschen Tiere *203*, 539–541
Hess, W.R.: Die Prüfung der motorischen Koordination der Augen im graphischen Verfahren. Laupen (bei Bern): Polygraphische Gesellschaft
Hess, W.R.: Méthode graphique pour l'examen de la motilité oculaire. Rev. Gen. Ophtalmol. *38*, 5–20
Hess, W.R.: Die Physiologie der Arbeit. Naturwissenschaften *12*, 1031–1039
Hess, W.R.: Diskussion zu Michaud, L.: Le rôle du système nerveux autonome dans la pathogénèse des maladies internes. Schweiz. Arch. Neurol. Psychiatr. *14*, 308–309
Hess, W.R., Neergaard, K. v.: Die Beziehungen der Acetylcholinverkürzung des Skelettmuskels zur Einzelzuckung und zum Tetanus. Pflügers Arch. Ges. Physiol. Menschen Tiere *205*, 506–517

1925 *Hess, W.R.: *Über die Wechselbeziehungen zwischen psychischen und vegetativen Funktionen*. Schweiz. Arch. Neurol. Psychiatr. *15*, 260–277 (1924); *16*, 36–55, 285–306 [p. 71]
Hess, W.R.: Über die Bedeutung der Fusion für die motorische Koordination der Augen. Klin. Monatsbl. Augenheilkd. *75*, 289–294
Hess, W.R.: Die physiologische Bedeutung des Sportes für die Jugenderziehung. Schweiz. Z. Gesundheitspflege *5*, 1–6
Hess, W.R.: Die Wirkung von Ergotamin auf das Auge. Klin. Monatsbl. Augenheilkd. *75*, 295–300
Hess, W.R., Messerle, N.: Untersuchungen über die motorische Koordination der Augen. Pflügers Arch. Ges. Physiol. Menschen Tiere *210*, 708–720

1926 Hess, W.R.: Klima und Schlaf. In: Verhandlungen der klimatologischen Tagung in Davos 1925, pp. 1–8. Basel: Schwabe
Hess, W.R.: Untersuchungen über den Einfluß der Peripherie auf die Regulierung der Herztätigkeit. Schweiz. Med. Wochenschr. *56*, 690–692
Hess, W.R.: Untersuchungen über den Mechanismus der Kreislaufregulierung. Pflügers Arch. Ges. Physiol. Menschen Tiere *213*, 163–170
Hess, W.R.: Funktionsgesetze des vegetativen Nervensystems. Klin. Wochenschr. *5*, 1353–1354
Hess, W.R.: Die Temperaturregulierung im Bienenvolk. Z. Vergl. Physiol. *4*, 465–487
Hess, W.R.: Experimentelles zur Dynamik der Augenmuskeln. Arch. Augenheilkd. *97*, 460–466
Hess, W.R.: Vegetatives Nervensystem und Auge. Klin. Monatsbl. Augenheilkd. *77*, 708
Hess, W.R.: Praktische Erfahrungen in der Diagnostik von Augenmuskelstörungen. Klin. Monatsbl. Augenheilkd. *77*, 708–709
Hess, W.R.: Photogrammatlanten von Stammganglien und Zwischenhirn der Katze, Standardserie. Physiologisches Institut Zürich
Hess, W.R., Lehmann, F.E.: Der Einfluß vegetativer Reizstoffe auf den Adaptierungszustand der Retina. Pflügers Arch. Ges. Physiol. Menschen Tiere *211*, 603–611
Hess, W.R., Rehsteiner, R.: Die Wirkung von Acetylcholin auf den Zuckungsablauf des Froschmuskels. Pflügers Arch. Ges. Physiol. Menschen Tiere *214*, 463–470

1927 Hess, W.R.: Die innere Reibung des Blutes. In: Handbuch der physikalischen und technischen Mechanik, Vol. V. Auerbach, F., Hort, W. (eds.), pp. 471–472. Leipzig: Barth

Hess, W.R.: Das Löschphänomen der Acetylcholin-Contraktur. Pflügers Arch. Ges. Physiol. Menschen Tiere *217*, 511–515

Hess, W.R.: Methodisches zur Untersuchung von Störungen der Augenmotilität. Klin. Monatsbl. Augenheilkd. *79*, 289–292

Hess, W.R.: *Die Gesetze der Hydrostatik und Hydrodynamik*. In: Handbuch der normalen und pathologischen Physiologie, Vol. VII/2: Blutzirkulation II. Bethe, A. et al. (eds.), pp. 889–903. Berlin: Springer

Hess, W.R.: *Die Verteilung von Querschnitt, Widerstand, Druckgefälle und Strömungsgeschwindigkeit im Blutkreislauf*. In: Handbuch der normalen und pathologischen Physiologie, Vol. VII/2: Blutzirkulation II. Bethe, A. et al. (eds.), pp. 904–933. Berlin: Springer

Hess, W.R.: Stammganglien-Reizversuche. Ber. Ges. Physiol. Exp. Pharmakol. *42*, 554–555

Hess, W.R.: Bewegungsanalytische Filme. Ber. Ges. Physiol. Exp. Pharmakol. *42*, 574–575

Hess, W.R.: Das Schlafzentrum. Verh. Schweiz. Naturforsch. Ges. Basel 108. Vers., 247–248

Hess, W.R., Büsch, J.: Der Einfluß von Atropin auf die Reaktion des Skelettmuskels bei direkter Reizung. Pflügers Arch. Ges. Physiol. Menschen Tiere *216*, 644–650

1928 Hess, W.R.: Hirnreizversuche über den Mechanismus des Schlafes. Arch. Psychiatr. Nervenkr. *86*, 287–292

Hess, W.R.: Praktische Ernährungslehre auf wissenschaftlicher Grundlage. Führer durch die Wanderausstellung für Ernährungslehre. Zürich: Schweizer. Zentralstelle für Gesundheitspflege

1929 Hess, W.R.: Über den Schlaf. Vierteljahrsschr. Naturforsch. Ges. Zürich *74*, XIX–XX

Hess, W.R.: Versteckte Feinde – Wissen gibt Macht – Von den Vitaminen – Gewissenskonflikt – Tierschutz im Tierversuch. In: Aussprache zwischen Vertretern der experimentellen Medizin und Vivisektionsgegnern über die Zweckdienlichkeit und Zulässigkeit des wissenschaftlichen Tierversuches. Tierfreund *55*, 32–36

Hess, W.R.: The Alpine research station of Switzerland. J. State Med. *37*, 11–14

Hess, W.R.: The mechanism of sleep. Am. J. Physiol. *90*, 386–387

Hess, W.R.: Lokalisatorische Ergebnisse der Hirnreizversuche mit Schlafeffekt. Arch. Psychiatr. Nervenkr. *88*, 813–816 [p. 131]

1930 Hess, W.R.: Die Funktionen des vegetativen Nervensystems. Klin. Wochenschr. *9*, 1009–1012

*Hess, W.R.: Kritik der Hering-Breuer'schen Lehre von der Selbststeuerung der Atmung. Pflügers Arch. Ges. Physiol. Menschen Tiere *226*, 198–211 [p. 211]

Hess, W.R.: *Die Regulierung des Blutkreislaufes, gleichzeitig ein Beitrag zur Physiologie des vegetativen Nervensystems*, pp. 1–162. Leipzig: Thieme

Hess, W.R., Arthus, M.: Traité pratique d'alimentation rationnelle à base scientifique. Guide de l'exposition itinérante d'alimentation rationnelle. Zürich: Centrale Suisse pour la Propagande d'Hygiène

1931 Hess, W.R.: Die Verwendung des Schmalfilms für biologische Zwecke. In: Handbuch der biologischen Arbeitsmethoden, Abt. II: Physikalische Methoden, Vol. 2/II. Abderhalden, E. (ed.), pp. 2317–2336. Berlin: Urban & Schwarzenberg

Hess, W.R.: Biologie und Universitas. Vierteljahrsschr. Naturforsch. Ges. Zürich *76*, 7–15
Hess, W.R.: Le sommeil. C.R. Soc. Biol. (Paris) *107*, 1333–1364
Hess, W.R.: Der Schlaf. Schweiz. Med. Wochenschr. *61*, 849
Hess, W.R.: *Die Regulierung der Atmung, gleichzeitig ein Beitrag zur Physiologie des vegetativen Nervensystems*, pp. 1–137. Leipzig: Thieme
Hess, W.R.: Entstehungsgeschichte und Ziele des Forschungsinstitutes. In: Hochalpine Forschungsstation Jungfraujoch. Internationale Stiftung Bern, pp. 11–16. Zürich: Orell Füssli
Hess, W.R.: Die Grundlagen der Arbeitsphysiologie. Tech. Rundschau (Bern) *23/45*, 1–2
Hess, W.R.: The autonomic nervous system. Lancet *223*, 1199–1201

1932 Hess, W.R.: Hirnstamm-Reizversuche. Dtsch. Z. Nervenheilkd. *124*, 132–136
Hess, W.R.: *Beiträge zur Physiologie des Hirnstammes. I. Die Methodik der lokalisierten Reizung und Ausschaltung subkortikaler Hirnabschnitte*, pp. 1–122. Leipzig: Thieme
Hess, W.R.: Demonstration: Der Hering-Breuer'sche „Schaltreflex" in neuer Deutung. Schweiz. Med. Wochenschr. *62*, 263
Hess, W.R.: Demonstration der Methodik zur lokalisierten Reizung und Ausschaltung subkortikaler Hirnabschnitte. Schweiz. Med. Wochenschr. *62*, 1042

1933 Hess, W.R.: Motorische Zentren im Hypothalamus. Schweiz. Med. Wochenschr. *63*, 457
Hess, W.R.: Über den Mechanismus der Bewegungsinnervation. Schweiz. Med. Wochenschr. *63*, 1030
*Hess, W.R.: Der Schlaf. Klin. Wochenschr. *28*, 129–134 [p. 115]
Hess, W.R.: Teil und Ganzes im Organismus. Schweiz. Lehrerztg. [Suppl. to No. 44] *78*, 21–25

1934 Hess, W.R.: Modell-Demonstrationen zum Zentrenproblem. Schweiz. Med. Wochenschr. *64*, 583
Hess, W.R.: Plastizitätslehre und Lokalisationsfrage. Verh. Dtsch. Ges. Inn. Med. 46. Kongr. 212–218
Hess, W.R.: Diskussionsvotum zum Vortrag Muralt, A. v.: Über die Blutregulation im Höhenklima. Schweiz. Med. Wochenschr. *64*, 1174
Hess, W.R.: Die Struktur der starren und der gesteuerten Motorik. Schweiz. Med. Wochenschr. *64*, 1176
Hess, W.R.: Medizinisch-biologische Fragen und Allgemeines über den Betrieb des Internationalen Forschungsinstitutes. Vierteljahrsschr. Naturforsch. Ges. Zürich *79*, XIX–XX
Hess, W.R.: Die hochalpine Forschungsstation Jungfraujoch. Vierteljahrsschr. Naturf. Ges. *79*, 1–26

1935 Hess, W.R.: Beispiele funktionellen Denkens. In: Festschrift Heinrich Zangger, pp. 769–775. Zürich: Rascher
Hess, W.R.: Das physiologische Zusammenspiel von Kreislauf und Atmung. Verh. Dtsch. Ges. Kreislaufforsch. 8. Tagung, 9–31
Hess, W.R.: Diskussionsvotum zum Vortrag Sauerbruch, F.: Physiologie und Klinik des Zwerchfells. Schweiz. Med. Wochenschr. *65*, 567
Hess, W.R.: Physiologische Varia. Schweiz. Med. Wochenschr. *65*, 648
Hess, W.R.: Aktuelle Probleme der Physiologie. Schweiz. Med. Wochenschr. *65*, 852–854

Hess, W.R.: Filmdemonstrationen zur Physiologie des Zwischenhirns. Verh. Schweiz. Physiol. (Juni) pp. 18–19. Basel: Schwabe
Hess, W.R.: Zentrale Vertretung von Hilfsfunktionen des vegetativen Systems. Schweiz. Med. Wochenschr. *65*, 729
Hess, W.R.: Diskussionsvotum zu Leuthardt, F.: Allgemeines über die Regulierung der Reaktion und ihre physiologische Schwankungsbreite. Schweiz. Med. Wochenschr. *65*, 1035
*Hess, W.R.: Kritisches zum Zentrenbegriff. In: Problèmes de biologie et de médecine. Volume jubilaire dédié au Prof. Lina Stern, pp. 43–49. Leningrad, Moscow: Governmental Publications in Science and Medicine [p. 205]

1936 Hess, W.R.: Hypothalamus und die Zentren des autonomen Nervensystems: Physiologie. Arch. Psychiatr. Nervenkr. *104*, 548–557
Hess, W.R.: Die Rolle des Vagus in der Selbststeuerung der Atmung. Pflügers Arch. Ges. Physiol. Menschen Tiere *237*, 24–39
Hess, W.R.: Ziel- und Fertigkeitsbewegungen in der Zeitlupe. Verh. Ver. Schweiz. Physiol. (June), pp. 6–7. Basel: Schwabe
Hess, W.R.: Diskussion zu Ruzicka, L.: Über die Sexualhormone, unter besonderer Berücksichtigung der männlichen. Schweiz. Med. Wochenschr. *66*, 1020
Hess, W.R.: Zentrale Regulierung von Kreislauf und Atmung. Schweiz. Med. Wochenschr. *66*, 1078
Hess, W.R., Wyss, O.A.M.: Die Analyse der physikalischen Atmungsregulierung an Hand der Aktionsstrombilder des Phrenicus. Pflügers Arch. Ges. Physiol. Menschen Tiere *237*, 761–770

1937 Hess, W.R.: Die Lokomotion von Tier und Mensch in der Zeitlupe. Verh. Ver. Schweiz. Physiol. (January), p. 6. Basel: Schwabe
Hess, W.R.: I. Reflexbeziehungen zwischen afferente Vagus und Innervation der respiratorischen Thoraxmuskulatur. II. Beziehungen zwischen der Innervation synergistischer Muskeln. Verh. Ver. Schweiz. Physiol. (June), pp. 5–6. Basel: Schwabe
Hess, W.R.: O pokusném vyšetřování vegetativního nervového systému. Biol. Listy *22*, 77–82

1938 Hess, W.R.: Über experimentelle Untersuchungen des vegetativen Nervensystems. Med. Klin. *34*, 323–326
Hess, W.R.: Inverser Vaguseffekt auf das Zwerchfell. Verh. Ver. Schweiz. Physiol. (January) p. 7. Basel: B. Schwabe
Hess, W.R.: Diskussion zu Wagner, H.: Der heutige Stand der operativen Therapie der Netzhautablösung. Schweiz. Med. Wochenschr. *68*, 703
Hess, W.R.: Wo steht heute die Physiologie und welches sind ihre aktuellen Probleme? Schweiz. Med. Wochenschr. *68*, 945–947
Hess, W.R.: Eröffnungs-Ansprache des Kongreß-Präsidenten. 16. Int. Physiol. Kongr. Zürich *3*, 96–103
Hess, W.R.: Rückblicke auf den XVI. Internationalen Physiologen-Kongreß. Schweiz. Med. Wochenschr. *68*, 1331
Hess, W.R.: *Beitrag zur Physiologie des Hirnstammes*. II. Das Zwischenhirn und die Regulation von Kreislauf und Atmung, pp. 1–127. Leipzig: Thieme

1939 Hess, W.R.: Beziehungen zwischen Winterschlaf und Außentemperatur beim Siebenschläfer. Z. Vgl. Physiol. *26*, 529–536
Hess, W.R.: Die funktionelle Organisation des Nervensystems. Vierteljahrsschr. Naturforsch. Ges. Zürich *84*, 197–210
Hess, W.R.: Diskussion zu Rossier, P.: Die Bedeutung des Carotis-Sinus für die Klinik. Schweiz. Med. Wochenschr. *69*, 329

Hess, W.R.: Kollaboration der Hirnforschung durch Standardisierung der Hilfsgeräte. Schweiz. Arch. Neurol. Psychiatr. *43*, 221–224
Hess, W.R.: La coordination des mécanismes régulateurs du système respiratoire. J. Physiol. Pathol. Gén. *37*, 1233–1238
Hess, W.R.: Eine Stellungnahme der Schweizerischen Medizinisch-biologischen Gesellschaft zum Tierversuch. Schweiz. Med. Wochenschr. *69*, 892–894
Hess, W.R.: Pupille und Zwischenhirn. Klin. Monatsbl. Augenheilkd. *103*, 407–413
Hess, W.R.: Zentrogene Einflüsse auf Iris und Augenmuskeln. Schweiz. Med. Wochenschr. *69*, 843
Hess, W.R.: Prof. Justus Gaule† 1849–1939. Verh. Ver. Schweiz. Physiol. (July), pp. 3–4. Basel: Schwabe
Hess, W.R.: Beziehungen des Hypothalamus zu Trieb- und Stimmungsäußerungen. Verh. Ver. Schweiz. Physiol. (July), pp. 5–9. Basel: Schwabe
Hess, W.R.: Georg Justus Gaule (1849–1939); Mitglied der Gesellschaft seit 1887. Vierteljahrsschr. Naturforsch. Ges. Zürich *84*, 375–377

1940 Hess, W.R.: Ergebnisse von Reizversuchen in Zwischenhirn und Nachbargebieten. Pflügers Arch. Ges. Physiol. Menschen Tiere *243*, 409–430
Hess, W.R.: Beitrag zur Technik des zentralen Reizversuches. Pflügers Arch. Ges. Physiol. Menschen Tiere *243*, 431–438
Hess, W.R.: Die Ernährung der Zivilbevölkerung. Schweiz. Z. Gemeinnützigkeit *79*, 190–199
Hess, W.R.: Statistische Analyse zentraler Reizeffekte. Verh. Ver. Schweiz. Physiol. (June), pp. 15–16. Basel: Schwabe
Hess, W.R.: Zwischenhirn und Motorik. Pflügers Arch. Ges. Physiol. Menschen Tiere *243*, 634–650
Hess, W.R.: Motorische Symptome im Gesicht bei Reizung im Zwischenhirn. Pflügers Arch. Ges. Physiol. Menschen Tiere *243*, 678–685
Hess, W.R.: Diencephale Reizsymptome am Körperstamm; Beziehungen derselben zu Bewegungseffekten im Gesicht und an den Extremitäten. Pflügers Arch. Ges. Physiol. Menschen Tiere *243*, 741–747
Hess, W.R.: Aufbau und Aufbauordnung physiologischer Funktionen. Umschau *47*, 737–739
Hess, W.R.: Der Reflextonus des Zwerchfells und der Einfluß der Lungenentfaltung auf den Atmungstypus. Veröff. der Reichsstelle für den Unterrichtsfilm zu dem Archivfilm Nr. B 438/1940
Hess, W.R.: Zwischenhirn und Gleichgewichtsregulierung. Verh. Schweiz. Naturforsch. Ges. Locarno, 120. Vers., 212

1941 Hess, W.R.: Weitere Beobachtungen über den tonischen Vaguseinfluß bei verschiedenem konstantem Lungenvolum. Pflügers Arch. Ges. Physiol. Menschen Tiere *244*, 360–364
Hess, W.R.: Zwischenhirn und Gleichgewichtsregulierung. Schweiz. Med. Wochenschr. *71*, 893–895
Hess, W.R.: Beziehung von Zwischenhirn zur Organisation der Bewegung. Verh. Ver. Schweiz. Physiol. (July), p. 4. Basel: Schwabe
Hess, W.R.: Experimenteller Beitrag zur Frage der extrapyramidalen Motorik. Z. Ges. Neurol. Psychiatr. *172*, 639–642
Hess, W.R.: Die Motorik als Organisationsproblem. Biol. Zentralbl. *61*, 545–572
*Hess, W.R.: Charakter der im Zwischenhirn ausgelösten Bewegungseffekte. Pflügers Arch. Ges. Physiol. Menschen Tiere *244*, 767–786 [p. 223]
Löffler, W., Hess, W.R.: Über Isthmusstenose der Aorta als pathologisch-physiologisches Problem. Schweiz. Med. Wochenschr. *71*, 636–647

1942 Hess, W.R.: Das vegetative Funktionssystem. In: Schweizer Medizinisches Jahrbuch, pp. 51–67. Basel: Schwabe

*Hess, W.R.: *Biomotorik als Organisationsproblem.* Naturwissenschaften *30*, 441–448, 537–541 [p. 243]

Hess, W.R.: Der naturwissenschaftliche Film für Forschung und Unterricht. Vierteljahrsschr. Naturforsch. Ges. Zürich *86*, XXVI–XXVIII

Hess, W.R.: Die S.E.D.-Methode in der Hirnforschung. Verh. Ver. Schweiz. Physiol. (June), pp. 22–23. Basel: Schwabe

Hess, W.R.: Vegetatives Nervensystem. Fragen der Organisation der Begriffe und Bezeichnungen. Schweiz. Arch. Neurol. Psychiatr. *50*, 1–5

Hess, W.R.: Physiologische Aspekte der extrapyramidalen Motorik. Nervenarzt *15*, 457–466

Hess, W.R., Bartorelli, C., Bucher, V.: Diencephale Motorik und Marchi-Degeneration. Z. Ges. Neurol. Psychiatr. *174*, 340–352

1943 Hess, W.R.: Die Regulation der Laktogenese als Beispiel allgemein-physiologischer Gesetzlichkeit. Schweiz. Med. Wochenschr. *73*, 53–60

Hess, W.R.: Induzierte Störungen der optischen Wahrnehmung. Nervenarzt *16*, 57–66

Hess, W.R.: Alte und neue Ziele der Physiologie. Vierteljahrsschr. Naturforsch. Ges. Zürich *88*, 67–76

Hess, W.R.: Von den Gefühlen und Stimmungen eines Hundes. Ein Beitrag zur vergleichenden Psychophysiologie. Z. Vgl. Physiol. *30*, 123–138

Hess, W.R.: Symptomatik des durch elektrischen Reiz ausgelösten Schlafes und die Topographie des Schlafzentrums (Film) Helv. Physiol. Acta *1*, C 61–C 62

*Hess, W.R.: Teleokinetisches und ereismatisches Kräftesystem in der Biomotorik. Helv. Physiol. Acta *1*, C 62–C 63 [p. 265]

Hess, W.R.: Das Zwischenhirn als Koordinationsorgan. Helv. Physiol. Acta *1*, 549–565

*Hess, W.R., Brügger, M.: Das subkortikale Zentrum der affektiven Abwehrreaktion. Helv. Physiol. Acta *1*, 33–52 [p. 183]

Hess, W.R., Brügger, M.: Der Miktions- und der Defäkationsakt als Erfolg zentraler Reizung. Helv. Physiol. Acta *1*, 511–532

Hess, W.R., Magnus, W.O.C.: Leck- und Kau-Automatismen bei elektrischer Reizung. Helv. Physiol. Acta *1*, 533–547

1944 Hess, W.R.: Diskussion zu Sulzer, W.: Arbeitsschutz und Arbeitsmedizin. Schweiz. Med. Wochenschr. *74*, 270

Hess, W.R.: Wandernde Tonusstörungen nach zentraler Reizung. Helv. Physiol. Acta *2*, 131–135

*Hess, W.R.: Hypothalamische Adynamie. Helv. Physiol. Acta *2*, 137–147 [p. 171]

*Hess, W.R.: Das Schlafsyndrom als Folge dienzephaler Reizung. Helv. Physiol. Acta *2*, 305–344 [p. 131]

Hess, W.R.: Alfred Vogt †. Von seinem Leben und Schaffen. Vierteljahrsschr. Naturforsch. Ges. Zürich *89*, 136–140

*Hess, W.R.: *Kollektive Ordnung in biologischem Aspekt.* In: Festschrift Max Huber – Vom Krieg und vom Frieden, pp. 151–172. Zürich: Schulthess, [p. 3]

Hess, W.R., Stoll, W.A.: Experimenteller Beitrag betreffend die Regulierung der Körpertemperatur. Helv. Physiol. Acta *2*, 461–480

1945 Hess, W.R.: Der extrapyramidal-motorische Innervationsapparat. Helv. Physiol. Acta *3*, C 3–C 4

Hess, W.R.: Von den höheren Zentren des vegetativen Funktionssystems. Bull. Schweiz. Akad. Med. Wiss. *1*, 138–166

1946 Hess, W.R.: Experimentelle Darstellung des mesencephalen Blickzentrums, seiner Struktur und Funktionsweise. Helv. Physiol. Acta *4*, C3–C4
Hess, W.R.: Demonstrationen und Erläuterungen zur Funktion des Tectum opticum. Bull. Schweiz. Akad. Med. Wiss. *2*, 51–54
Hess, W.R.: Statistik betreffend die Ernährungsverhältnisse während der Kriegszeit. Bull. Eidgen. Gesundheitsamtes Nr. *45*, 1–8
Hess, W.R., Müller, H.R.: Schnupperbewegungen als zentrale Reizeffekte. Helv. Physiol. Acta *4*, 339–345
Hess, W.R., Müller, H.R.: Einflüsse des Mittel- und Zwischenhirns auf die Atmung. Helv. Physiol. Acta *4*, 347–358
Hess, W.R., Bürgi, S., Bucher, V.: Motorische Funktion des Tektal- und Tegmentalgebietes. Monatsschr. Psychiatr. Neurol. *112*, 1–52 [p. 269]
Hess, W.R., Brügger, M., Bucher, V.: Zur Physiologie von Hypothalamus, Area praeoptica und Septum, sowie angrenzender Balken- und Stirnhirnbereiche. Monatsschr. Psychiatr. Neurol. *111*, 17–59

1947 Hess, W.R.: Die Relativität biologischer Reaktionen. Schweiz. Med. Wochenschr. *77*, 3–9
Hess, W.R.: Stufen im Aufbau des vegetativen Nervensystems. Schweiz. Med. Wochenschr. *77*, 659–665
Hess, W.R.: *Vegetative Funktionen und Zwischenhirn*. Helv. Physiol. Acta [Suppl.] *IV*, 1–65
Hess, W.R.: Das endophylaktisch-trophotropische System und seine Beziehung zur Symptomatik der Überempfindlichkeit. Schweiz. Med. Wochenschr. *77*, 1151–1154

1948 Hess, W.R.: *Die funktionelle Organisation des vegetativen Nervensystems*, pp. 1–226. Basel: Schwabe
Hess, W.R.: *Zwischenhirn und Motorik*. Helv. Physiol. Acta [Suppl.] *V*, 1–55
Hess, W.R.: Korrespondierende Symptome aus Stirnhirn, innerer Kapsel und vorderem Thalamus. Helv. Physiol. Acta *6*, 731–738

1949 Hess, W.R.: Kleine Beiträge zur Biologie der Wirkstoffe. Helv. Physiol. Acta *7*, 178–188
Hess, W.R.: *Das Zwischenhirn. Syndrome, Lokalisationen, Funktionen*, pp. 1–187. Basel: Schwabe
Hess, W.R.: Die Bedeutung des Hypothalamus für die Regulierung vegetativer Funktionen. Dtsch. Arch. Klin. Med. *195*, 55–57
Hess, W.R.: Das Zwischenhirn als Organisator vegetativer Funktionen. Nervenarzt *20*, 533–536
Hess, W.R.: Le sommeil comme fonction physiologique. J. Physiol. (Paris) *41*, 61-A–67-A
Hess, W.R.: Motorik und Zwischenhirn. Arch. Psychiatr. Nervenkr. *184*, 267–270
Hess, W.R.: Funktionen des Zwischenhirns im Rahmen der extrapyramidalen Motorik. Bull. Schweiz. Akad. Med. Wiss. *5*, 221–226
Hess, W.R., Weisschedel, E.: Die höheren Reflexzentren der regulierten Körperhaltung. Helv. Physiol. Acta *7*, 451–469
Hess, W.R., Koella, W., Szabo, Th.: Experimentelle Studien über die antagonistische Innervation. Z. Ges. Exp. Med. *115*, 481–490

1950 Hess, W.R.: Physiologie du thalamus. Rapport présenté au IVe Congrès Neurologique International, Paris 1949. Paris: Masson
Hess, W.R.: Symposion über das Zwischenhirn mit einleitenden Referaten und Demonstrationen. Helv. Physiol. Acta [Suppl.] *VI*, 5–80

Hess, W.R.: Die zentrale Regulation der Tätigkeit innerer Organe. In: Le Prix Nobel en 1949, pp. 115–123. Stockholm: Norstedt

Hess, W.R.: Die Kontrolle von Wirkstoffen am Irispräparat im isometrischen Verfahren. Helv. Physiol. Acta *8*, C49–C50

Hess, W.R.: Überblick über die Dokumente von 25 Jahren Hirnforschung. Helv. Physiol. Acta *8*, C50–C51

*Hess, W.R.: Funktion und nervöse Regulation der inneren Organe. Vierteljahrsschr. Naturforsch. Ges. Zürich *95*, 249–264 [p. 17]

Hess, W.R., Akert, K.: Die Koordination morphologischer und physiologischer Untersuchungsmethoden in der Hirnforschung. Folia Psychiatr. Neurol. Neerl. *53*, 268–274

Hess, W.R., Koella, W.: Experimentelle Studien über die antagonistische Innervation II. Z. Ges. Exp. Med. *116*, 431–443

Hess, W.R., McDonald, D.: Film "Basilar artery". Helv. Physiol. Acta *8*, C51

1951 Akert, K., Hess, W.R., McDonald, D.A.: Autonomic and somatic activity evoked by stimulation of frontal cortex and thalamus in cats. (Film). J. Physiol. (London) *113*, 19P–20P

Hess, W.R.: Die automatische Regulierung der Körperhaltung. Experientia *7*, 51–58

Hess, W.R.: Prinzipien organischer Ordnung am Beispiel des vegetativen Nervensystems. Klin. Wochenschr. *29*, 105–111

Hess, W.R.: Beitrag zur experimentellen Analyse des Stirnhirnes. Bull. Schweiz. Akad. Med. Wiss. *7*, 295–306

Hess, W.R.: Methodische Erfahrungen über die experimentelle Erforschung des Zentralnervensystems. Helv. Physiol. Acta *9*, 367–377

Hess, W.R., Akert, K.: Die corticale Repräsentation von Gesicht, Oral- und Pharyngealsphäre bei der Katze. Helv. Physiol. Acta *9*, 269–289

Hess, W.R., Weisschedel, E.: Aus dem Zwischenhirn ausgelöste motorische Symptome an den Extremitäten und im Gesicht. Nervenarzt *22*, 14–22

Hess, W.R., Akert, K., McDonald, D.A.: Beziehungen des Stirnhirnes zum vegetativen System. Helv. Physiol. Acta *9*, 101–124

Hess, W.R., McDonald, D.A., Livingston, R.B.: Head-turning and eye deviation evoked by cortical excitation in the freely moving cat. Fed. Proc. *10*, 63

1952 Hess, W.R.: Experimentalphysiologie und Psychologie. Helv. Physiol. Acta *10*, 85–92

Hess, W.R.: Fundamentos de la regulacion vegetativa. Folia Clin. Int. *2*, 1–6

Hess, W.R.: Aktuelle Aufgaben der Physiologie. Wien. Klin. Wochenschr. *64*, 621–623

Hess, W.R.: Vom Lichtreiz zur bildhaften Wahrnehmung. Helv. Physiol. Acta *10*, 395–402

Hess, W.R.: Physiologische Grundlagen der Ästhetik. Helv. Physiol. Acta *10*, 462–468

Hess, W.R., Akert, K., McDonald, D.A.: Functions of the orbital Gyri of cats. Brain *75*, 244–258

1953 Hess, W.R.: Vergleichende Physiologie und pharmakologische Untersuchungen der Iris-Muskulatur. Arch. Exp. Pathol. Pharmakol. *219*, 284–290

Hess, W.R.: Die physiologische Interpretation pathologischer Symptome aus dem Hirnstamm. Schweiz. Arch. Neurol. Psychiatr. *72*, 405–406

Rüegg, J.C., Hess, W.R.: Die Wirkung von Adrenalin, Nor-Adrenalin und Acetylcholin auf die isolierten Irismuskeln. Helv. Physiol. Acta *11*, 216–230

1954 Hassler, R., Hess, W.R.: Experimentelle und anatomische Befunde über die Drehbewegungen und ihre nervösen Apparate. Arch. Psychiatr. Nervenkr. *192*, 488–526
Hess, W.R.: *Das Zwischenhirn, Syndrome, Lokalisationen, Funktionen*, 2nd ed. pp. 1–218. Basel: Schwabe
Hess, W.R.: Das vegetative Funktionssystem im Lichte der experimentellen Forschung. Neuralmed. *2*, 240–251
Hess, W.R.: *Diencephalon. Autonomic and extrapyramidal functions*, pp. 1–79. New York: Grune & Stratton

1955 Hess, W.R.: Übergeordnete corticale Zentren vegetativer Regulationen? Ber. Ges. Physiol. Exp. Pharmakol. *172*, 116
Hess, W.R., Akert, K.: Experimental data on role of hypothalamus in mechanism of emotional behavior. Arch. Neurol. Psychiatr. (Chicago) *73*, 127–129

1956 Hess, W.R.: *Hypothalamus and Thalamus. Documentary pictures – Hypothalamus und Thalamus-Experimental-Dokumente*. Atlas with German and English legends, pp. 1–70. Stuttgart: Thieme
Hess, W.R.: Beziehungen zwischen psychischen Vorgängen und Organisation des Gehirnes I. Stud. Gen. *9*, 467–479
Hess, W.R., Meyer, A.E.: Triebhafte Fellreinigung der Katze als Symptom diencephaler Reizung. Helv. Physiol. Acta *14*, 397–410

1957 Hess, W.R.: Beziehungen zwischen psychischen Vorgängen und Organisation des Gehirnes II. Stud. Gen. *10*, 327–339
Hess, W.R.: Experimentelle Medizin der Gegenwart. Schweiz. Med. Wochenschr. *87*, 1177–1188
Hess, W.R.: Die Formatio reticularis des Hirnstammes im verhaltensphysiologischen Aspekt. Arch. Psychiatr. Z. Neurol. Nervenkr. *196*, 329–336
Hess, W.R.: Physiologische Aspekte extrapyramidaler Motorik. Schweiz. Arch. Neurol. Psychiatr. *80*, 299–305
Hess, W.R.: Die vegetative Innervation des Kortex. Acta Neuroveg. *16*, 5–8
Meyer, A.E., Hess, W.R.: Diencephal ausgelöstes Sexualverhalten und Schmeicheln bei der Katze. Helv. Physiol. Acta *15*, 401–407

1958 Hess, W.R.: Von den integrierenden Funktionen des Diencephalon. In: Pathophysiologia Diencephalica. Internationales Symposium Mailand, May 1956. Curri, S.B., Martini, L. (eds.), pp. 210–216. Wien: Springer

1959 Hess, W.R.: Objektive und subjektive Aspekte des Lärms. Z. Präventivmed. *4*, 1–2
Hess, W.R.: Kritische Bemerkungen zum Hirnreizversuch. Pflügers Arch. Ges. Physiol. Menschen Tiere *270*, 9–11

1960 Hess, W.R.: Zentrale Einwirkungen auf die Muskeltätigkeit. Bull. Schweiz. Akad. Med. Wiss. *16*, 398–401; 421–422

1961 Hess, W.R.: Schlaf, sein Mechanismus und seine physiologische Funktion. Naturwiss. Rundschau *14*, 1–7
Hess, W.R.: Die physiologische Grundlage der Psychosomatik. Dtsch. Med. Wochenschr. *86*, 3–8
Hess, W.R.: Psychophysiology and Psychopharmacology. In: Neuropsychopharmacology, Vol. 2. Rothlin, E. (ed.), pp. 3–8. Amsterdam: Elsevier

1962 Akert, K., Hess, W.R.: Über die neurobiologischen Grundlagen akuter affektiver Erregungszustände. Schweiz. Med. Wochenschr. *92*, 1524–1530
Hess, W.R.: *Psychologie in biologischer Sicht*, pp. 1–120. Stuttgart: Thieme

Hess, W.R.: „Kausalität“ als psychophysiologischer Begriff. Schweiz. Arch. Neurol. Neurochir. Psychiatr. *91*, 59–64
Hess, W.R.: Etude des corrélations entre l'activité psychique et le cerveau. J. Psychol. Norm. Pathol. *59*, 369–378

1963 Hess, W.R.: From medical practice to theoretical medicine: An autobiographic sketch. Perspect. Biol. Med. *6*, 400–423 [p. 309]
Hess, W.R.: Prinzipien cerebraler Organisationen. Experientia *19*, 511–513
Hess, W.R.: *Psicologia y biologia.* La psicologia desde el punto de vista biologico. Morata: Filosofia, Psicologia, Pedagogica. Madrid (Registro No. 3112/63)

1964 Hess, W.R.: Problemas planteados en los límites de la psicofisiologia. Folia Clin. Int. *14*, 3–6
Hess, W.R.: Die Motivierung bewußten Verhaltens. Schweiz. Arch. Neurol. Neurochir. Psychiatr. *93*, 338–341
Hess, W.R.: *The biology of mind*, pp. 1–203. Chicago: University of Chicago Press
Hess, W.R.: Psychische Manifestation cerebraler Organisation. Arch. Psychiatr. Nervenkr. *205*, 333–348
Hess, W.R.: Der Schlaf als Phänomen des integralen Organismus. Praxis *53*, 1355–1357

1965 Hess, W.R.: Gehirn und psychische Begabung. Bull. Schweiz. Akad. Med. Wiss. *21*, 43–50
Hess, W.R.: Von Beziehungen zwischen Psychologie und Psychiatrie. Schweiz. Med. Wochenschr. *95*, 317–319
Hess, W.R.: Sleep as a phenomenon of the integral organism. In: Sleep mechanisms. Progr. Brain Res. *18*, 3–8. Akert, K., Bally, C., Schadé, J.P. (Eds.). Amsterdam: Elsevier
Hess, W.R.: Cerebrale Organisation somatomotorischer Leistungen. I. Physikalische Vorbemerkungen und Analyse konkreter Beispiele. Arch. Psychiatr. Nervenkr. *207*, 33–44

1966 Hess, W.R.: Cerebrale Organisation somatomotorischer Leistungen. II. Experimentelle Untersuchungen über die zentrale Koordination der Motorik. Arch. Psychiatr. Nervenkr. *208*, 209–233

1967 Hess, W.R.: Causality, consciousness, and cerebral organization. Science *158*, 1279–1283

1968 Hess, W.R.: *Hypothalamus und Thalamus.* Experimental-Dokumente. 2nd ed. pp. 7–77. Stuttgart: G. Thieme
Hess, W.R.: *Psychologie in biologischer Sicht.* Mit kritischen Beiträgen von K. Akert, G. Baumgartner, R. Hess, jr., R. Hunsperger, R. Jung, P. G. Wasser, G. Weber. 2nd ed. pp. 1–132. Stuttgart: Thieme

1969 Hess, W.R.: *Hypothalamus and thalamus.* Experimental documentation. 2nd ed. pp. 1–77, enlarged edition. Stuttgart: G. Thieme

1970 Hess, W.R.: Polarität regulierender Funktionen und zerebrale Organisation. Dtsch. Med. Wochenschr. *95*, 1672–1675
Hess, W.R.: Interdisciplinary discussion of selected problems with reference to "The biology of mind". Perspect. Biol. Med. *13*, 267–293
Hess, W.R.: Ein Beitrag zur Physiologie des Kleinhirns. Zentrale Organisation somatomotorischer Leistungen. Biol. Zbl. *89*, 457–466

1973 Hess, W.R., Fischer, H.: Brain and consciousness. A discussion about the function of the brain. Perspect. Biol. Med. *17*, 109–118

Subject Index

MIX
Papier aus verantwortungsvollen Quellen
Paper from responsible sources
FSC® C105338

If you have any concerns about our products,
you can contact us on
ProductSafety@springernature.com

In case Publisher is established outside the EU,
the EU authorized representative is:
Springer Nature Customer Service Center GmbH
Europaplatz 3, 69115 Heidelberg, Germany

Printed by Libri Plureos GmbH
in Hamburg, Germany